SPECTROSCOPIC ANALYSES - DEVELOPMENTS AND APPLICATIONS

Edited by **Eram Sharmin** and **Fahmina Zafar**

Spectroscopic Analyses - Developments and Applications
http://dx.doi.org/10.5772/65617
Edited by Eram Sharmin and Fahmina Zafar

Contributors

Fritea Luminita, Jurca Tünde, Marian Eleonora, Vicas Laura Gratiela, Muresan Mariana Eugenia, Maria Ines Tavares, Marwa Elazazy, Mihaela Ilie, Marinela Florea, Antonio Martinez-Richa, Regan Silvestri, Martin Jaeger, Robin Legner, Hayriye Eda Şatana Kara, Nusret Ertaş, Anna Timoszyk, Nivia Coelho, Luciana Coelho, Bruno Elias S. Costa, Thais S. Neri, Henrique P. Rezende, Priscila A.R. Souza, Liliam Q. Tavares, Uttam Singh Baghel, Deeksha Singh, Atamjit Singh, Manish Sinha, Fabian Dayrit, Angel De Dios

Published by InTech
Janeza Trdine 9, 51000 Rijeka, Croatia

Publishing Process Manager Romina Rovan
Technical Editor SPi Global
Cover InTech Design team

First published December, 2017
Printed in Croatia
Legal deposit, Croatia: National and University Library in Zagreb

Additional hard copies can be obtained from orders@intechopen.com

Spectroscopic Analyses - Developments and Applications, Edited by Eram Sharmin and Fahmina Zafar
p. cm.
Print ISBN 978-953-51-3627-9
Online ISBN 978-953-51-3628-6

For EU product safety concerns:
IN TECH d.o.o., Prolaz Marije Krucifikse Kozulić 3, 51000 Rijeka, Croatia,
info@intechopen.com or visit our website at intechopen.com.

Meet the editor

Dr. Eram Sharmin is an assistant professor in the Department of Pharmaceutical Chemistry, College of Pharmacy, Umm Al-Qura University, Makkah, Saudi Arabia. She obtained her PhD degree in Chemistry from Jamia Millia Islamia (JMI)—A Central University, New Delhi, India, in 2007. She received her MSc degree in Organic Chemistry, in 2000, and BSc degree in Chemistry, in 1998, from Aligarh Muslim University (AMU), India. She has previously worked as a senior research associate (under Scientists' Pool Scheme, Council of Scientific and Industrial Research [CSIR], New Delhi, India), research associate (CSIR, New Delhi), and senior Research fellow (CSIR, New Delhi) at Materials Research Laboratory, Department of Chemistry, JMI. Dr. Sharmin has more than 50 publications in peer-reviewed journals and books and has presented more than 30 research papers in national and international conferences. Her research interests include the development of "green" materials with applications as antimicrobial and corrosion-resistant films, coatings, and packaging materials.

Dr. Fahmina Zafar is a senior researcher working at the Department of Chemistry, JMI, New Delhi, India, under the Women Scientists Scheme for Research in Basic/Applied Sciences, DST, India. Dr. Zafar has received her PhD degree in Chemistry from JMI in 2006. She has worked as a postdoctoral fellow under UGC Kothari Postdoctoral Fellowship, as a scientist pool, research associate, and senior research fellow (CSIR) at the same Department. She has more than 50 publications in peer-reviewed journals and books, and has presented more than 40 research papers in national and international conferences. Her research work involves the development of bio-based materials for green environment in different fields including adsorption, antimicrobial, and corrosion-protective applications.

Contents

Preface

Spectroscopic methods hold ubiquitous importance in analyses of materials, both qualitative and quantitative. They are being actively used as characterization and diagnostic tools in plethora of analyses as they are simple, sensitive, accurate, cost-efficient, reproducible, and quick. Spectroscopic and spectrophotometric methods have gained considerable importance in pharmaceutical and biomedical applications, as diagnostic tools, in drug quality control, for analyses and estimation of major/active constituents in drugs and natural products, and many other applications.

The book presents developments and applications of these methods, such as NMR, mass, and others, including their applications in pharmaceutical and biomedical analyses. The book is divided into two sections. The first section covers spectroscopic methods, their applications, and their significance as characterization tools; the second section is dedicated to the applications of spectrophotometric methods in pharmaceutical and biomedical analyses. The first four chapters in the first section deal with the recent advancements and applications of NMR spectroscopy, followed by a couple of chapters on the applications of NMR and Mass spectroscopy in profiling of natural products, and other pharmaceutical applications, while the last two chapters highlight the use of these methods as characterization tools. The second section focuses particularly on the pharmaceutical and biomedical applications of spectrophotometric methods such as ion-pair spectrophotometry and flow-injection spectrophotometry.

This book would be useful for students, scholars, and scientists engaged in synthesis, analyses, and applications of materials/polymers.

Due to great efforts of the authors, contributors, and technical staff of InTech Open Access Publisher, the book project is finally accomplished. We appreciate Ms. Romina Rovan, Publishing Process Manager, InTechOpen for her patience and prompt responses, during the whole process. Dr. Fahmina Zafar is thankful to the Department of Science and Technology, New Delhi, India, for the award of fellowship under the Women Scientists Scheme (WOS) for Research in Basic/ Applied Sciences (Ref. No. SR/WOS-A/CS-97/2016).

We would like to give special thanks to our family members for their great support during the entire process of compilation of the book project *Spectroscopic Analyses - Developments and Applications*.

Dr. Eram Sharmin
Department of Pharmaceutical Chemistry,
College of Pharmacy, Umm Al-Qura University,
Makkah Al-Mukarramah, Saudi Arabia

Dr. Fahmina Zafar
Inorganic Materials Research Laboratory,
Department of Chemistry, Jamia Millia Islamia,
New Delhi, India

Section 1

Spectroscopic Techniques Such as Proton and Carbon NMR and Mass and Their Applications

Chapter 1

Solid State NMR

Maria Ines Bruno Tavares

Additional information is available at the end of the chapter

http://dx.doi.org/10.5772/intechopen.71004

Abstract

The nuclear magnetic resonance (NMR) spectroscopy is a very powerful tool in the chemical characterization, both in solution and in solid state. With the development of NMR spectrometers more potent field, employing radio frequency pulse, provided the development of studies on materials, especially amorphous materials. Thus, there was a need to develop techniques to obtain spectra in solid state with high resolution in comparison to those obtained in solution. Therefore, the study of polymers and polymeric materials could be developed quickly as a result a lot of information about the structure-property could be obtained with more details. The use of NMR in the solid state has become particularly important in the study of amorphous materials, as well as in the study of crystal structures, and permits us to detect different constituents present in material. This chapter covers the basic solid-state NMR techniques that provide important information on sample molecular behavior because they are powerful and versatile tools to evaluate polymer and complex materials like nanomaterials.

Keywords: NMR, solid-state, polymer, nanomaterials, relaxation times

1. Introduction

Solid-state nuclear magnetic resonance (NMR) consists of several techniques, which are distinguished by different pulse sequences and generate different responses on the sample, allowing obtaining data on different time scales. This makes the development of new analytical methods for the study of polymer materials interesting. The solid-state analyses differentiate the solution by two main factors; the first is signal width. In the solid state, the signals are broader than solution, concerning to polymers due to the high-molecular weight and monomer ordination; among other factors, the signals are wider. The second point concerns the type of response obtained; in the solid state, the number of information obtained is greater than solution. When the material is insoluble or has soluble difficulties, the study of structure-property relation is of great interest because the search for responses with respect to homogeneity, phase dispersion, and intermolecular interaction between components is of great importance.

1.1. Nuclear magnetic resonance experiment

In the NMR experiment, a sample is placed in probe in a strong magnetic field, denominated Bo; if this sample presents magnetic moment (μ), its nuclear spins magnetize and at the excess of spin population, called magnetization (Mo), is applied for a radio frequency (RF) pulse by a radio frequency emission (B_1) field, the magnetization is excited by transfer to the *xy* plane. When the RF emission is turned off, nuclear spins tend to return to the steady state since this state has less energy, occurring in a process called relaxation. In this process, two types of relaxation occur simultaneously: one denominated transverse or spin-spin relaxation, which occurs in the *xy* plane and has a time constant—T_2, and the other called longitudinal or spin-lattice relaxation that occurs along the axis *z* and is characterized by the time constant—T_1. The NMR signal is detected after the withdrawal of radio frequency, with a process of free induction decay (FID) of the radio frequency, which results from the free precession of the nuclear spins, upon its return to steady state.

The NMR relaxation processes with time constants T_1 and T_2 are governed by fluctuating magnetic fields associated with molecular motion. The relaxation process that determines the T_1 values involving energy absorption, since this process is enthalpic. The temporal evolution of the transverse relaxation is fundamentally different from longitudinal, and it corresponds to a loss of phase coherence between the individual magnetic moment process in it, and, thus an increase of entropy. In many cases, solid sample loss phase coherence initially created by B_1 is due to direct interactions between the spin moments of individual [1–10].

1.2. High-field NMR

1.2.1. Solid-state NMR analyses

The solid-state NMR is constituted by several distinct pulse sequences, which generate different responses on the sample, allowing obtaining data on different time scales. This makes possible to develop new analytical methods for the study of complex solid materials [1–5], as polymer nanocomposites.

The analysis of the complex materials in the solid state differentiates the analysis in solution by two main factors: the first is the signal width. In the solid state, the signals are broader than in solution and especially for polymers due to their high molecular weight and mere ordination, among other factors, the signals become even wider. The second point concerns the type of response to be obtained in the solid state the number of information to be obtained is greater than in solution [11–14]. However, when the material is insoluble or has soluble difficulties, the study of the structure-property relation is of great interest because the search for responses with respect to homogeneity, phase dispersion, and interaction between the components is of great importance.

a. Nuclear magnetic resonance signal line width

Generally, the spectra obtained in solution generate narrow signals best resolved comparing to solid-state signals, due to the isotropy of the chemical shift, since all interactions as shielding, dipolar coupling, and indirect coupling depend on the orientation of the local environment in nuclear magnetic field B_o and when the samples are in solution, these effects are compensated. However, they are dependent on the nature of the sample and the external magnetic field strength applied [7–10].

In solids, there is usually little movement relative to the liquid. However, most samples (except single crystals) have a substantial molecular orientation range of line width. This fact comes from the anisotropy of the chemical shift as well as the strong dipolar interaction between the hydrogen nuclei and carbon-13. The nature of the sample and the type of nucleus to be observed are also two points of fundamental importance to the spectral resolution in solid state.

b. Solid-state nuclear magnetic resonance response

The type of answer you want to get on a specific material or on a polymeric system may be a reason why those must be analyzed by solid-state NMR. Information on the molecular dynamics is of great interest for answers about the correlation structure-molecular-dynamic property.

The problem of signal line width in NMR solid-state spectra led to the development of techniques that allow them to obtain signals in the solid state the narrowest possible, like liquids. Along with the information to be obtained on the material, different techniques are performed to analyze more different polymer systems.

1.2.1.1. High-resolution solid-state NMR basic theory

The Hamiltonian that governs the analysis involves a solid sum of different Hamiltonians, according to expression 1.

$$\mathbf{H}_{\mathrm{NMR}} = \mathbf{H}_{\mathrm{Z}} + \mathbf{H}_{\mathrm{RF}} + \mathbf{H}_{\mathrm{CSA}} + \mathbf{H}_{\mathrm{D}} + \mathbf{H}_{\mathrm{J}} + \mathbf{H}_{\mathrm{Q}} \quad (1)$$

where $\mathbf{H}_Z$ is the Zeeman effect; $\mathbf{H}_{RF}$ is the radio frequency effect; $\mathbf{H}_{CSA}$ is the chemical shift anisotropy; $\mathbf{H}_D$ is the dipolar interaction between hydrogen nucleus and the carbon-13 nucleus; $\mathbf{H}_J$ is the coupling constant; and $\mathbf{H}_Q$ is the quadrupolar moment.

When one observes spin nuclei ½, as carbon-13 (^{13}C), for example, the Hamiltonian that promotes more interference in the signal enlargement are H_{CSA} e H_D. Improving the resolution of the signals in the NMR spectra obtained in the solid state requires techniques to eliminate factors that cause this signal enlargement [10–14]. Thus, techniques were developed using methods that mathematically eliminate these effects.

1.2.1.1.1. Magic angle spinning (MAS)

The strong dipolar interactions between hydrogen nuclei and carbon-13, facilitated by the internuclear distance between them and the restricted mobility of the chains and the anisotropy of the chemical shift generate signals in very wide solid, with line width of 20 kHz order. The elimination of the dipole-dipole interaction generates a decrease in the signal line width of 5 kHz, and the elimination of the anisotropy of chemical shift width of the signals decreases to 100 Hz, making possible the detection of signals. Both the dipolar interaction and the anisotropy of the chemical shift dependence have with the term $3\cos^2\theta-1$. The elimination of these two effects occurs when the solid analyzes are performed by rotating the sample at high rotational speeds (each nucleus suitable for a given magnetic field) in a sample introduction probe angle corresponding to the amount of 54.74°, able to eliminate the term $3\cos^2\theta-1$, aligned with a strong decoupling of the hydrogen nucleus generating a significant narrowing of the line width in spectrum.

The employed pulse sequence is simple:

$$\begin{array}{ll} \text{hydrogen} & \text{decoupling} \\ \text{nucleus observed} & [90^{\circ}x \rightarrow \text{FID} - t]n \end{array} \quad (2)$$

where t is the time interval between 90° pulses (delay) and n is the number of scans.

The time t is variable and it is directly related to the relaxation time of different types of nuclei that are analyzed. Thus, variations in this parameter allow studies that provide information about the molecular mobility of the sample and the time of spin-lattice relaxation.

All nuclei that undergo the phenomenon of resonance can be analyzed by this technique. However, for the observation of nuclei that have quadrupole moments, line widths are so large that the signals have no resolution. However, for nuclei having dipole moment, this technique generates high-resolution spectra. It should be considered that for high-molecular weight materials such as polymer, for example, the chemical structure can be defined by this technique. However, a fine or detailed microstructure structure cannot be observed as they are well resolved by the NMR solution techniques. Because in the carbon-13 solid state analysis the signals broadened comes from the dipolar interactions and the chemical shift anisotropy, generating large signals that contain all molecular information's.

Note that using the MAS technique can obtain quantitative spectra solid. However, the long analysis time, comes from the high values of the spin-lattice relaxation times of the different nuclei, mainly rare spin. It makes this type of spectrum replaced by spectra expressing or representing only a portion of the sample. Therefore, variation in spectral parameters of this pulse sequence provides information about the increased mobility of a sample region, for example, a mixture of polymers, polymers, composites, amorphous, and nanocomposite materials. Thus, a greater number of applications of this technique can be obtained, when seeking information about the homogeneity, compatibility, and purity of polymers or any material samples.

The analysis of materials by the MAS technique using a small interval between pulses (ms) can detect only one region or the region that has the highest mobility. This variation in the MAS technique enables, in the case of amorphous polymers, that is, ethylene-*co*-vinyl acetate (EVA), identifies the region of increased molecular mobility, or distinguishes the mobility of different areas, which cause changes in the properties of the material [6–10].

http://dx.doi.org/10.5772/intechopen.71004

Poly(ethylene-*co*-vinyl acetate) (EVA) is a random copolymer that has a distinct percentage of vinyl acetate, which promotes changes in their mechanical and thermal properties and consequently changes the processing conditions and materials. The monomer sequence of the random copolymer is shown in **Figure 1**.

Analyzing EVA containing 28% of acetate by carbon-13 (C-13) solid-state NMR basic techniques will be shown as an example of how useful is the application of solid-state techniques.

Figure 2 exhibits the powder EVA C-13 MAS NMR spectrum. Showing the highest signal located at 30.2 ppm referred to CH_2 (the methylene group) long chains and a small signal detected at 14.3 ppm attributed to the methyl group of the acetate part. Two small signals were detected at about 21 and 25 ppm, which were assigned as CH_2 from the ethylene branching [11, 12].

1.2.1.1.2. Cross-polarization and magic angle spinning angle (CPMAS)

The cross-polarization technique was developed aimed at detection of rare nuclei spins with the aim to minimize the analysis time because of the long relaxation times of these nuclei. This method relies on the transfer of polarization of a nucleus spin abundant, hydrogen nucleus (^1H), for example, to rare spin nuclei (i.e., ^{13}C), the cores ^{13}C and ^1H are in thermal contact for a stipulated period of time, called during the cross-polarization contact time at that time the nuclei are kept in contact due to precession frequencies of both nuclei are kept identical, in this case the nuclei are in a condition called condition Hartman-Hahn [1–5], which is an equality where the frequency precession of hydrogen nucleus versus magnetic field of hydrogen are equal to precession of carbon-13 nucleus versus magnetic field of carbon-13, in a period of time: $\omega_H \mathbf{B}_H = \omega_C \mathbf{B}_C$.

The combined cross-polarization technique with the rotation of the sample at the magic angle and strong hydrogen decoupling (CPMAS), generating NMR spectra of solid high-resolution rare spin nuclei with increasing signal strength in a shorter analysis time than the MAS, considering that the hydrogen nucleus controls the relaxation process [1–10, 13].

The pulse sequence used to obtain the spectra through CPMAS is the same for MAS, but with the inclusion of the condition Hartman-Hahn, which is inserted a contact time between the two nuclei for transferring the polarization between them. Thus, the combination of cross-polarization technique, magic angle spinning process, and strong hydrogen decoupling of carbon-13 nucleus technique informs about the compatibility of polymer blends at the molecular level. The changes in the widths of the lines of NMR and the values of the chemical shifts provide information about changes in mobility at the molecular level.

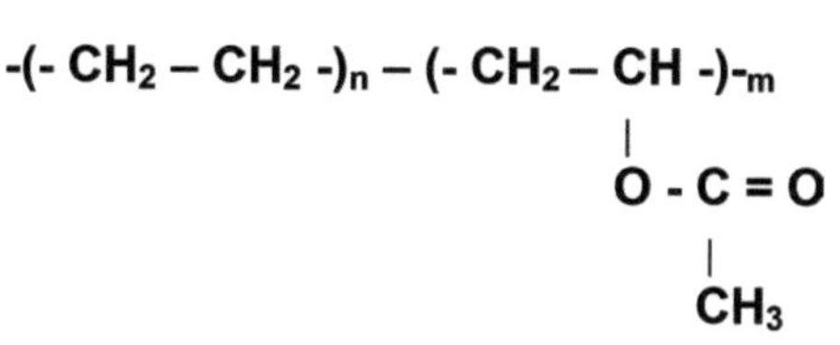

Figure 1. Poly(ethylene-*co*-vinyl acetate) (EVA) chemical structure.

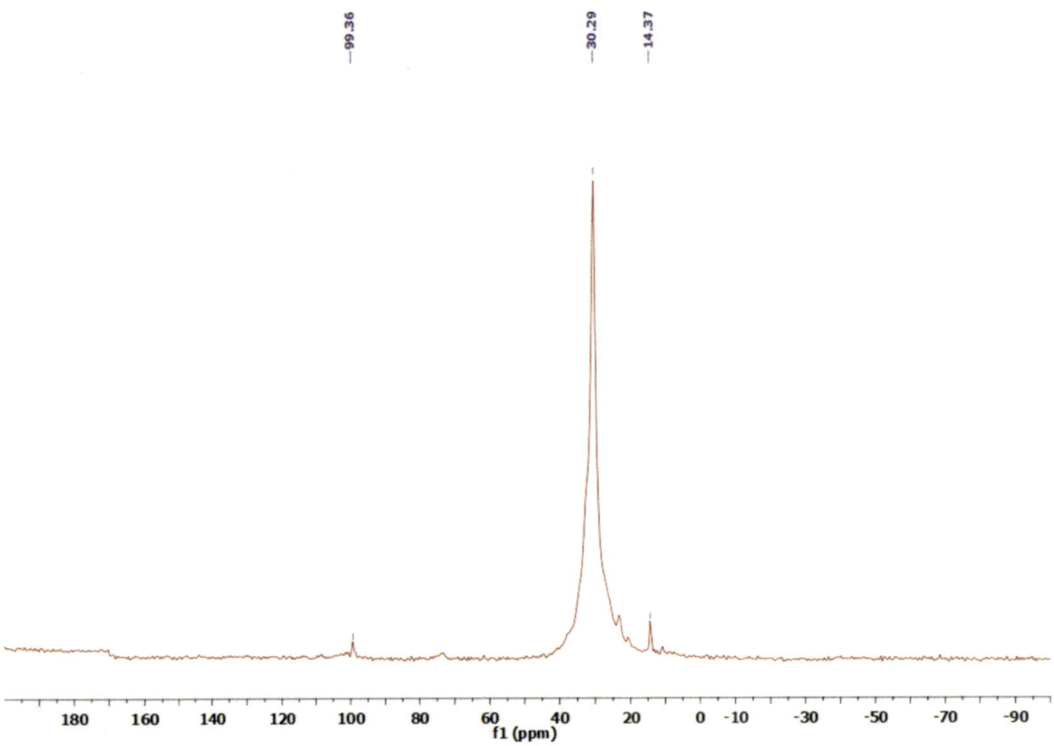

Figure 2. Solid-state NMR C-13 MAS spectrum of powder EVA.

Figure 3 shows the powder EVA CPMAS C-13 NMR spectrum, with 1 ms of contact time. It already showed two signals: one located at short chemical shift, centered at 30.7 ppm referring to mobile region and the other one located at 32.3 ppm due to the segments of rigid region.

One type of solid-state NMR studies is to use a comparison between ^{13}C NMR spectra obtained by techniques MAS and CPMAS, which can first show the different regions of the samples. One example was showed in the literature, which exhibits the MAS and CPMAS solid state NMR spectra of seed flour bourbon mango spectra [15, 16]. It shows that in these spectra have at least two segment areas of different molecular mobilities, and it may also a third due to the interaction of these two domains that may not be detected in this type of measurement.

1.2.1.1.3. Variable contact time (VCT) during polarization transfer

This technique generates a variation of contact times during the cross-polarization experiment, leading to a series of ^{13}C CPMAS spectra with different contact times, and through this experiment, one can obtain some important information, such as heterogeneity of the sample, material stiffness, different types of domains, and the value of the hydrogen spin-lattice relaxation time in the rotating frame ($T_1\rho H$). This parameter can be obtained from the intensities decay of carbon-13 nucleus during the cross-polarization transfer experiment, according to the changes in the contact time, since the hydrogen nucleus is the one that controls this relaxation process. **Figure 4** exhibits the variable contact-time experiment for powder EVA.

http://dx.doi.org/10.5772/intechopen.71004

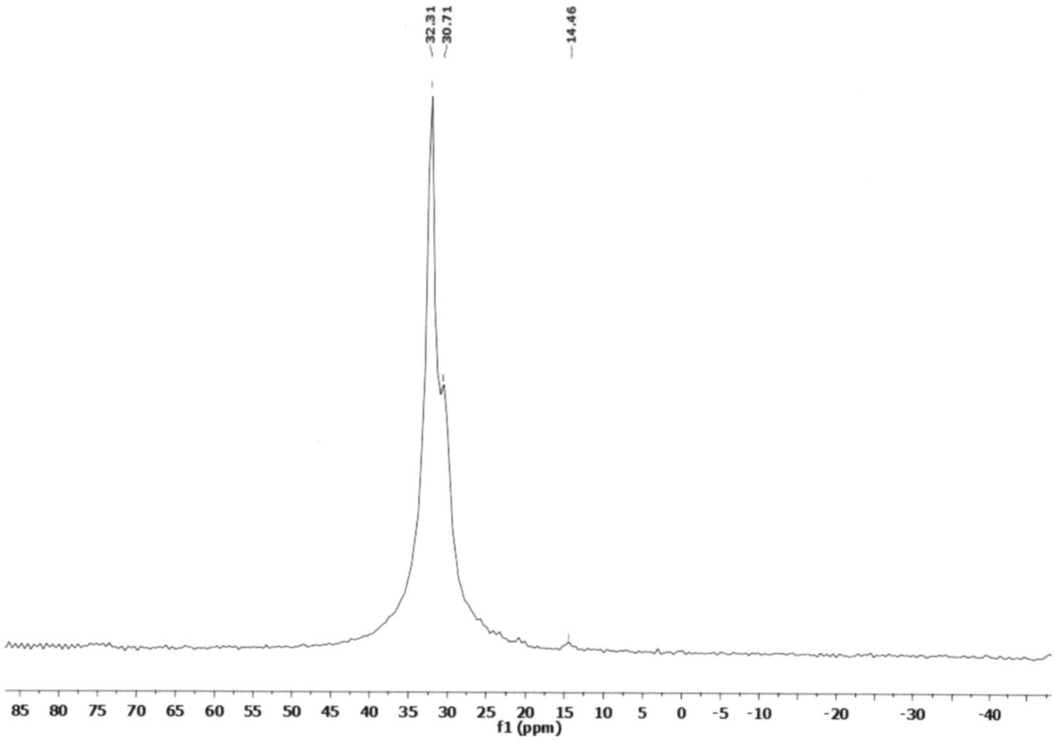

Figure 3. Solid-state NMR CPMAS C-13 NMR spectrum of powder EVA.

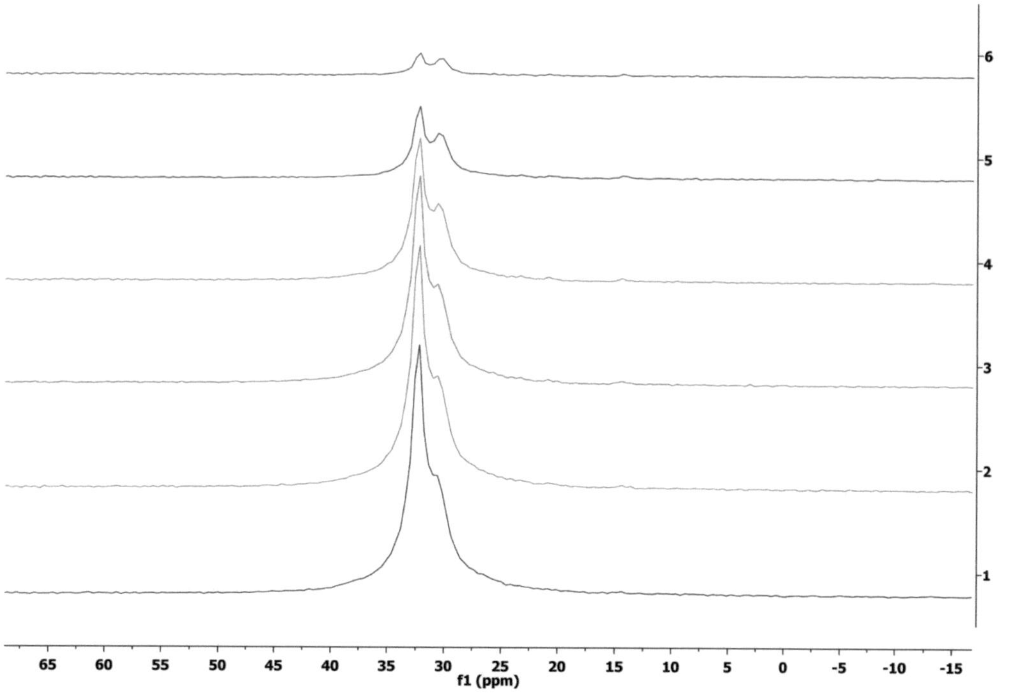

Figure 4. Variable contact time experiment of powder EVA.

This experiment shows the decay of both resolved carbon types detected from CPMAS with contact-time variation, showing the rigid part of the sample and part of the mobile one with the increase in the signal related to the CH_2 of the nonrigid phase.

2. Case study

2.1. Example of the determination of $T_1\rho H$ for polymer nanocomposites based on poly(3-hydroxy butyrate) (PHB)

In this study, it was evaluated the $T_1\rho H$ for the PHB/silica (PHB/S) systems contain different proportions of silica. All samples were obtained through solution casting, and the films after being dried were out into the rotor, the analyses were carried out at 30°C, and the values of this parameter are listed in **Table 1**.

From the data listed in **Table 1**, just a small proportion of silica affects the polymer, promoting a formation of a new material. Therefore, the relaxation data for the samples containing 0.5 or more silica exhibit a similar behavior, promoting an increase in this parameter comparing to pure PHB indicating that a new material with good dispersion and distribution of the nanoparticle in the polymer was obtained. The limit of silica is 0.5%, and no more is needed since no change in the $T_1\rho H$ for all carbons was detected. The evaluation of this parameter is not very much used for polymer nanocomposites yet. Therefore, studies have been already published by the group for other systems, as blends and composites [24], and show the behavior of these polymer materials [25].

Sample	$T_1\rho H$ (ms)			
	C = O	CH_2	CH	CH_3
PHB	13	26	31	17
PHB/S 0.2	15	19	27	19
PHB/S 0.5	33	32	34	32
PHB/S 0.75	31	32	34	30
PHB/S 1.0	31	31	34	30

Table 1. $T_1\rho H$ data for the PHB/S systems, containing different proportions.

3. Low-field NMR

In low-field NMR equipment due to low intensity and homogeneity of the magnetic field, the chemical shift cannot be used to discriminate between different molecules. The nuclear relaxation processes occurring in the nuclear magnetic resonance phenomenon inherent to spectroscopy are spin-lattice and spin-spin relaxation times; however, they provide detailed information about the molecular dynamics. These relaxation times influence from the structural and microstructural quantitative determination by the study of molecular dynamics. The relaxation processes are associated to the time constant for these processes: T_1, spin-lattice relaxation time, and T_2, spin-spin relaxation time. In the low-field NMR, the relaxation parameters T_1 and T_2 for the hydrogen nuclei that constitute the samples can be measured directly, using the pulse sequences for such experiments and are widely used to characterize the types of molecular segments present in the samples and the interactions between them [15–20]. The time T_1 is associated with the return of the nucleus excited by the absorption of radiation to the equilibrium. While T_2 relaxation is related to the inverse of the half-width of the signal and occurs due to the loss of phase coherence in the precession of the nucleus excited about the direction of the applied magnetic field [1–3].

The relaxation times determined by NMR offer detailed information on the molecular mobility of a material. Thus, one could detect the formation of rigid and flexible segments, plasticization or antiplasticization process, and any other change in the molecular dynamics of the sample or comparing the variation of the molecular mobility of structures from the same type of material, but of different origin. The time of relaxation times T_1H and T_2H can be measured in a wide temperature range and identifying small differences in similar structures.

3.1. Determination of T_1

The inversion-recovery pulse sequence is the most accurate technique for measuring the relaxation time spin–lattice process. The T_1H relaxation time é measured in the frequency of magnetic field generated by the external magnetic field. This relaxation must do with the return of the excited nuclei to its ground state after removal of the excitation frequency, and therefore, it allows to evaluate the molecular mobility of a material, global compatibility and homogeneity, as well as processes and plasticization [1–3, 10–15]. In this process, the excess energy is emitted to the lattice in the form of dipole interaction since there is a lowering of the enthalpy of the nuclear spin system. The return to the magnetization to the equilibrium state is usually exponential, and is a first order process with constant speed, R_1, and time constant T_1.

The applied pulse sequence is described below:

$$\text{Observed nuclei}: \left[180^{\circ}{}_{X} - \tau - 90^{\circ}{}_{X}\right]_{n} \tag{3}$$

where τ is a time interval between 90 pulses.

The relaxation time T_1H reports on the molecular dynamics of materials. In the case of starches and starch flour, for example, this parameter is sensitive to the formation of domains or segments and structural changes in the range 15–50 nm.

3.2. Determination of T_2

The proton spin-spin relaxation time, T_2H, in principle, can be obtained by measuring the width of the NMR signal at half height. However, the inhomogeneity of the magnetic field causes the magnetization of the different nuclei processes at different rates when removing the RF and a time tau (τ) is needed to wait, which is chosen depending on the mobility of the sample; however, this inhomogeneity is refocused after an application of a 180° pulse, generating a single magnetization [1, 15, 16]. The pulse sequence normally used is Carr-Purcell-Meiboom-Gill (CPMG), which contains a train of pulses [1–5].

This relaxation time, and the time constant of spin-lattice relaxation reports about molecular mobility of materials in the molecular level, thereby assess overall molecular dynamics and segmental. The T_2H parameter corroborates data obtained by T_1H and sometimes can inform more detail of the material behavior under observation, if this have some regions or segments containing high-molecular mobility.

In several studies that involve the evaluation of molecular mobility of materials, specially using the NMR relaxometry, it was normally employed the measurements of relaxation times spin-lattice and spin-spin and evaluate the T_1 and T_2 times constants of both relaxation phenomena, respectively. This comes from the fact that these parameters are sensitive to dynamic processes that occur at different frequency ranges. Thus, T_1 parameter measures the relaxation of the magnetization component parallel to the external magnetic field, being sensible to fast movements that are sensible to the movements of first order (MHz). The $T_{1\rho}$ relaxation parameter is measured in low frequencies in the range of tens of kilohertz [1, 7, 17–23].

4. Final comments

In the solids, there is a restricted molecular movement comparing to liquids. However, most of the samples range have a substantial molecular orientation of the line width. This fact stems from the anisotropy of the chemical shift as well as the strong dipolar interaction between the hydrogen and carbon-13. The nature of the sample and the type of nuclei to be observed are two points of fundamental importance to the spectral resolution. The type of answer you want to get on a specific material may be a reason why it must be analyzed by solid-state NMR. Information on the molecular dynamics is of great interest for answers about the correlation structure-molecular-dynamic property. The line width in NMR solid state spectra led to the development of techniques that allows obtaining signals in the solid state most narrow possible, like liquids. Along with the information to be obtained, different techniques are performed to analyze the more different polymer systems.

Author details

Maria Ines Bruno Tavares

Address all correspondence to: mibt@ima.ufrj.br

Federal University of Rio de Janeiro, Professor Eloisa Mano Macromolecules Institute, Rio de Janeiro, Brazil

References

[1] Komoroski RA. High Resolution NMR Spectroscopy of Synthetic Polymers in Bulk. Deerfield Beach: VCH Publishers; 1986

[2] McBrierty VJ, Packer KJ. Nuclear Magnetic Resonance in Solid Polymers. Cambridge: Cambridge University Press; 1993

[3] Schmidt-Rohr K, Spiess HW. Multidimensional Solid-State NMR and Polymers. New York: Academic Press; 1994

[4] Stejskal EO, Memory JD. High Resolution NMR in the Solid State. New York: Oxford University Press; 1994

[5] Bovey FA, Mirau PA. NMR of Polymers. New York: Academic Press; 1996

[6] Sandres JKM, Hunter BK. Modern NMR Spectroscopy: A Guide for Chemists. 2nd ed. Oxford: Oxford University Press; 1996

[7] Harris RK. NMR studies of solid polymer. In: Fawcell AH, editor. Polymer Spectroscopy. England: John Wiley & Sons; 1996

[8] Silva NM, Tavares MIB. Journal of Applied Polymer Science. 1996;**60**:663

[9] Costa DA, Oliveira CMF, Tavares MIB. Journal of Applied Polymer Science. 1998;**69**:129

[10] Harris RK. Recent advances in solid state NMR. The Fifth International Conference on Applications of Magnetic Resonance in Food Science, Aveiro, Portugal. 2000;**I**:1-11

[11] Tavares MIB. Polymer Testing. 2000;**19**:899

[12] Silva EO, Tavares MIB, Bathista ALBS, Filho NP, Nogueira JS. Journal of Applied Polymer Science. 2002;**86**:1848

[13] Souza CMG, Tavares MIB. Journal of Applied Polymer Science. 2002;**86**:116

[14] Tavares MIB. Journal of Applied Polymer Science. 2003;**87**:473

[15] Tavares MIB, Bathista ALBS, Silva EO, Filho NP, Nogueira JS. Carbohydrate Polymers. 2003;**53**:213

[16] Bathista ALBS, Silva EO, Tavares MIB, Prad RJ. Journal of Applied Polymer Science. 2012;**126**(S1):E132-E126

[17] Cheng HN. Modern Methods of Polymer Characterization. In: Barth HG, Mays JW, editors. Série Chemical Analysis. Vol. 113. 1991. p. 409-493

[18] Ebdon JR. In: Dawking JV, editor. Developments in Polymer Characterisation-2. London: Applied Science Published Ltd; 1980

[19] Silva MBR, Tavares MIB, Junior AWM, Cucinelli-Neto RP. Journal of Nanoscience and Nanotechnology. 2016;**16**:7606

[20] Sebastião PJO, Monteiro MSSB, Brito LM, Rodrigues E, Chávez FV, Tavares MIB. Journal of Nanoscience and Nanotechnology. 2016;**16**:7539

[21] Tavares MR, Menezes LR, Nascimento DF, Souza DHS, Reynaud F, Marques MFV, Tavares MIB. European Physical Journal Special Topics. 2016;**225**:779

[22] Monteiro MSSB, Tavares MIB, Sebastião PJO. Materials Scince and Applications. 2016;**7**:575

[23] Cunha APCB, Tavares MIB, Silva EO. Materials Sciences and Applications. 2016;**07**:380

[24] Preto M, Tavares MIB, Silva EP d. Polymer Testing. 2007;**26**:501

[25] Tavares MIB, Nogueira RF, San Gil RAS, Preto M, Silva EO, e Silva MBR, Miguez E. Polymer Testing. 2007;**26**:1102

Chapter 2

Developments in Solid-State NMR Spectroscopy of Polymer Systems

Antonio Martínez-Richa and Regan L. Silvestri

Additional information is available at the end of the chapter

http://dx.doi.org/10.5772/intechopen.70116

Abstract

Solid-state nuclear magnetic resonance (NMR) has long emerged as a valuable technique for characterizing the molecular structure, conformation, and dynamics of polymer chains in various polymer systems. The principles of the solid-state ^{13}C NMR cross-polarization experiment are described along with corresponding relaxation measurements. The ensuing recent applications of these techniques to different polymer systems are reviewed, with selected examples that have appeared in the recent literature. All of these applications of solid-state NMR to polymers have one feature in common: the interpretation of spectroscopic observations as related to the structural features and physical properties of the polymer.

Keywords: polymers, solid-state NMR, structure-property relationship, polymer morphology, polymer dynamics, NMR relaxation time

1. Introduction

Solid-state nuclear magnetic resonance (NMR) spectroscopy is at this time well-established as a valuable technique for characterizing a variety of polymer systems. A multitude of NMR experiments can be used to gain valuable practical information about the molecular structure, conformation, and dynamics of polymer chains in various polymer systems. Such information is useful in the design of polymer properties, and therefore the technique of solid-state NMR has been widely applied to numerous polymer systems.

The use of solid-state ^{13}C NMR is well established for elucidation of cross-linking structures in elastomers and bonding structures in adhesives [1], and the technique continues to be applied to new elastomer and adhesive systems. These same techniques are now being applied to new polymer systems such as self-assembled polymers, advanced functional polymers, electro-conducting polymers, microporous materials, and proteins [2].

2. High-resolution ^{13}C NMR spectroscopy of solid polymers

Nuclei with a spin quantum number of $I = ½$ such as ^{1}H, ^{13}C, ^{19}F, ^{29}Si, ^{15}N, and ^{31}P yield high-resolution NMR spectra and are therefore particularly informative in the study of polymer systems. These nuclei display spectra with unique peaks for each magnetically inequivalent nuclei in the chemical structure, essentially enabling the study of individual atomic positions within the chemical structure of a polymer [3]. As such, localized information can be ascertained about individual atomic sites within the chemical structure of a polymer.

In solution, local magnetic fields experienced by nuclei are averaged by rapid isotropic motions resulting in the observation of relatively sharp NMR peaks. Polymers however are largely used not in the solution state but as structural engineered materials such as plastics and elastomers in the solid state. Therefore, there is a need to study them as solids, to characterize their properties in the solid state, which is the state that they will be used as structural materials. Observation of the ^{13}C nucleus in the solid state is complicated by line broadening caused by strong dipolar interactions with the abundant ^{1}H isotope. This line broadening is reduced by heteronuclear dipolar decoupling (DD), a high powered radio frequency (rf) pulse at the ^{1}H frequency during the time in which the ^{13}C signal is observed [4].

The collection of solid-state NMR spectra is also complicated by high chemical shift anisotropy, which is motionally averaged in the liquid state. The line broadening caused by chemical shift anisotropy in the solid state is reduced by a technique termed magic angle spinning (MAS). The broad chemical shift anisotropy pattern of a solid is reduced to a single peak at the isotropic chemical shift by spinning the solid sample, typically at a rate of a few thousand hertz, at an angle of precisely 54.74° relative to the static magnetic field [5].

The ^{13}C isotope being only 1.1% naturally abundant, and the heteronuclear dipolar coupling and chemical shift anisotropy of solids being only partially reduced by DD and MAS, respectively, results in an inherently low signal-to-noise ratio for solid-state spectra. The ^{13}C signal for solid samples is enhanced by a technique termed cross-polarization (CP). Cross-polarization is achieved by simultaneously applying spin locking rf pulses to both ^{13}C and ^{1}H nuclei. The spin locking rf pulses are adjusted to reach a state where both nuclei process at the same frequency, a special condition referred to as the Hartmann-Hahn match. Under these conditions, magnetization is transferred from the naturally abundant ^{1}H spin reservoir to the more dilute ^{13}C spin reservoir, resulting in signal enhancement of the ^{13}C spin reservoir. As a result, an enhanced ^{13}C signal can be observed for solid samples [6].

The combination of these three techniques (DD, MAS, and CP) into one experiment provides a method for the collection of high-resolution ^{13}C NMR spectra of solids. The results observed by applying these techniques, in combination progressively and ultimately, simultaneously in one experiment, are shown in **Figure 1** [7]. The method of simultaneously combining DD,

MAS, and CP has become routine for solid samples, and applications to polymer systems are sufficiently plentiful to fill entire professional reference textbooks [8]. Particularly, application of the solid-state NMR technique has historically proven to be highly informative in elucidation on the conformational structure of solid polymers via chemical shift analysis and determination of cross-linking structures of cross-linked solid polymers.

Beyond the DD, MAS, and CP experiment to collect high-resolution solid-state spectra, sophisticated and elegant rf pulse sequences are now used to perturb the magnetization in specific ways. As such, observation of the magnetization as it processes back to equilibrium via various cleaver rf pulse sequences allows the collection of information far beyond the simple spectrum.

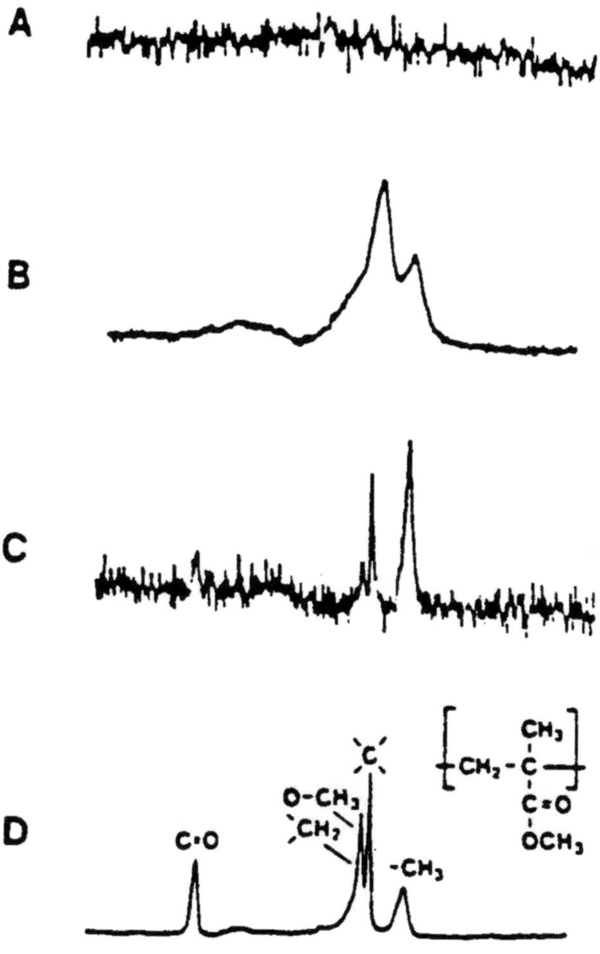

Figure 1. The ^{13}C NMR spectra of poly(methyl methacrylate) in the solid state under various experimental conditions: (a) using the experimental conditions for solution spectra, no discernible peaks are observed for the solid sample; (b) using dipolar decoupling with cross-polarization permits the observation of broad chemical shift anisotropy patterns; (c) using dipolar decoupling with magic angle spinning reduces the broad chemical shift anisotropy patterns to resolvable peaks; and (d) the combination of all three techniques, dipolar decoupling with magic angle spinning and cross-polarization, facilitates the observation of a high-resolution spectrum for the solid sample. Reprinted from Ref. [7].

For example, the study of selectively pulsed magnetization as it relaxes back to equilibrium yields insight into not only the molecular structure, but also the molecular dynamics of a polymer system. Relaxation of a nucleus back to equilibrium is modulated by fluctuations in the local magnetic field which the nucleus experiences, and the local field is modulated due to the local environment, including not only the physical-chemical structure around a nucleus, but also molecular motion that the nucleus is involved in. Various pulse sequences allow the measurement of relaxation parameters including T_1, T_2, and $T_{1\rho}$, which yield information about molecular motions in different frequency ranges, which occur over different scales of distance [9]. Beyond the direct measurement of these relaxation parameters, other pulse sequences exploit these relaxation phenomena to yield data which can be interpreted in terms of polymer structure and properties.

Two-dimensional (2D) NMR techniques enhance the resolution of a traditional one-dimensional spectrum by spreading the peaks out in a second dimension. Two-dimensional NMR techniques also facilitate the observation of through bond or through space interactions, where through space interactions being particularly related to the physical structure of polymers such as conformation [10]. While 2D NMR techniques are routine for solutions, they are still considered by some to be experimentally cumbersome in the solid state. Nonetheless, there has been considerable progress recently in the development of experimentally practical pulse sequences for 2D NMR of solid samples [11].

3. Applications to various polymer systems

3.1. Carbohydrates

Wood is a complex heterogeneous material composed mainly of hemicellulose, cellulose, and lignin. Solid-state NMR spectroscopy can discriminate wood samples based upon their provenance. In that regard, the traceability of wood samples can be undertaken by analyzing the solid-state NMR peak patterns in their MAS and CP-MAS spectra.

This technique has been used to characterize the chemical composition of wood, and the effects of aging, decomposition, and some physical or chemical treatments on the polymer structure. Recently, this technique was used to analyze maple samples from Norway [12] and spruce samples from Finland, Poland, and Italy [13]. The chemical structures of various components in soft wood are shown in **Figure 2**. Carbon-13 CP-MAS NMR spectra for spruce wood from Finland, Poland, and Italy are shown in **Figure 3** with resonance peaks labeled to correspond with the structures shown in **Figure 2**. While the same peak pattern is present in all three samples, small differences (mainly in peak intensity) can be distinguished between the three spectra.

In the spectral region 110–160 ppm, differences in the lignin aromatic components can be distinguished, whereas in the spectral region from 15 to 110 ppm, differences in the signals for cellulose and hemicellulose can be observed. In that regard, the most intense peaks in the spectra 12 and 13 are due to the C-2, C-3, and C-5 carbons of the carbohydrates. The two peaks 10 and 11 are assigned to C-4, in crystalline and amorphous (or less ordered surface) cellulose,

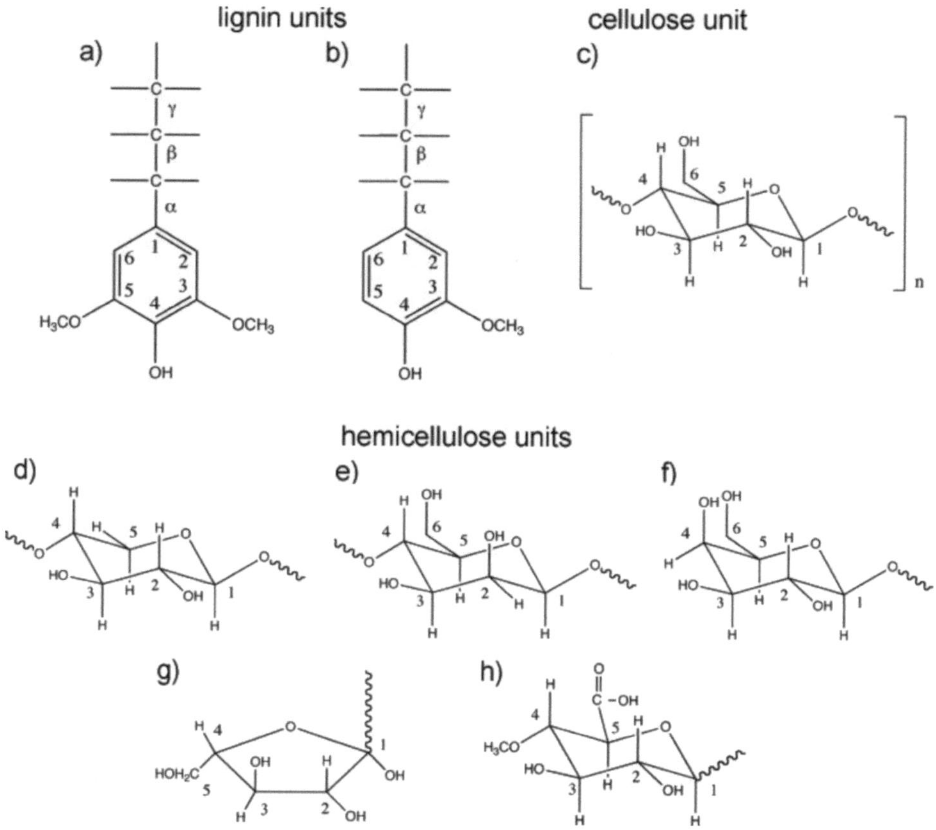

Figure 2. Chemical structures of various wood components in softwood. The numbering scheme corresponds to the resonance peaks as labeled in **Figure 3**. Reprinted from Ref. [13].

respectively. Peak 15 is related to C-6 in cellulose and Cγ in lignin. Peak 8 represents the C-1 of cellulose, with a high-field shoulder 9 attributed to hemicellulose (102 ppm). Only two signals can be undoubtedly attributed to hemicellulose: the methyl carbon peak 18 and the carboxylic carbon peak 1; both of these are rather weak in intensity. Three groups of signals in the range 160–105 ppm are attributed to the three aromatic units constituting the lignin lattice. Finally, the small peak 16 is assigned to the lignin methoxyl group.

As differences in peak pattern intensities for the three samples are small, ^{1}H $T_{1\varrho}$ measurements were obtained using a variable CP contact time experiment. The results suggest that higher polymer mobility and a higher homogeneity are observed in spruce wood from Finland, relative to a lower homogeneity measured in samples from both Italy and Poland.

3.2. Protein systems

Solid-state NMR spectroscopy has been widely applied to study a variety of protein systems. Various modern solid-state NMR techniques can be applied to gain insights into biophysics and structural biology in proteins.

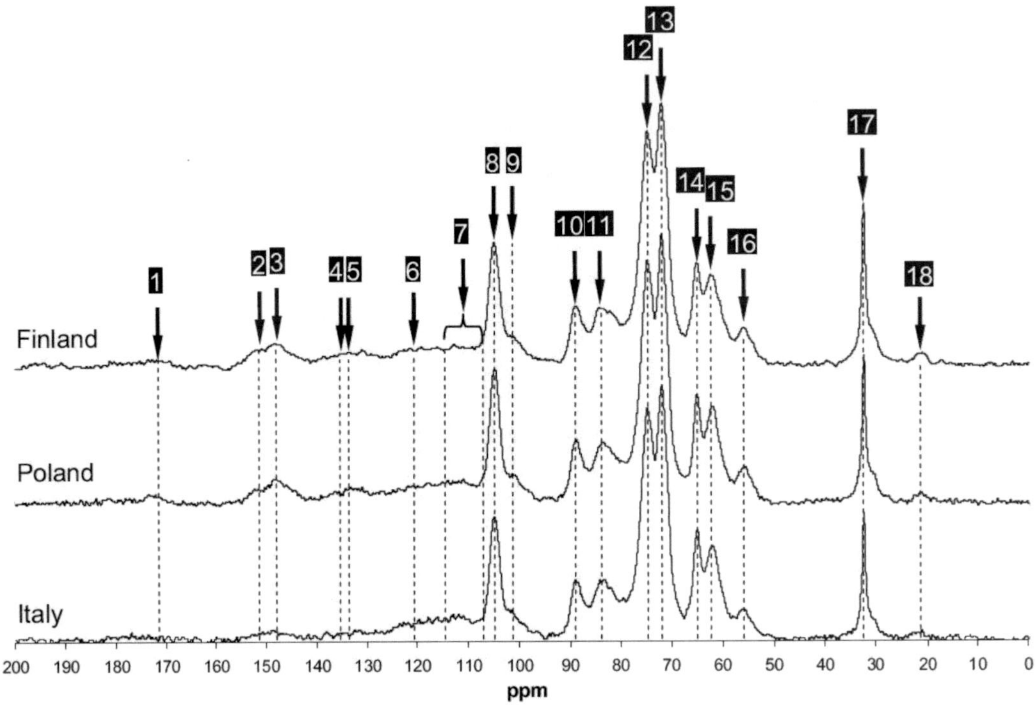

Figure 3. Solid-state ^{13}C NMR spectra of spruce woods differing in provenance and their peaks assignments. Differences in peak intensities can be used to distinguish between the geographic origins of woods. Reprinted from Ref. [13].

Peptide nanoassemblies have been studied by solid-state NMR. Using the recent development of high-field MAS dynamic nuclear polarization (MAS-DNP), an increase in sensitivity for the peaks of crystalline components of the polymer system can be achieved [14]. Using this approach, self-assembled structures of systems based on diphenylalanine dipeptide (FF) have been studied. These proteins have applications as organic semiconductors, and have been reported as a core motif of Alzheimer's amyloid-β. Supramolecular structural information such as hydrogen bonding and π-π stacking can be improved using dynamic nuclear polarization. Ultimately, an increase in sensitivity by an overall factor of 320 is achieved by using DNP.

Self-assembly peptide nanostructure can be studied by solid-state NMR. MAX8 is an amphiphilic peptide composed of 20 amino acids, having the amino acid sequence VKVKVKVKVDPPTKVEVKVKVNH2, where K and E are hydrophilic residues and V hydrophobic residues [15]. The self-assembled peptide contains a β-strand hairpin structure aligned into antiparallel β-sheets. As is commonly observed, changes in supramolecular structure can be induced by changes in temperature, pH, and ionic strength. The CP-MAS spectra of labeled and unlabeled MAX8 nanofibers are shown in **Figure 4**. The observed chemical shifts for the amino acid residues 1–8 and 13–20 indicates that they form part of a β-hairpin conformation. Polymorphism of K residues occurs uniformly across the MAX8 amino acid system. Calculations of ϕ and φ backbone torsion angles, derived from peak positions using commercial software, were consistent with β-strand secondary structures for residues 1–8 and

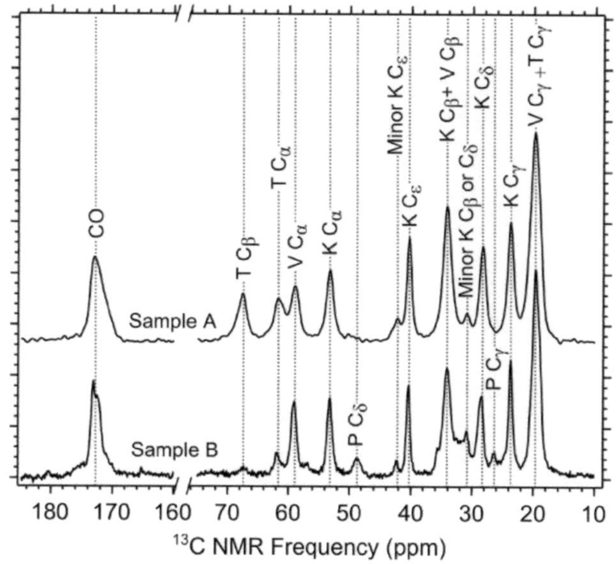

Figure 4. Solid-state ^{13}C CP-MAS spectra of labeled and unlabeled MAX8 nanofiber, a peptide composed of 20 amino acids. Observed chemical shifts for the amino acid residues allow determination of the three dimensional conformational structure of the peptide. Reprinted from Ref. [15].

13–20. Further evidence of the existence of a close β-hairpin conformation was obtained by (a) the analysis of cross-peaks in a 2D DARR ^{13}C—^{13}C NMR experiment, and (b) ^{13}C—^{13}C dipolar recoupling NMR experiments using the PITHIRDS-CT technique. The PITHIRDS-CT technique is a constant-time dipolar recoupling sequence in solid-state NMR with MAS, which yields experimental data that is insensitive to rf pulse imperfections and nuclear spin relaxation processes. This sequence has been widely used to determine intermolecular distances and molecular conformations in solid ^{13}C and ^{15}N labeled compounds, and as such one of the most important applications of this technique has been the study of biological samples such as amino acids and amyloid fibrils.

In a similar approach, β-sheet nanocrystalline domains in phosphorylated serine-rich motifs in caddisfly larval silk were studied by ^{13}C and ^{31}P solid-state NMR [16]. ^{13}C NMR data from isotopically enriched caddisfly silk, packed in its natural hydrated environment, are shown in **Figure 5**. ^{13}C chemical shifts were identified using ^{13}C—^{13}C DARR and $^{1}H \rightarrow {}^{31}P \rightarrow {}^{13}C$ DCP NMR experiments. Differences between CP-MAS and DD-MAS spectra are due to the aqueous environment. Water-solvated residues with a short T_1 exhibit enhanced signals in the DD-MAS spectrum, whereas carbons located in more rigid environments are better observed in the CP-MAS spectrum. An enhanced peak for the unmodified serine β carbon is seen in the DD-MAS spectrum, suggesting higher mobility. On the other hand, the signal of the phosphorylated serine β carbon is broader, indicating that these moieties mainly reside in the β-sheet regions. Similar conclusions have been derived for Valine residues. However, glycine and leucine residues, often seen in GGX repeats, exist predominantly in a random or disordered conformation.

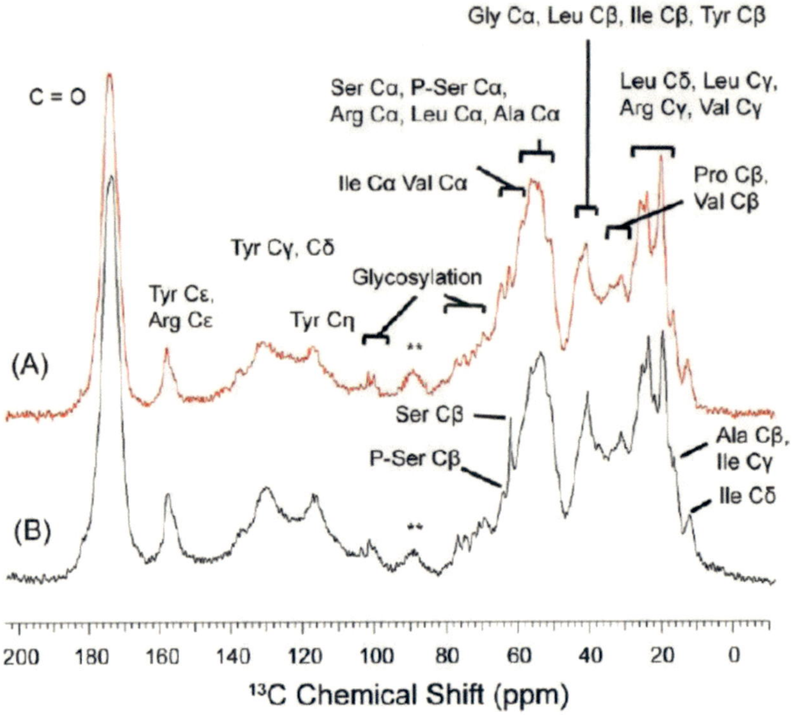

Figure 5. Solid-state ^{13}C NMR of isotopically enriched caddisfly silk from the species *Hyllisia consimilis* in its natural water-hydrated environment. (A) $^{1}H \rightarrow {}^{13}C$ CP-MAS NMR, and (B) ^{13}C DD-MAS NMR using a fast (1 s) repetition time. Peaks, which have a higher intensity without cross-polarization are in more mobile water-solvated regions, whereas peaks which have a higher intensity with cross-polarization are in more rigid environments in the β-sheet regions. Carbonyl spinning side bands are marked with a double asterisks. Reprinted from Ref. [16].

3.3. Conducting polymers

Advances in conducting and semiconducting materials have led to enhancements in the performance of organic thin transistors, organic photovoltaics, and other devices. High molar mass poly(3-hexylthiophene) (P3HT), an organic semiconductor, has been studied by solid-state ^{13}C CP-MAS NMR. Information at the subnanometer length scale can be obtained via solid-state NMR.

Recently, a modified approach to this technique has facilitated a correlation between peak pattern observed in CP-MAS spectra and the degree of crystallinity [17]. Measurements of order in semicrystalline, high molar mass poly(3-hexylthiophene) (P3HT) were made via solid-state ^{13}C CP-MAS. The relative degrees of order were compared between two films under different drying conditions: one slow-dried and one fast-dried. Ordered and disordered fractions within the polymer were separated using a $T_{1\varrho}$ filtered CP-MAS experiment. The spectrum for the ordered P3HT component is then obtained by spectral subtraction, as shown in **Figure 6**. The peak pattern of the crystalline component for slow-dried P3HT is narrower than that observed for the fast-dried sample. NMR does not measure long-range order, but instead is sensitive to order on a subnanometer scale. Line shape analysis shows that chains in noncrystalline regions

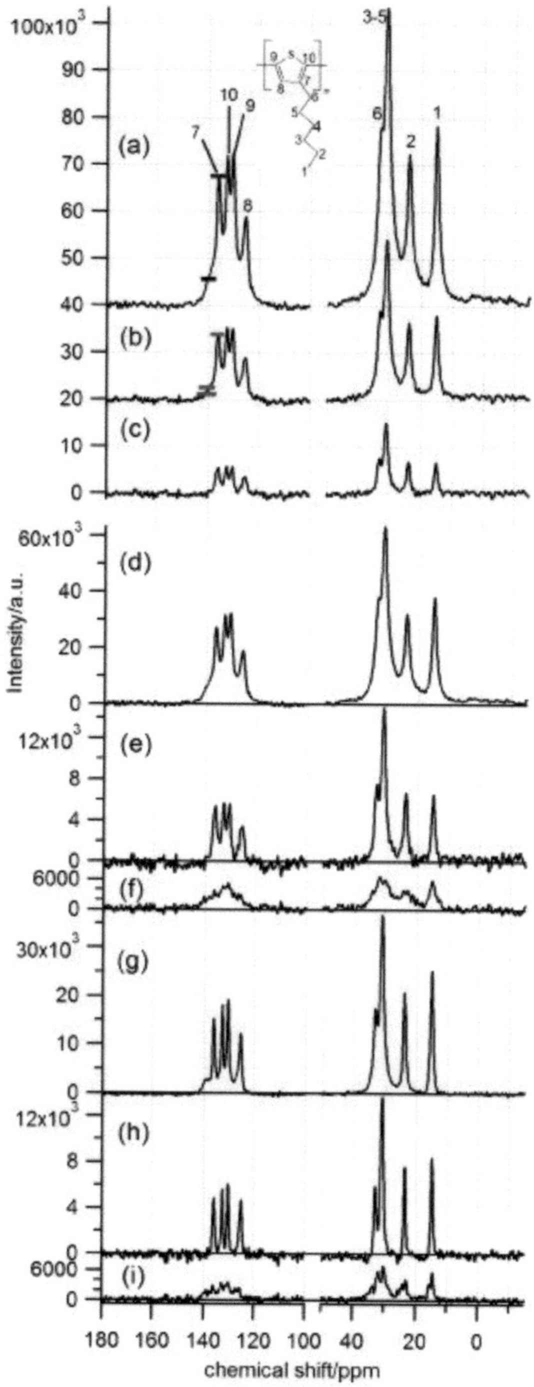

Figure 6. Solid-state ^{13}C CP-MAS spectra of the fast-dried organic semiconductor poly(3-hexylthiophene) (P3HT) (a) without and (b) with a $T_{1\rho H}$ spectral editing (spin-lock) pulse prior to cross-polarization, as well as the difference spectrum: (c) = (b) – (0.35 × (a)). The (d) CP-MAS spectrum for fast-dried P3HT, and (e) its ordered and (f) disordered fractions. The (g) CP-MAS spectrum for slow-dried P3HT, and (h) its ordered and (i) disordered fractions. Reprinted from Ref. [17].

can exhibit uniform local packing, which is presumed to be the result of uniform molecular conformations. Ultimately, observation of a narrower peak pattern of the crystalline component for slow-dried than for fast-dried is interpreted to mean that the quality of order is different, and that P3HT may be classified as a conformationally disordered crystal.

The solid-state NMR technique provides information about the conducting polymer system, which is close to the natural state and natural local conformation and packing arrangement. As such, information from solid-state NMR experiments may be regarded as favored to information derived by other techniques such as wide-angle X-ray scattering and differential scanning calorimetry.

4. NMR relaxation

The magnetic properties of a nucleus are modulated by the magnetic fields of neighboring nuclei. As molecular motions occur in the local environment changes resulting in altered interactions between nuclear spins. The study of NMR relaxation parameters yields information about the molecular motions in polymer systems.

The NMR relaxation parameters, which are most commonly measured are T_1, T_2, and $T_{1\varrho}$. T_1 is the so-called spin-lattice relaxation time, T_2 is the spin-spin relaxation time, and $T_{1\varrho}$ is the spin-lattice relaxation time in the rotating frame. These various relaxation parameters are related to molecular motions on varied time scales, and the correlation time of a molecular motion is related to the scale or distance over which a molecular motion occurs. As such, each of the relaxation parameters T_1, T_2, and $T_{1\varrho}$ is characteristic of molecular motions involving different frequency ranges, which occur accordingly over different distance scales as shown in **Figure 7**. The relaxation parameter T_1 probes fast motions with frequencies in the MHz regime; such fast motions are small-scale short-range motions, which are typically internal to a molecule. The relaxation parameter $T_{1\varrho}$ probes slower motions with frequencies in the kHz regime; these relatively slower motions occur over a larger distance and correspond to long chain motions in polymers. The theoretical relationship between the ^{13}C NMR relaxation parameters T_1, T_2, and $T_{1\varrho}$ in the solid state and correlation time τ_c are shown in **Figure 8** [18].

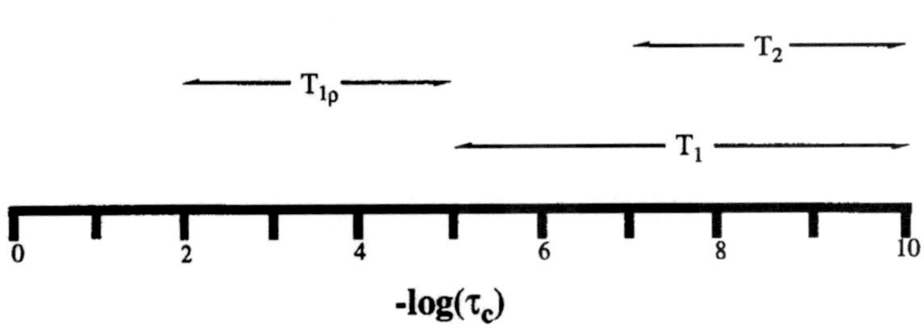

Figure 7. The solid-state ^{13}C NMR relaxation parameters T_1, T_2 and $T_{1\varrho}$ (in seconds) and the corresponding correlation times τ_c (in seconds) for the motions which they are sensitive to. Each of the relaxation parameters T_1, T_2 and $T_{1\varrho}$ are characteristic of molecular motions in different frequency ranges which occur over different distance scales.

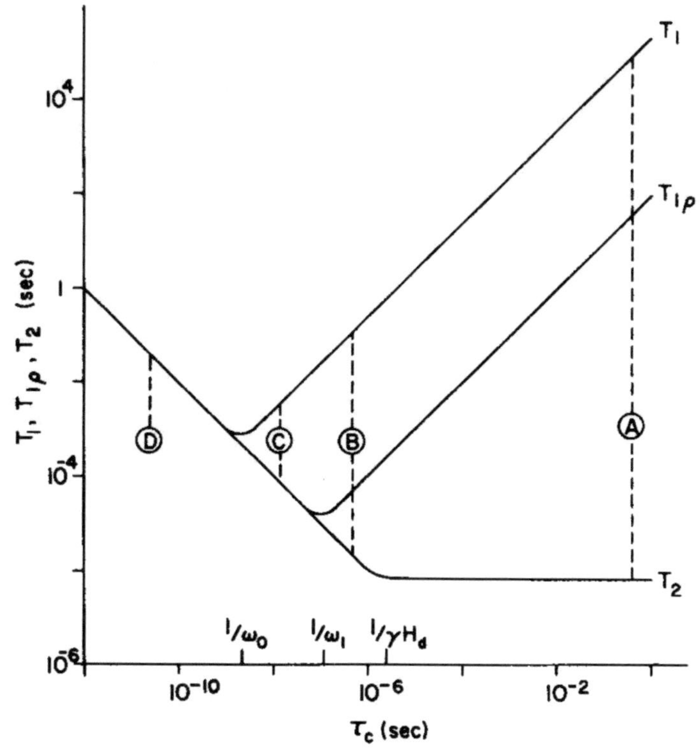

Figure 8. Theoretical dependence of the relaxation times T_1, T_2, and $T_{1\varrho}$ on the correlation time τ_c of the molecular motions responsible for the relaxation, as predicted by molecular motions that result in changes in dipole-dipole interactions. Region A is characteristic of a rigid lattice, region B of a nonrigid solid, region C of a viscous liquid, and region D of a nonviscous liquid. Reprinted from Ref. [18].

Semi-crystalline polymers are composed of amorphous (noncrystalline) bulk material, which contains crystalline domains. As NMR relaxation is modulated not only by physical-chemical structure, but also by molecular motions, the study of NMR relaxation yield insight into phases and phase structure in such multiphase polymer systems. The different domains present in such polymer materials display differences in molecular mobility, with the motion in crystalline domains being more restricted. Nuclei in amorphous domains show a higher mobility than those of nuclei in more rigid crystalline domains. Insight into polymer phase structure can be gained by the measurement of the NMR relaxation parameters T_1, T_2, and $T_{1\varrho}$ [19].

There are numerous examples of how such NMR relaxation studies have been exploited to yield valuable information about the phase structure in polymer blends. In the interest of blending thermoplastics with biodegradable polymers, the degree of crystallinity of microbial poly(ε-L-lysine) has been estimated via ^{13}C T_1 relaxation experiments [20]. In another study, the domain size of poly(ε-L-lysine) blended with poly(vinyl isobutyl ether) was estimated via ^1H spin-lattice relaxation experiments, and further miscibility of blends prepared under various processing conditions was explained in terms of crystallinity observations from the relaxation experiments [21]. In a similar study, phase separation in starch/polycaprolactone blends was investigated and the length scale over which phase separation occurs was determined via ^{13}C T_1, ^1H T_1, and ^1H $T_{1\varrho}$

relaxation experiments [22]. Further, the effect of the concentration of poly(ethylene glycol) in poly(ethylene glycol)/silica blends has been investigated via ^{13}C T_1 and ^1H $T_{1\varrho}$ relaxation experiments yield information about inhomogeneous separated phases [23]. As a final example, blends of poly(ethylene terephthalate) toughened by natural rubber have been studied by ^{13}C chemical shift and ^1H $T_{1\varrho}$ relaxation experiments, yielding evidence of interactions between the carbonyl groups of the poly(ethylene terephthalate) with some functionality in the natural rubber [24].

Likewise, there are numerous examples of how such NMR relaxation studies have been exploited to yield practical and useful information about copolymer systems. For example, poly(styrene-butadiene-styrene) has been found to compatibilize otherwise incompatible blends of polystyrene/polybutadiene, and ^1H T_1, ^1H T_2 and ^1H $T_{1\varrho}$ relaxation experiments have been used to explain this observation in terms of preferential localization of the copolymers at the polystyrene/polybutadiene interface [25]. Likewise, the content and length of soft/hard segments and microphase-separated morphology of poly(ether-block-amide) copolymers have been elucidated by and ^{13}C T_1, ^{13}C $T_{1\varrho}$, and ^1H $T_{1\varrho}$ relaxation experiments [26]. As a final example, segmented copolymers of poly(N-isopropyl acryl amide) and poly(tetrahydrofuran) have been studied by ^1H T_1 and ^1H $T_{1\varrho}$ relaxation experiments, to monitor the multiphase characteristics of the segmented copolymer networks as the polymerizable end group of the copolymer was varied [27].

Further, there are numerous examples of how NMR relaxation studies have been exploited to yield practical and useful information about polymer composites. For example, a ^1H T_1 and ^1H $T_{1\varrho}$ relaxation study of poly(p-phenylene benzobisoxazole) fibers demonstrated the observation of crystalline and noncrystalline regions, and further allowed determination of the crystal size [28]. As another example, the dispersion of organomodified clay fillers in nanocomposites with various thermoplastics was investigated by ^1H T_1 relaxation. The comparison in **Figure 9** of the ^1H T_1 saturation-recovery curve for polyamide 6 with that of a nanocomposite of polyamide 6

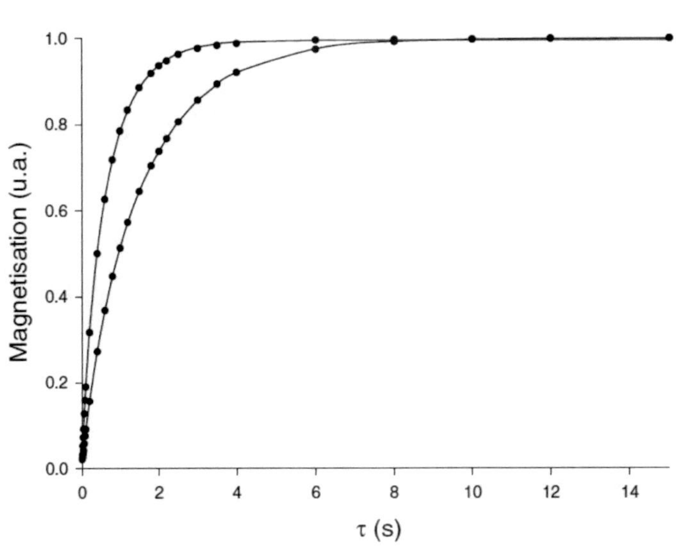

Figure 9. Solid-state ^1H T_1 saturation-recovery curves for polyamide 6 and a nanocomposite of polyamide 6 with an organomodified clay. The presence of clay shortens the relaxation time, indicating nanodispersion of the clay in the composite. Reprinted from Ref. [29].

with an organomodified clay, shows that the presence of clay shortens the relaxation time. The nanocomposites have a shorter relaxation time due to paramagnetically induced relaxation at the polymer-clay interface, indicating nanodispersion of the clay. Through a detailed and systematic study thereby, quantitative measurements of the degree of nanodispersion have been made [29].

Finally, in an example involving the combination of copolymers and composites, cotton fibers for composites were modified by surface copolymerization of ethyl acrylate followed by styrene. The graft copolymer-encapsulated cotton fibers were studied by ^{1}H T_1 relaxation experiments, yielding the observation of a heterogeneous morphology of the grafted skin [30].

5. Concluding remarks

Solid-state NMR continues to advance as a valuable technique for characterizing the molecular structure, conformation, and dynamics of polymers. There is already a rich history of applications of various solid-state NMR experiments to numerous polymer systems. Not only are advances occurring in the development of new solid-state NMR techniques, but also techniques which are now considered traditional are finding application to newly developed modern polymer materials. Ultimately, the application of solid-state NMR to polymers provides for the interpretation of spectroscopic observations as related to the structural features and physical properties of the polymer in the solid state, the state in which they are predominately used in as materials.

Author details

Antonio Martínez-Richa[1]* and Regan L. Silvestri[2]

*Address all correspondence to: richa@ugto.mx

1 Universidad de Guanajuato, Guanajuato, Mexico

2 Lorain County Community College, Elyria, USA

References

[1] Koenig JL. Spectroscopy of Polymers. 2nd ed. New York: Elsevier; 1999. 491 p

[2] Saalwachter K, Spiess HW. Solid-state NMR of polymers. In: Moeller M, Matyjaszewski K, editors. Polymer Science: A Comprehensive Reference, Volume 2. Amsterdam: Elsevier; 2012. pp. 185-219. DOI: 10.1016/B978-0-444-53349-4.00025-X

[3] Randall JC. Characterization of long-chain branching in polyethylenes using high-field carbon-13 NMR. In: Woodward AE, Bovey FA, editors. Polymer Characterization by ESR and NMR. Washington DC: American Chemical Society; 1980. pp. 93-118

[4] Sarles LR, Cotts RM. Double nuclear magnetic resonance and the dipole interaction in solids. Physical Review. 1958;**111**(3):853-859. DOI: 10.1103/PhysRev.111.853

[5] Andrew ER, Bradbury A, Eades RG. Nuclear magnetic resonance spectra from a crystal rotated at high speed. Nature. 1958;**182**:1659. DOI: 10.1038/1821659a0

[6] Hartmann SR, Hahn EL. Double nuclear resonance in the rotating frame. Physical Review. 1962;**128**(5):2042-2053. DOI: 10.1103/PhysRev.128.2042

[7] Fyfe CA. Solid State NMR for Chemists. Ontario: CFC Press; 1983. 593 p

[8] Mirau PA. A Practical Guide to Understanding the NMR of Polymers. New York: Wiley; 2004. 418 p

[9] Silvestri RL, Koenig JL. Applications of nuclear magnetic resonance spectrometry to solid polymers. Analytica Chimica Acta. 1993;**283**(3):997-1005. DOI: 10.1016/0003-2670(93)80261-I

[10] Ernst RR, Bodenhausen G, Wokaun A. Principles of Nuclear Magnetic Resonance in One and Two Dimensions. Oxford: Clarendon Press; 1987. 640 p

[11] Schmidt-Rohr K, Clauss J, Spiess HW. Correlation of structure, mobility, and morphological information in heterogeneous polymer materials by two-dimensional wideline-separation NMR spectroscopy. Macromolecules. 1992;**25**(12):3273-3277. DOI: 10.1021/ma00038a037

[12] Cao X, Pignatello JJ, Li Y, Lattao C, Chappell MA, Chen N, Miller LF, Mao J. Characterization of wood chars produced at different temperatures using advanced solid-state 13C NMR spectroscopic techniques. Energy Fuels. 2012;**26**(9):5983-5991. DOI: 10.1021/ef300947s

[13] Santonia I, Calloneb E, Sandaka A, Sandaka J, Dirè S. Solid state NMR and IR characterization of wood polymer structure in relation to tree provenance. Carbohydrate Polymers. 2015;**117**:710-721. DOI: 10.1016/j.carbpol.2014.10.057

[14] Takahashi H, Viverge B, Lee D, Rannou P, DePaepe G. Towards structure determination of self-assembled peptides using dynamic nuclear polarization enhanced solid-state NMR spectroscopy. Angewandte Chemie International Edition. 2013;**52**(27):6979-6982. DOI: 10.1002/anie.201210093

[15] Leonard SR, Cormier AR, Pang X, Zimmerman MI, Zhou HX, Paravastu AK. Solid-state NMR evidence for b-hairpin structure within MAX8 designer peptide nanofibers. Biophysical Journal. 2013;**105**(1):222-230. DOI: 10.1016/j.bpj.2013.05.047

[16] Addison JB, Ashton NN, Weber WS, Stewart RJ, Holland GP, Yarger JL. β-sheet nanocrystalline domains formed from phosphorylated serine-rich motifs in caddisfly larval silk: A solid state NMR and XRD study. Biomacromolecules. 2013;**14**(4):1140-1148. DOI: 10.1021/bm400019d

[17] Nieuwendaal RC, Snyder CR, DeLongchamp DM. Measuring order in regioregular poly(3-hexylthiophene) with solid-state 13C CPMAS NMR. ACS Macro Letters. 2014;**3**(2):130-135. DOI: 10.1021/mz4005343

[18] Farrar TC, Becker ED. Pulse and Fourier Transform NMR. New York: Academic Press; 1971. 118 p

[19] Martinez-Richa A, Silvestri RL. Solid-state NMR spectroscopy of multiphase polymer systems. In: Boudenne A, Ibos L, Candau Y, Thomas S, editors. Handbook of Multiphase Polymer Systems. 1st ed. New York: John Wiley and Sons; 2011. pp. 519-549. DOI: 10.1002/9781119972020.ch13

[20] Maeda S, Kunimoto K-K, Sasaki C, Kuwae A, Hanai K. Characterization of microbial poly (ε-L-lysine) by FT-IR, Raman and solid state 13C NMR spectroscopies. Journal of Molecular Structure. 2003;**655**(1):149-155. DOI: 10.1016/S0022-2860(03)00218-7

[21] Asano A, Murata Y, Kurotsu T. Crystallinity and miscibility of poly(vinyl isobutyl ether)/poly(ε-L-lysine) blends by solid state 13C NMR study. e-Journal of Soft Materials. 2007;**3**:1-8. DOI: 10.2324/ejsm.3.1

[22] Spevacek J, Brus J, Divers T, Grohens Y. Solid-state NMR study of biodegradable starch/polycaprolactone blends. European Polymer Journal. 2007;**43**(5):1866-1875. DOI: 10.1016/j.eurpolymj.2007.02.021

[23] Chen W, Feng H, He D, Ye C. High resolution solid-state NMR and DSC study of poly(ethylene glycol)-silicate hybrid materials via sol-gel process. Journal of Applied Polymer Science. 1998;**67**(1):139-147.DOI:10.1002/(SICI)1097-4628(19980103)67:1<139::AID-APP16>3.0.CO;2-X

[24] Phinyocheep P, Saelao J, Buzare JY. Mechanical properties, morphology and molecular characteristics of poly(ethylene terephthalate) toughened by natural rubber. Polymer. 2007;**48**(19):5702-5712. DOI: 10.1016/j.polymer.2007.07.016

[25] Joseph S, Laupretre F, Negrell C, Thomas S. Compatibilising action of random and triblock copolymers of poly(styrene-butadiene) in polystyrene/polybutadiene blends: A study by electron microscopy, solid-state NMR spectroscopy and mechanical measurements. Polymer. 2005;**46**(22):9385-9395. DOI: 10.1016/j.polymer.2005.07.053

[26] Hucher C, Eustache R-P, Beaume F, Tekely P. Motional heterogeneity in poly(ether-block-amine) copolymers as revealed by solid-state NMR. Macromolecules. 2005;**38**(22):9200-9209. DOI: 10.1021/ma051625q

[27] Lequieu W, Van De Velde P, Du Prez FE, Adriaensens P, Storme L, Gelan J. Solid-state NMR study of segmented polymer networks: fine-tuning of phase morphology via their molecular design. Polymer. 2004;**45**(23):7943-7951. DOI: 10.1016/j.polymer.2004.09.024

[28] Bourbigot S, Flambard X, Revel B. Characterisation of Poly(p-phenylenebenzobisoxazole) fibres by solid state NMR. European Polymer Journal. 2002;**38**(8):1645-1651. DOI: 10.1016/S0014-3057(02)00049-6

[29] Samyn F, Bourbigot S, Jama C, Bellayer S, Nazare S, Hull R, Castrovinci A, Fina A, Camino G. Crossed characterization of polymer-layered silicate (PLS) nanocomposite morphology: TEM, X-ray diffraction, rheology and solid-state nuclear magnetic resonance measurements. European Polymer Journal. 2008;**44**(6):1642-1653. DOI: 10.1016/j.eurpolymj.2008.03.021

[30] Castelvetro V, Geppi M, Giaiacopi S, Mollica G. Cotton fibers encapsulated with homo- and block copolymers: Synthesis by the atom transfer radical polymerization grafting-from technique and solid-state NMR dynamic investigations. Biomacromolecules. 2007;**8**(2):498-508. DOI: 10.1021/bm060602w

Chapter 3

Dynamics of Model Membranes by NMR

Anna Timoszyk

Additional information is available at the end of the chapter

http://dx.doi.org/10.5772/intechopen.69866

Abstract

Amphiphilic molecules can create various aggregates in water. Concern about exploring such structures has been unabated for several decades due to the wide range of possible applications of lipid aggregates, from food technology to the pharmaceutical industry. The form of self-assembled structures depends on many factors, such as the type of amphiphilic molecule, the concentration, the level of hydration, the temperature, and the pH. Liposomes and micelles are the most widely known types of closed structures. Liposomes are more often used in the fields of medicine and pharmacy because they consist of nontoxic compounds and their composition and size can be controlled. Nuclear magnetic resonance (NMR) is one of the methods, which is most commonly used to study liposome properties. It can be used to observe changes in the structure, dynamics, and phase transition of lipid membranes. The membrane properties are changed under the influence of external factors, such as temperature, pH, and the presence of ions or drugs. The chapter aims to introduce and discuss the possibilities of the most useful NMR methods, ^{31}P and ^{1}H, to study the liposome properties. It also aims to show how various changes in the structure or dynamics of lipid molecules are visible in the NMR spectra.

Keywords: lipid aggregates, model membrane, liposome, dynamics, splitting, half-width of signal, ^{1}H NMR, ^{31}P NMR

1. Introduction

The phenomenon of creating various types of aggregates by lipids in water is of particular interest to professionals in the fields of biophysics, biochemistry, medicine, and pharmacy. The reason for this broad interest is the similarity between formed aggregates and subcellular structures, such as lysosomes and biological membranes. Thus, the structures, particularly those such as bilayers, have been used as models of biological membranes for many years.

The main advantage of this model is its ability to decide the composition, both in terms of lipid and protein content and the environment in which they are located (one in which various types of ions are present). Model membranes enable the study of their thermotropic properties, the transport of ions through them, and the phenomenon of vesicle fusion. Currently, model biological membranes (liposomes) are widely used as drug delivery systems and in various kinds of therapies.

The use of nuclear magnetic resonance (NMR) to study these structures poses several challenges. For instance, the technique uses a method of sample preparation that differs from the standard methods used for measurement in a liquid. An additional problem is the formation of lipid aggregates in water, which exhibits differ in the NMR time scale. This is due to the fact that NMR spectrometer "sees" the hydrated surface of the lipid aggregate as a substance similar to liquid, and the hydrophobic core of the aggregate as something similar to a solid. In fact, biological membranes are in the liquid crystalline phase (L_α); therefore, model membranes are usually studied in this phase. This shows that, in the case of the dispersion of lipids in water, other parameters typical for this type of structure will have a significant impact on the appearance of the NMR spectrum. The main features of lipid aggregates affecting the NMR spectrum include the type of lipid aggregate, its size, the degree of lipid hydration, and the thermodynamic state of the membrane. These four parameters determine the dynamics of lipid molecules and individual chemical groups trapped in a complex structure.

1.1. Amphiphilic molecules self-assembly and critical micelle concentration

Amphiphilic molecules (e.g., phospholipids) have lipophilic parts (hydrophobic) and polar parts (hydrophilic). In aqueous environments, these kinds of molecules undergo two basic effects [1]: the adsorption of water molecules on the surface of lipids and self-association. The result of both effects is that when dispersed in water, amphiphiles spontaneously aggregate. It is precisely this property of phospholipids that makes them the basic material from which the cell membrane is formed. Lipids, due to their ability to self-assemble, are divided by the properties of their polar headgroups; thus, the characteristics of the lipids' headgroups (i.e., non-ionic, zwitterionic, anionic, cationic, and catatonic) are emphasized, while the hydrophobic parts of lipids differ in their number of hydrocarbon chains, in their length (number of carbons in the chain), and in their degree of saturation [1–3]. Therefore, the self-assembly phenomenon occurs as the result of two opposite forces: first, that connected with the hydrophobic effect (the energetically unfavorable contact between fatty acid chains and water) and second, that connected to hydrophilic interactions with water molecules [1]. The result of these effects is the formation of micelles, bilayers, or other aggregates, because only the hydrophilic headgroups of lipids can be exposed to water.

Micelles are the simplest structures that can form amphiphilic lipids in water. The Gibbs phase rule says that at a certain temperature and under certain pressure, lipid molecules and micelles can be in equilibrium only at a fixed value of the mole fraction of hydrocarbon in water [4]. This value of lipid concentration in water is known as critical micelle concentration (CMC). Below CMC value, lipid molecules exist as monomers dispersed in water, whereas above CMC value lipid molecules tend to self-associate, forming micelles, bilayers, or other aggregates. The CMC, which is greater for charged molecules than for uncharged molecules [1], decreases significantly in conjunction with decreases in the length of the fatty acid chains [5]. Lipids with relatively weak headgroups (i.e., with weak opposing forces) form bilayer-like aggregates, such as vesicles, and disc-like micelles [6].

Formed aggregates are usually large in size, mainly because they must take forms in which the fatty acid chains are not exposed to direct contact with water. Moreover, the type of lipid structure depends on the energetic of the lipid-water interface and on the shape of the lipid molecule [2, 3, 6]. Further consideration of this issue must take into account factors that influence the formation of the free energy of micellization and the micelle size. The formation of the free energy of micellization and its dependence on the aggregate size involves the bulk term, surface term, curvature term, and packing term [2–4]. The major driving force in the formation of aggregates is its hydrophobic effect and the contribution to the bulk term; however, it is not associated with the size of the formation, which depends on the free energy of micellization [4]. The hydrophobic interaction between fatty acid chains exposed to water and the different repulsive interactions between headgroups (steric, electrostatic, and hydration) contribute to the surface term. The repulsive interactions increase the surface area, whereas the hydrophobic interactions decrease the surface area [4, 7]; these effects are known as opposite interactions. The molecular conformation and motional properties of polar headgroups and the formed membrane surface, which identify the lipid type, are well known. However, information about the structure of lipid molecules packed at aggregates, especially if more than one kind of lipid is present, is very limited. The study of lipid-lipid interactions at the membrane polar-apolar interface is important because the membrane surface is the most probable site of selective electrostatic or steric associations [2–4]. The existing opposite forces have an influence on the curvature of formed agglomerates and restrict the packing of lipid molecules, contribute to the curvature and packing terms, and result in an optimum aggregate size [4]. All terms involved in the free energy of micellization determine favorable molecular packing, which is directly connected to the formed favorable aggregate structures of a specific type of lipid. Moreover, in the bilayer-like structures, the lipid molecules manifest an asymmetric transmembrane lipid-packing geometry [2, 3]. It follows, then, that the average area per lipid headgroup and the effective length of the lipid molecule are greater in the membrane's outer layer than in its inner layer [4]. Therefore, the packing and curvature terms are closely connected to the lipids' molecular shape and configuration. As a consequence, the micelles of single-chain lipids may be formed favorably as a result of weak packing restrictions, whereas those of double-chain lipids, due to stronger packing restrictions, are favorably formed bilayer-like structures [2].

1.2. Lipid hydration

Amphiphilic lipid self-assembly is a specific equilibrium between hydrophobic and hydrophilic interactions, but the bilayer hydration is determined mainly by interactions between the hydrophilic headgroups and the solvent [2, 3]. The level of hydration affects the self-assembly, curvature, shape and size of the aggregate, and the phase behavior. The hydration of the lipid aggregate depends on the specifics of the lipids (i.e., their headgroups). The hydration of hydrocarbon chains is much smaller and is restricted by the hydrophobic interactions [6]. The hydration process is also connected to the thermodynamic state of the lipid membrane. When the fluidity of fatty acid chains increases, the lipid molecules occupy a larger area, which increases hydration as a result of the increase of exposure of the headgroups and hydrocarbon chains to the water molecules [1]. The properties of interbilayer water differ from those of free bulk water. Thus, each lipid molecule, with its water of hydration, should be treated as a separate thermodynamic and physicochemical entity [3, 4]. In general, it seems that the steric density fluctuations have only a slight influence on hydration parameters; however, they play an extremely important role in the surface hydration [1].

Of particular interest in this area is the use of lipids which build, in nature, biological membranes. The most common amphiphilic glycerophospholipids contained in their polar headgroups include phosphate, carboxyl, carbonyl, and choline residues, all of which take part in creating the hydrogen bonds [1, 2]. The availability of the headgroups for hydrogen bonding with the water molecules is the most important factor in the hydration of the bilayer. The coulombic charges of the lipid molecules participate less in the hydration process, probably because of an insufficient concentration of water molecules [1, 3].

1.2.1. Influence the hydration on lipid dynamics

The motion of phospholipid molecules within the lipid bilayer has been characterized as lateral diffusion, axis rotational Brownian motion of the headgroup, or glycerol backbone, wobbling, and flip-flopping [7–9]. The collective properties are different from those at the local molecular level. The phenomena accompanying the local molecular motion in the lipid aggregate include phase transitions, a morphological change of the lipid membrane as a whole (e.g., fusion/fission), pore formation, and the formation of heterogeneous structures, such as phase separation/domain formation [10]. Thus, the motion of the lipid aggregate as a whole cannot be explained on the basis of the lipid molecule motions (i.e., at the molecular level).

Thus, studies of membrane dynamics are concerned with the molecular motion of lipids. As mentioned previously, the rotational motion of the headgroups relates to hydrophobic interactions, hydrogen bonding, and the curvature of the lipid membrane; in this way, an increase in the curvature of, for example, liposomes, induce the level of hydration [11–16]. The reorientation of the headgroups is also restricted by the intermolecular force between them [1, 19]. Thus, the reorientation of phospholipid headgroups is restricted by breaking the intermolecular bonds (hydrogen bond and/or $PO_4^-/N^+(CH_3)_3$ bond) [10, 16–18]. Thus, the hydration of the polar headgroups weakens the strength of the $PO_4^-/N^+(CH_3)_3$ bond [10, 16, 19]. The dynamics of the lipid membrane interface is also connected with the mobility of the glycerol backbone of phospholipid ($PO_4^-–(CH_2)_2–N^+(CH_3)_3$) [7, 16, 20]. The mobility of the glycerol backbone is associated with the lateral diffusion of lipid molecules because it promotes hydration and, consequently, the reorientation of the headgroups. The reorientation of the headgroups (i.e., high hydration) causes the hydrocarbon chains to be more greatly exposed to water molecules, which indicates that the increased mobility of the phospholipid headgroups make the membrane polar-apolar surface more hydrophobic [6, 10, 16, 19].

1.3. Model membrane phase transitions

Membranes composed of one type of phospholipid have a clearly defined phase transition, which is caused by temperature variations. The phase transition temperature is primarily dependent upon the type of phospholipid (number and length of hydrocarbon chains) and the level of lipid molecule hydration. The most frequently described phase of self-assembling aggregates is the liquid crystalline phase as a characteristic of cell membranes. Therefore, the most commonly studied phases are lamellar and nonlamellar, such as the hexagonal and cubic phases (normal and inversed).

The typical phase for lipid membranes at low temperatures is the lamellar crystalline phase L_C. As the temperature increases, the van der Waals' interactions decrease, which maintains the order of the hydrocarbon chains in the crystalline phase. The rotational motion of the hydrocarbon chains is then activated by the temperature. The phase transition between crystalline L_C and gel L_β phase occurs at T_s temperature [4]. In the lamellar L_β phase, the lipid molecules take up a larger area and are more hydrated than in the L_C phase [21]. The correlation time of the rotational motion of the acyl chains is about 10^{-5} s, which is about 100 times slower than the isomerization of the hydrocarbon chains [22]. Further increases in temperature cause a rise in the quasilamellar segments within the membrane in the intermediate periodic phase P_β. In this phase, the surface of the membrane is usually rippled, which occurs at T_p temperature [23]. After the intermediate phase, above T_p temperature, the hydrocarbon chains start to melt and form *trans-gauche* isomers. Internal reorientations, which have stochastic characteristics, are transferred along the acyl chains at times ranging from microseconds to milliseconds [24]. The presence of *trans-gauche* isomers determines an increase of distance between the lipid molecules in the membrane and decrease van der Waals' interactions. The increase of movement in the hydrophobic and hydrophilic parts of the lipid bilayers is characteristic of the melting state (i.e., the liquid crystalline phase) L_α [25]. The temperature of transition from the intermediate pleated phase to the liquid crystalline phase is called the melting temperature or the temperature of main phase transition T_t. The value of T_t thus depends on the length and level of the saturation of the hydrocarbon chains as well as on the level of hydration of the lipid molecules. The most important factor in the determination of this value is a hydration level between 0 and 30%. Hydration levels in this range have a major impact on decreases in the temperature T_t. However, because a 30% hydration level mainly increases the amount of water molecules not associated with the membrane (bulk water), changes of the hydration level between 30 and 50% are less likely to influence decreases in the temperature T_t [4].

When CMC values are extremely high, which results in low volumes of water, some kinds of lipids form nonlamellar phases, such as hexagonal (H_I) and cubic (C_I). Inversed hexagonal (H_{II}) and cubic (C_{II}) phases may be formed in trace amounts of water. Not all lipids can form these phases: their ability to form hexagonal and cubic phases depends on the stereochemical structure of the molecules. Lysophospholipids tend to form normal nonlamellar phases, whereas phosphatidylethanolamine (PE), cholesterol (Ch), cardiolipin (CL), and phosphatidic acid (PA) tend to form inversed nonlamellar phases [26].

However, preferred aggregation structures depend not only on the type of lipid, temperature, and hydration level but also on the pH. Under neutral pH conditions, the phosphatidylcholine (PC) and PE headgroups are electrically neutral, whereas the phosphatidylserine (PS), phosphatidylglycerol (PG), and phosphatidylinositol (PI) headgroups have net negative charges. The mixture of the lipids in the membrane transmits a surface charge density that has an effect on the membrane permeability to ions and charged molecules, on the membrane protein function, and on the thermodynamic phase of the membrane [2–4]. For example, in the case of PE bilayers, which interact with fewer water molecules, the main gel-to-liquid crystalline phase transition (L_β/L_α) temperature increases by as much as 30°C compared to their counterpart PC bilayers (about 20–25°C) [4].

2. NMR spectra of model membranes

Amphiphilic lipids form aggregates of many different shapes and sizes; these aggregates can be at different phases. Moreover, minor changes in the concentration, temperature, or chemical structure of the lipid molecules may induce phase transitions between states. Additionally, the effects of molecular interactions and dynamics on macroscopic properties are evident in self-assembly systems [27]. NMR studies of self-assembly systems therefore begin by observing the dynamic parameters, which results in a better understanding of the static properties of the system [1]. Certainly, NMR is the most powerful technique with which to quantify the molecular dynamic in solution; however, in the case of lipid aggregates, it has some limitations.

NMR relaxation studies provide information about local dynamics and the conformational state of lipid hydrocarbon chains. This method is used to study aggregate properties (e.g., the size of micelles) [1]. The reorientation dynamics of aggregated lipid molecules is characterized by a locally preferred orientation; that is, lipid molecules undergo rapid internal motions, such as *trans-gauche* isomerizations, which are slightly anisotropic. NMR spectra from the lipids in micelles and bilayers are generally in the motional narrowing regime, which is caused by a time scale of lipid reorientation of 10^{-9} s or less [1–3, 7]. Thus, at conventional magnetic field strengths, essential information about the lipid aggregates is stored in the transverse relaxation rate T_2.

Polar headgroups and hydrocarbon chains (typically 12–16 carbons) can be studied using ^1H, ^{13}C, and ^{31}P NMR (e.g., dipole-dipole interactions, scalar interactions, and chemical shift anisotropy (CSA)) [27]. The ^{31}P has relatively large CSA. A complication revealed during the analysis is a change in the degree of the solvent's protonation, which is caused by direct dependency of CSA from the pH. In fact, the relaxation time T_2 is also sensitive to changes in micelle/liposome size because the rotational correlation time is proportional to the cube of the radius [1, 27]. Electrostatic and hydrodynamic intermolecular interactions are independent on the rotational diffusion of lipid molecules.

2.1. NMR time scale

The motional model of aggregated lipid molecules concerns the time scale separation of fast local and slow overall motion. The reorientational motion of lipid molecules divides into [4]

-fast local motion, which is slightly anisotropic, and

-slow isotropic motion (i.e., aggregate tumbling and the lateral diffusion of lipid molecules within the membrane surface).

These two motions occur on different time scales. Therefore, the special density can be written as the sum of the fast and slow components [28] as follows:

$$j(\omega_0) = (1 - S^2) j_f(\omega_0) + S^2 j_s(\omega_0) \tag{1}$$

where $j_f(\omega_0)$ and $j_s(\omega_0)$ are reduced spectral density functions that describe the fast and slow motions, respectively; ω_0 is the resonance frequency; and S is the order parameter.

The order parameter can be described as the average [28]

$$S = \frac{1}{2}\left(3\cos^2\theta - 1_f\right) \tag{2}$$

where θ is the angle between the axis of the maximum component of the electric field gradient tensor and the director axis.

For spherical aggregates such as micelles and liposomes, the slow motion (tumbling and lateral diffusion) is described by the Lorentzian spectral density function [1]:

$$j_{s(\omega_0)} = \frac{2\tau_s}{1 + (\omega_0 \tau_s)^2} \tag{3}$$

where τ_s is the correlation time, and the correlation function is [1]

$$g_s = g_t g_d = e^{\frac{-t}{\tau_t}} e^{\frac{-t}{\tau_d}} = e^{\frac{-t}{\tau_s}}, \tag{4}$$

where subscripts t and d correspond to the tumbling and lateral diffusion motions, respectively.

The correlation time of tumbling and lateral diffusion of the sphere of radius R can be written as [1]

$$\tau_t = \frac{4\pi\eta R^3}{3k_B T} \wedge \tau_d = \frac{R^2}{6D} \tag{5}$$

where D is the lateral diffusion coefficient.

Taking into consideration the above equations, it is possible to write the relations for relaxation times [28] as follows:

$$T_1 = \frac{3\pi^2}{40}\chi^2 \left[(1 - S^2) 20\tau_f + S^2 \left(\frac{4\tau_s}{1 + (\omega_0\tau_s)^2} + \frac{16\tau_s}{1 + (2\omega_0\tau_s)^2}\right)\right] \tag{6}$$

and

$$T_2 = \frac{3\pi^2}{40}\chi^2 \left[(1 - S^2) 20\tau_f + S^2 \left(6\tau_s + \frac{10\tau_s}{1 + (\omega_0\tau_s)^2} + \frac{4\tau_s}{1 + (2\omega_0\tau_s)^2}\right)\right] \tag{7}$$

where subscripts s and f correspond to slow and fast motions, respectively; χ is the gyromagnetic ratio.

2.2. Preparation of the NMR sample

Liposomes are most often used in NMR studies as models of biological membranes. Liposomes, spherical structures consisting of one (large LUV or small SUV unilamellar vesicle) or more (multilamellar vesicles (MLV)) lipid bilayers, are divided by the number of bilayers as well as by size (**Figure 1**).

The classical preparation MLV method consists of hydrating the thin lipid film. Suitable amounts and types of lipids are dissolved in organic solvent (e.g., chloroform) and are predried under a stream of dry nitrogen. After the formation of a thin lipid film, the sample is allowed to continue drying in a vacuum evaporator for 1–12 h, after which it is hydrated [16]. In the preparation of the NMR sample, deuterated solvents (deuterated water) must be used. An appropriate amount of water should be added to the sample to obtain a final lipid concentration of not less than 20 mg/ml. The sample is then gently mixed, often in a water bath, at a temperature close to the main phase transition. After removing the thin lipid film from the glass walls, it is vortexed for 5–7 min. After completion of the procedure, a sample containing MLV is obtained [16]. When are need LUV or SUV, other methods (most commonly ultrasound disintegration and extrusion) are used. The sonication is carried out in an ice-water bath for 15–45 min, depending on the unit capacity and the expected size of the liposomes [16]. In the case of extrusion, special filters with a proper pore size and pressurized MLV are forced through the pores, thus depriving them of the unwanted bilayers. Combined methods are frequently used (e.g., sonication combined with extrusion). Extrusion is then used to calibrate the liposomes (i.e., to reduce deviations in the size distribution) [29]. Sonication is used more frequently than other methods due to the procedure's lower cost.

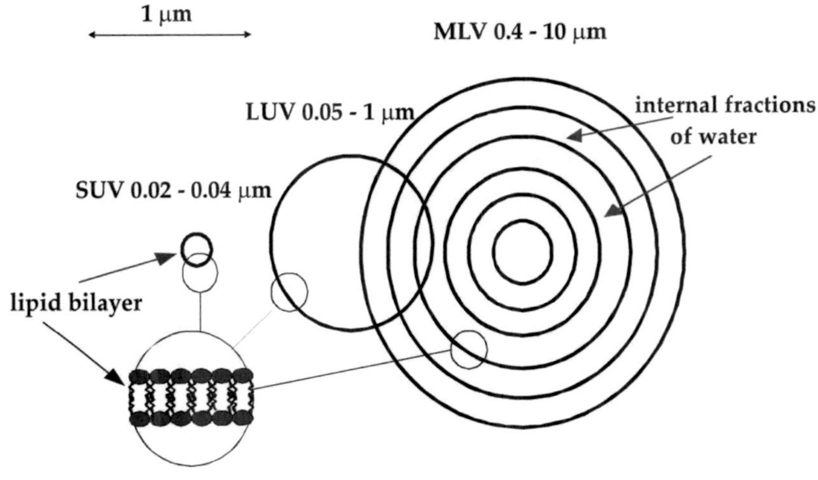

Figure 1. Types of the liposomes.

2.3. Lipid membrane dynamic studies

The ^{31}P NMR technique provides information about the local order, the mobility of the phosphate part of the lipid head, and the overall structure of the lipid aggregate [30]. The lineshape of ^{31}P NMR spectra is closely related with the CSA tensor and to the orientation of the lipid molecules relative to the applied magnetic field [31]. Thus, the CSA $\Delta\sigma$ depends on the phosphate group motion and on the temperature. The ^{31}P NMR spectrum shows the characteristic lamellar lineshape ($\sigma_\perp$—high-field maximum) and low-field shoulder ($\sigma_\parallel$). $\Delta\sigma$ can be calculated from the following expression [30]:

$$\Delta\sigma = 3\left(\sigma_\parallel + \sigma_\perp\right), \tag{8}$$

where $\sigma_\parallel$ and $\sigma_\perp$ are the values of ^{31}P shielding of the lipid molecules, oriented parallel or perpendicular relative to the magnetic field.

The spectral second moment is a measure of the shape of the ^{31}P signal related to the various motions of the headgroup. The second moment S_2 can be calculated from the following formula [30]:

$$S_2 = \frac{\int_{-\infty}^{+\infty} \nu^2 f(\nu) d\nu}{\int_{-\infty}^{+\infty} f(\nu) d\nu}, \tag{9}$$

where ν is frequency and $f(\nu)$ is the lineshape.

The CSA for lipid molecules ordered in water is about 100 ppm; for lipid aggregates, it decreases to about 50 ppm due to lateral diffusion of the lipid molecules, leading to further averaging; for liposomes larger than 200 nm, the CSA is reduced to 40 ppm [31]. Thus, the ^{31}P lineshape depends on the size of the lipid aggregate. The CSA value decreases as the liposome size decreases, that is, the liposome curvature increases. In the case of small aggregates (e.g., SUV), the CSA can be reduced to 10 ppm.

Analysis methods of ^{31}P NMR spectra most often are used to test functionalized liposomes that are used as drug carriers, the best known being PEGylated liposomes. The polyethylene glycol (PEG) caps the liposome, which gives it a longer circulation time in the blood, meaning that it decreases the uptake of the liposomes by the reticulum endothelium system (RES) and allows the drug to be released gradually [32]. In addition, PEG is biocompatible, which enables the possibility of further functionalization of the liposomes by attaching antibodies or ligands [33, 34]. The ^{31}P spectra analysis of PEGylated liposomes obtained using the thin film method revealed information about the amount of free lipids or building micelles (narrow line) and lipids building MLV (broad shoulder) [31, 35]. It is also possible to obtain more than one narrow signal (**Figure 2**).

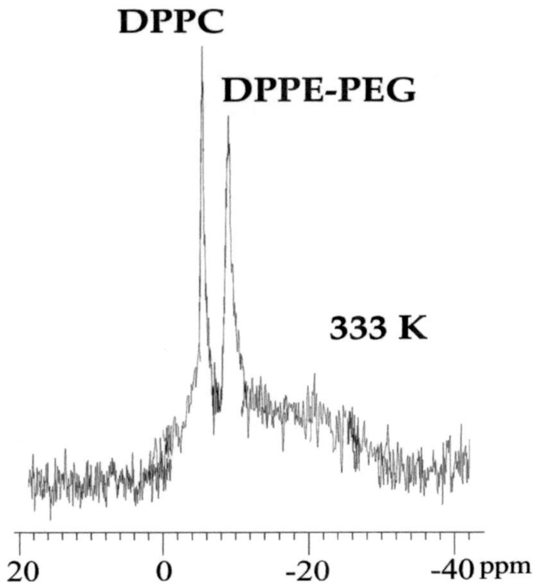

Figure 2. The ^{31}P NMR spectrum of the PEGylated DPPE/DPPC/Ch (phosphatidylethanolamine/dipalmitoylphosphocholine/cholesterol) multilamellar vesicles obtained at 333 K.

The number of narrow peaks depends on the number of phospholipid types used in the experiment. Then, each phosphate group is in a different chemical surrounding, which causes different chemical shifts. Splitting (δ) between narrow signals also depends on motional averaged CSA rather than on an isotropic chemical shift [31]. Moreover, it may affect the value of the chemical shift and cause difficulties in the assignment of peaks. As mentioned previously, the lineshape depends on the orientation and motion of the lipid molecules and all aggregates. Thus, the temperature of measure is very important. Additionally, in the case of lipid aggregates, a change in temperature causes a change in the phase/physical state. Temperature changes have a slight impact on narrow signals but a significant impact on the broad shoulder [31]. The ^{31}P NMR spectrum of MLV in the low-field shoulder is extremely broad below the temperature during the major phase transition.

The ^{31}P NMR studies of the influence of drugs on the organization and fluidity of a liposome membrane as a function of temperature have been previously reported [30]. The presence of the antibiotic azithromycin affects the thermotropic behavior of DOPC (1,2-dioleoyl-*sn*-glycero-3-phosphocholine) and DPPC (1,2-dipalmitoyl-*sn*-glycero-3-phosphocholine) membranes. Moreover, temperatures between 35 and 45°C affect the ^{31}P lineshape of DOPC and DPPC liposomes, as follows: the position of the narrow peak and CSA values does not change; however, the lineshape in a part of the broad shoulder does. The monitoring of the lineshape at the same range of temperatures of DPPC or DOPC liposomes with azithromycin revealed the following new information: the presence of the antibiotic causes smoothing of the line and a decrease in CSA. In fact, only above 40°C did the narrow line stay within the spectrum, and the CSA value averaged zero. The azithromycin molecules increased the membrane fluidity below the main phase transition temperature [30].

The example of hydrophobic molecules contained in the lipid membrane and its influence on the dynamics of the lipid bilayer were presented in Ref. [36]. β-carotene, a well-known hydrocarbon carotenoid, appeared in photosynthetic membranes. The presence of β-carotene in a DPPC liposome membrane fulfilled opposite roles in different membrane states. The ^{31}P spectra of β-carotene/DPPC MLV as a function of β-carotene concentration and temperature showed changes in the ^{31}P resonance lineshape [36]. In both cases, the CSA decreased and the spectral line smoothed. Thus, at temperatures above those in the main phase transition for DPPC, the β-carotene caused a fluidization effect on the membrane (in the liquid crystalline state). The effect is also connected to a decreased temperature in the main phase transition. However, at temperatures below those in the main phase transition, β-carotene rigidified the membrane (in the fluid state) [36]. This effect manifested itself as a broadening of the ^{31}P signal. The similar effect can be observed in the case of the PC/octadecylamine MLV contained lycopene (**Figure 3**).

In the ^{1}H NMR spectra of LUV/SUV, the most useful parameter for analyzing the spectrum is the half-width ($\Delta\nu_{1/2}$) of the signals, because the $\Delta\nu_{1/2}$ of the resonance signals is directly connected to the motional freedom of particular chemical groups. The increase of the $\Delta\nu_{1/2}$ of

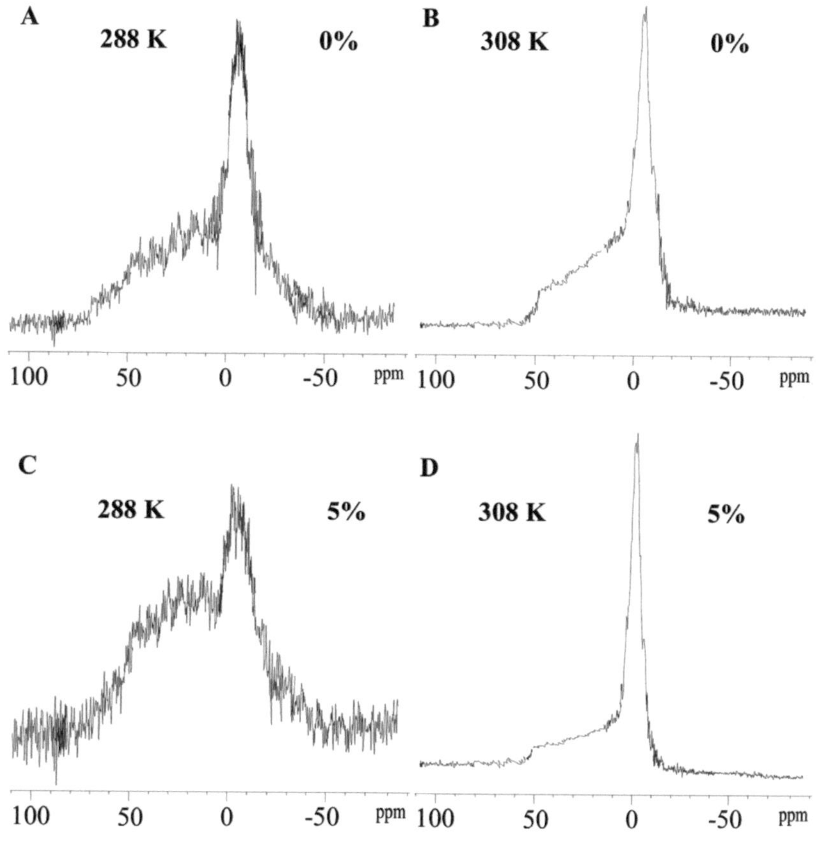

Figure 3. The ^{31}P NMR spectra of the positively charged PC/octadecylamine MLVs (containing 5 mol% of octadecylamine) obtained at 288 K (A and C) and at 308 K (B and D) with 5% of the lycopene (C and D).

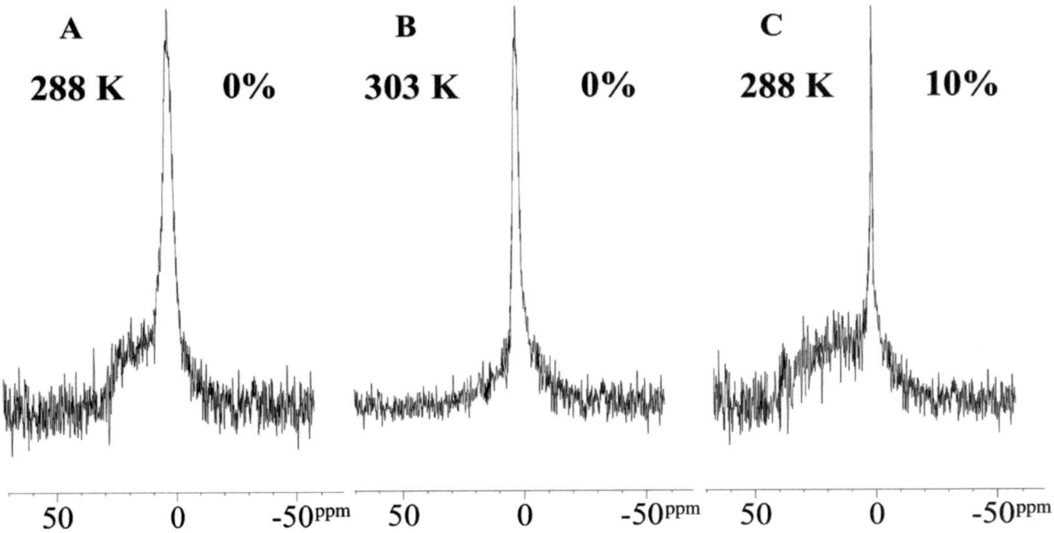

Figure 4. The ^{31}P NMR spectra of the PC MLVs (A). The temperature (B) and polysialic acid (C) effect of narrowing the ^{31}P NMR spectra of PC MLVs.

the ^{1}H resonance signal indicates a restriction of the motion. A change in the $\Delta\nu_{1/2}$ signal assigned to a choline group from the outer layer indicates that the studied drug interacts only with the liposome surface; in other words, it does not penetrate the hydrophobic core of the membrane. This effect was observed in the case of amphotericin B [37], which rigidified the hydrophilic surface of the PC membrane; this effect increased the fluidity of the hydrophobic part of the lipid bilayer. It is manifested as a decrease in the $\Delta\nu_{1/2}$ signals assigned to protons from the $-(CH_2)_n$ and $-CH_3$ groups of hydrocarbon chains [37].

Quite the opposite effect was observed in the case of polysialic acid (**Figure 4**). The ^{31}P NMR spectra of PC MLVs showed a slight narrowing of the isotropic part and a broadening of the anisotropic part of the resonance line at a temperature range of 10–30°C [38].

However, when the polysialic acid was involved in several cell processes in the external environment of the liposome, the effect was even more noticeable. The interaction between well-hydrated and anionic polysialic acid and the polar headgroups of PC increased the fluidity of the hydrophilic part (decrease of $\Delta\nu_{1/2}$) and simultaneously rigidified the hydrophobic core (increase broad shoulder) of the membrane [38].

While the soy isoflavone, genistein, reduces the hydration level of the phosphate groups (hydrophilic part), i.e. decreases its mobility, and rigidified the hydrophobic part of the asolectine liposomes [35]. It also was found that isoflavone prevents lipid molecules from peroxidation [35].

2.4. Phase transitions of lipid membrane studies

Phase transition studies via the use of NMR are based on the fact that residual couplings depend on the extension of the anisotropic domains in combination with the rate of molecular diffusion [39]. The isotropic phases are perfectly visible in ^{1}H and ^{13}C NMR spectra, but anisotropic phases from liquid crystals are much more difficult to observe. The broadening of ^{1}H and ^{13}C spectra can provide some information about non-isotropic phases but do not bring detailed information about phase transition. In this case, the ^{31}P NMR technique is the one that is most useful. If the phase is isotropic on the NMR time scale, static dipolar, quadrupolar, and shift anisotropy interactions are averaged to zero by molecular motion. When the phase is anisotropic, however, a static interaction effect should be seen on the spectrum [39].

Some of the first and most comprehensive studies on the physical properties of liposomes have been presented in Ref. [40]. The authors analyzed the changes of the ^{31}P spectra of DPPC and DPPE MLV membranes with the addition of a different concentration of piracetam. An additional narrow peak appeared on the spectrum assigned to piracetam. The intensity of the signal was dependent on the piracetam-to-lipid molar ratio [40]. Temperature had a significant influence on the lipid bilayer's physical properties. Fortunately, the possible effect of temperature on the main phase transition of the lipid membrane caused by the drug could be observed in the ^{31}P spectra. The presence of piracetam decreased CSA and increased the mobility of the lipid polar headgroup, manifested as a narrowing ^{31}P line and suggesting that piracetam molecules are inserted between lipid molecules at the hydrophilic part of the membrane [40]. Thus, the presence of hydrophilic piracetam molecules decreased the temperature of the main phase transition. In the H_{II} phase, which can be the transient form in the fusion process, the piracetam resonance signal was no longer distinguishable. Finally, it can be concluded that piracetam combines with lipid molecules, which is exhibited as one signal due to the isotropic motion of the entire complex [40].

The ^{31}P method, therefore, also may be used to fix the temperature of the MLV phase transition [41–44]. The temperature studies of PC/Ch MLV revealed the changes of lineshape ranging from 10 to 40°C (**Figure 5**). Temperature changes could be observed during the phase transition between L_{α} and H_{II} [41].

The monitoring of the ^{31}P lineshape every 2°C led to observations of intermediate lineshapes between those characteristic of L_{α} and H_{II} and allowed us to precisely fix the temperature of the phase transition [41].

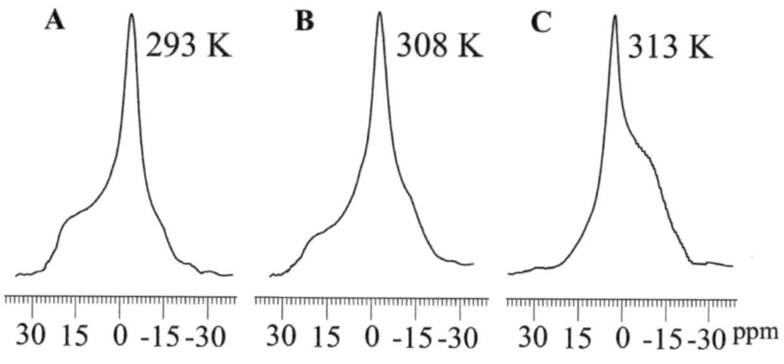

Figure 5. The ^{31}P NMR spectra of PC/Ch MLVs obtained in different temperatures. The characteristic ^{31}P lineshape for lamellar phase at 293 K (A), transient form at 308 K (B) and nonlamellar phase (inversed hexagonal) at 313 K (C).

2.5. Usage of paramagnetic ions as a chemical shift reagent and transport of ions through lipid bilayer

Paramagnetic ions are used to distinguish signals. The most frequently used are praseodymium (Pr^{3+}) ions or other ions from the lanthanides group [37]. The concentration of paramagnetic ions added to the external environment of the liposomes is important. The shift reagent could distinguish the signals within a few ppms; however, if the concentration is too great, it could drastically broaden the studied signal or even broaden the entire spectrum. In this case, the paramagnetic effect is dominant, and paramagnetic interactions may destroy the membrane structure [45]. It may be observed in the ^1H NMR spectra of DOPC SUV as broadened signals due to Eu^{3+} ions. In fact, this effect is associated with the properties of Eu^{3+} ions, which interact to the same extent with the hydrophobic and hydrophilic parts of the lipid bilayer. Moreover, the $\Delta\nu_{1/2}$ of the signal assigned to water also increased, indicating that the Eu^{3+} ions also interact with water molecules from the hydration shell of the liposome [45].

The Pr^{3+} ions enabled the distinguishing of signals assigned to the polar headgroups of lipids from the inner and outer layers of liposomes in the ^1H spectra. The splitting of choline signals of PC/Ch SUVs is showed in **Figure 6**. The signal corresponding to the protons from the $-N^+(CH_3)_3$ groups split due to interaction with paramagnetic ions in the external environment of LUV/SUV [46].

The signal shifted toward lower magnetic field values was assigned to protons from the outer layer, and the signal shifted to higher magnetic field values was assigned to protons from the inner layer [37]. The splitting of the $-N^+(CH_3)_3$ signal is strongly dependent on geometric

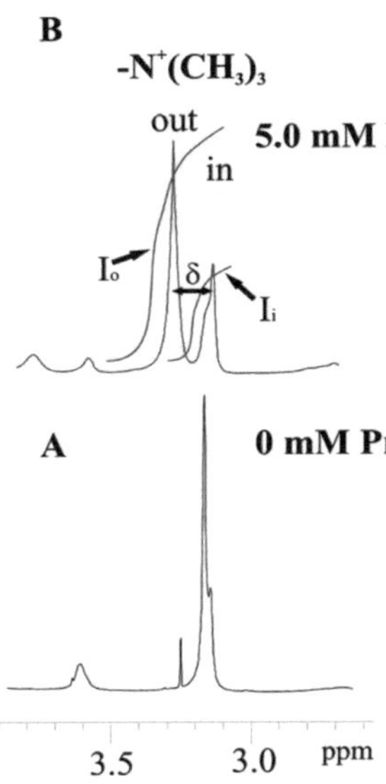

Figure 6. The ^1H NMR spectrum of choline groups of PC/Ch SUVs in the absence (A) and presence (B) of Pr^{3+} ions (5 mM).

conditions and axial symmetry at the lipid-lanthanide binding site [37, 46]. The ratio of the area under the signal corresponded to the outer layer and the area under the signal corresponded to the inner layer (I_o/I_i), thereby providing information about the size of the liposome [46]. Along with decreases in liposome size, the number of lipid molecules in the inner layer decreased. Thus, the areas under the signals from the inner and outer layers differed. The distinguishing of signals from the choline groups in the inner and outer layers of the membrane presents new possibilities for interpreting the results. For example, the addition of the amphotericin B to the external environment of the PC SUV liposome (after hydration of the lipid bilayer) did not have an effect on the size of the liposomes because there was no change in the I_o/I_i ratio. However, the Pr^{3+} ions could interact with the carboxylic group of amphotericin B due to increased splitting of the choline group signals [37].

The ^{31}P NMR spectra can reveal information about the transport through membrane or about ion competing binding sites in lipids and in other biomolecules. This is very important in the assay of biological membranes, since ion transport defects may cause various illnesses, such as manic depressive and neurodegenerative diseases [47]. The adenosine triphosphate (ATP) could be used as a model membrane ligand of metal cations [47]. In these types of studies, it is very important to keep a constant pH and temperature during the experiment. It is well known that the binding of metal cations is dependent on both parameters. The narrowing/broadening of the ^{31}P signal or changes in the distance (splitting) between the three phosphate signals could be the effect of complexes created by ATP and Mg^{2+}/Li^{+} ions [47]. The analysis of ^{31}P spectra led to the conclusion that the biochemical action of Li^{+} ions may be explained as their ability to compete with Mg^{2+} binding sites. Thus, the therapeutic role of Li^{+} in maniac-depressive illness is enabled by modulating the activity of G proteins in signal transduction [47]. It has been suspected that the Mn^{2+} ions also can complete Mg^{2+} binding sites. This ability of Mn^{2+} ions probably pays a significant role in course of neurodegenerative illnesses [48]. Thus, the same effect can be observed in the case of Mn^{2+} ions (**Figure 7**).

The ^{1}H and ^{31}P NMR may also be used to study the macroscopic rearrangement of liposome membrane as, for example, a fusion process [41]. A number of authors have suggested that the fusion process is associated with the development of transient forms related to the appearance of the H_{II} phase [41, 49, 50]. To induce the fusion process of PE/PS/PC liposomes, Ca^{2+} ions should be used as a fusogenic agent [41]. The ^{31}P spectra of MLV showed changes in resonance lineshapes with an increased molar ratio of Ca^{2+} ions to PS (**Figure 8**).

The characteristic lineshape for the nonlamellar phase, H_{II}, was revealed when the molar ratio of Ca^{2+}/PS was 2.0 [41]. The ^{1}H and ^{31}P NMR spectra of SUV, after the addition of Pr^{3+} ions, demonstrated splitting of the resonance signals. In both cases, there was an overall decrease in splitting and even in the intensity of the split signals [41]. Only the Pr^{3+} ions are associated with the outer layer of the membrane. During the fusion process, because the lipid molecules translocation from the outer layer to the inner layer due to transient form (inversed micelle) formation, the Pr^{3+} ions with the translocated lipids moved to the internal water portion of the liposome [41]. This effect determined the decrease in the splitting. In fact, during the fusion process, the internal and external chemical environments of the polar heads became identical, and the size of the liposomes increased [41].

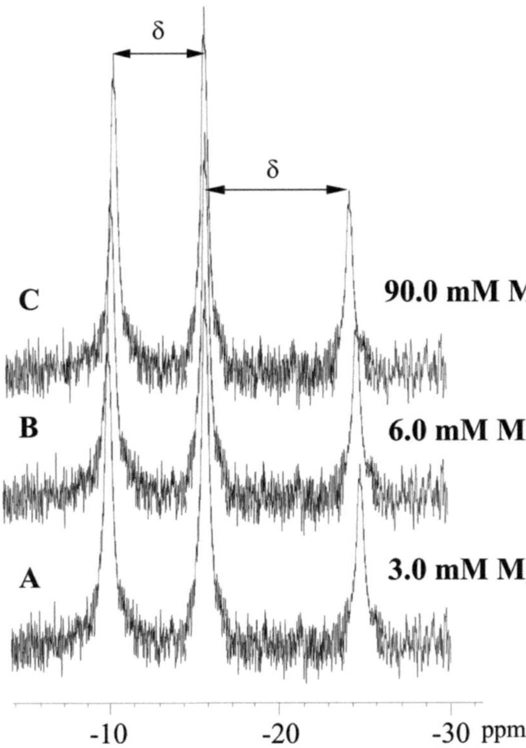

Figure 7. The ^{31}P NMR spectra of ATP in the presence of Mg^{2+} ions (A) and Mn^{2+} ions (B, C).

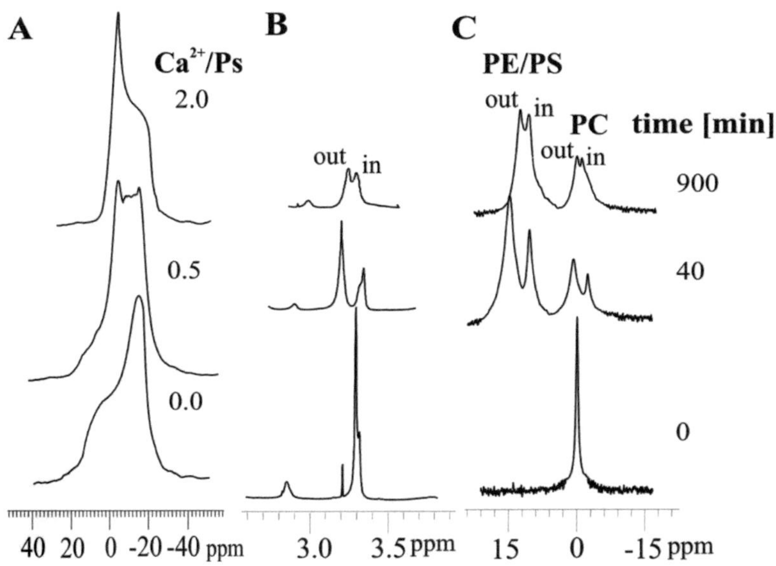

Figure 8. The changes of the ^{31}P resonance lineshape of PC/Ch MLVs caused by fusogenic reagent (Ca^{2+} ions) (A); the time changes of the ^{1}H NMR spectra (B) and the ^{31}P NMR spectra (C) of PE/PS/PC SUVs in the presence of Ca^{2+}/PS molar ratio 2.0 after addition of 0.5 mM Pr^{3+} ions.

2.6. Drug delivery study

The potential use of LUV/SUV as a drug carrier to cells is attractive as a therapy for increasing therapeutic effects and for reducing drug toxicity in normal cells. The concept of the application of temperature-sensitive liposomes is based on an increase in the permeability of the lipid bilayer at a proper temperature, due to rearrangement of the molecules from one stable state to another [51–53]. Temperature-sensitive liposomes may be used with local hyperthermia for the treatment of cancer via chemotherapy [51, 54, 55]. The release rate of the drug depends on the rate of change in temperature and by the serum compounds (i.e., lipoproteins) [56]. Thus, as drug carriers, the liposomes should be stable in the serum and release the drug slowly at a proper temperature [34, 57]. The ^{1}H spectra of PC and PC/octadecylamine (positively charged) LUV showed changes of the chemical shift of signals with increased temperatures ranging from 5 to 50°C (**Figure 9**).

For both types of liposomes, the values of chemical shift increased; that is, the spectral lines shifted toward the direction of the lower magnetic field [51, 57]. This effect is typical for lipid membranes. The ^{1}H spectra also revealed the narrowing of spectral lines (decrease in $\Delta\nu_{1/2}$) assigned to $-N^+(CH_3)_3$, $-(CH_2)_n$, and $-CH_3$. Moreover, the largest changes were observed in the $-(CH_2)_n$ signal from the hydrophobic core of the membrane [51]. Studies of the splitting of signals from the choline groups have demonstrated decreased splitting with increased temperatures, the result of increased liposome size. The PC liposome size changes from 20–30 nm to 1 μm. In fact, between temperatures 35 and 40°C, the structure of the PC liposomes becomes damaged and unstable. On the other hand, the size of the PC/octadecylamine liposomes changes slightly from 20 to 60 nm and seems to be stable at temperatures of 40–50°C [51]. Additionally, temperature-sensitive liposomes at higher temperatures may aggregate or fuse [49]. It is possible for temperature-sensitive PC/octadecylamine liposomes to transfer drug to cells by fusion or via an endocytosis process in moderate hyperthermia [51].

The ^{1}H NMR can also be used to study the permeability of lipid membrane. As mentioned before, the controlled release of drugs is very important. The PA/Ch/PEG-Ch (palmitic acid/cholesterol/PEGylated cholesterol) liposomes exhibit no permeability of drugs (calcein and doxorubicin) up to 20 mol% PEGylated cholesterol concentration, but in 10 mol% PEG-Ch concentration the permeability of membrane limited and can be controlled [34]. Moreover, the PA/Ch/PEG-Ch liposomes are stable in various pHs [34].

The release of drugs (cytosine 1-β-D-arabinofuranoside and 5-fluorouracil) from the DPPC liposomes was studied [57]. The ^{1}H spectra showed the shifting signals and the changes in splitting of signals dependent from temperature. The temperature-dependent controlled release of 5-fluorouracil was successfully provided [57].

The ^{31}P NMR technique also can be used to study liposomes used as drug carriers, such as in the case of cisplatin-loaded PEGylated LUV. The ^{31}P NMR technique has been used to measure the chemical shift of placebo (control) liposomes and cisplatin-loaded liposomes at room temperature and at 60°C [58]. The results revealed a characteristically broadened signal at temperatures below those in L_β to L_α phase transition (52.5°C for PC). At a temperature of 60°C, sharp signals were obtained in both cases. The analysis of spectra revealed some asymmetry in peaks on the

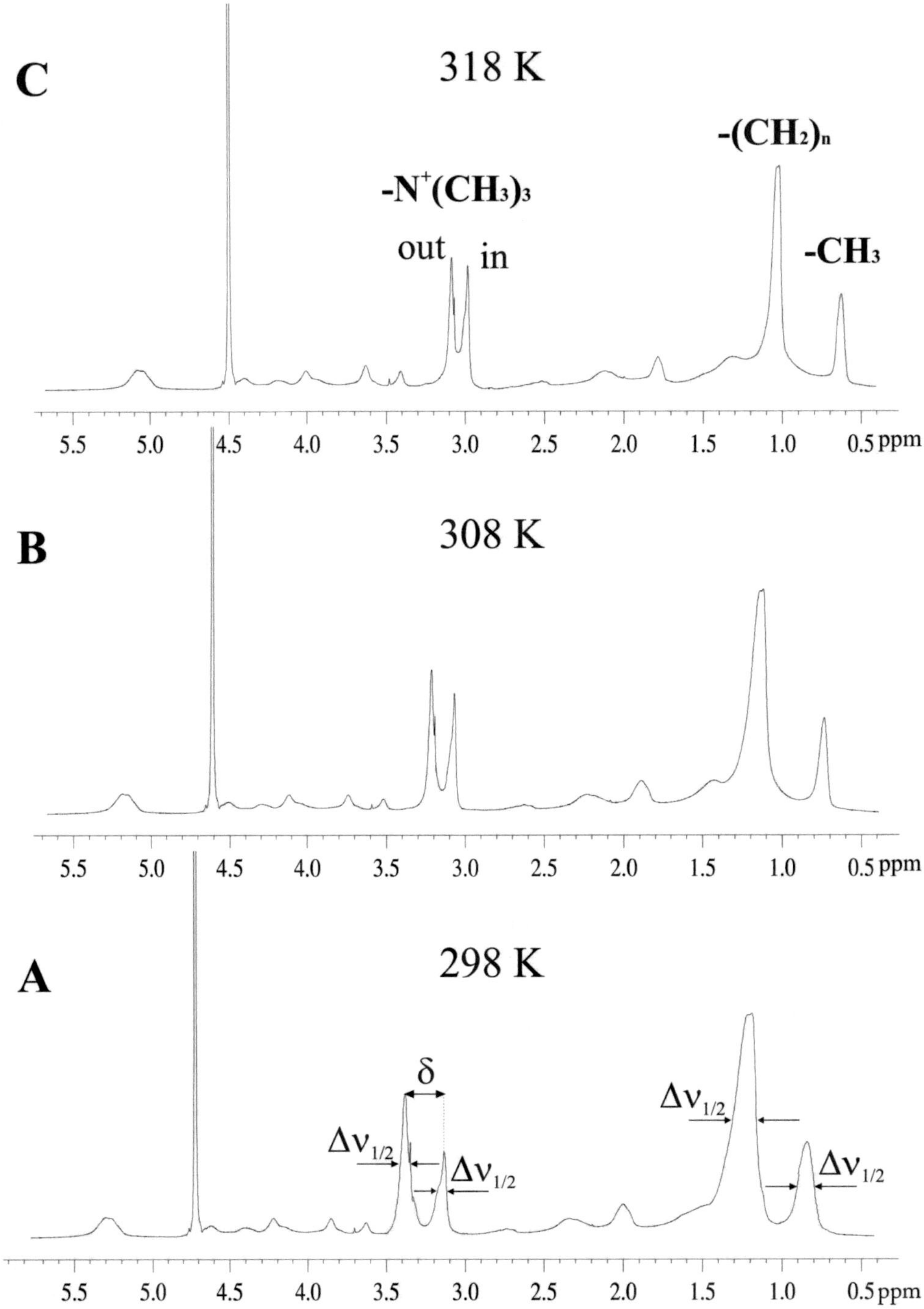

Figure 9. The effect of temperature on half-width and splitting of the ^{1}H resonance signals of PC/octadecylamine LUVs at 298 K (A), at 308 K (B), and at 318 K (C).

right side. This effect is expected in the case of phospholipid LUV. The monitoring of the $\Delta\nu_{1/2}$ signal can reveal information about interactions between cisplatin and phospholipid molecules. The $\Delta\nu_{1/2}$ signal has a lower value for cisplatin-loaded liposomes than that for control liposomes [58]. Because both types of LUV have the same lipid composition and concentration and similar size distributions, the implication is that differences in $\Delta\nu_{1/2}$ are caused by interactions between cisplatin and phospholipid molecules. This effect is probably a result of the hydration process [58].

3. Conclusions

The NMR techniques usually are used to determine the molecular structure but, in the case of lipid aggregates, are more important to know a nature of interactions between the molecules and their dynamics. Thus, the most often are used the ^{1}H and ^{31}P NMR techniques. The ^{1}H and ^{31}P spectra of liposomes led to observe the dynamics changes in the hydrophilic and hydrophobic part of membrane (half-width of signal). The changes of molecules/chemical groups dynamics can be caused by various substances added to liposome membrane, loaded to liposome, or coated a liposome. Moreover, the changes in splitting of signals can revealed the information about permeability of liposome membrane. These parameters are important to characterize the properties of liposome membrane. Additionally, the measurement can be provided in various physicochemical conditions. The pH, temperature, and concentration of added substances have significant influence on the physical state of membrane, the dynamics of molecules, the interactions between molecules, and the processes occurred on the membrane surface. Thus, the NMR technique is a proper tool to study the phase behavior, the transport of ions, the diffusion of drugs through the membrane, the membrane permeability, and the stability of membrane in various conditions.

Author details

Anna Timoszyk

Address all correspondence to: a.timoszyk@wnb.uz.zgora.pl

Faculty of Biological Sciences, University of Zielona Góra, Zielona Góra, Poland

References

[1] Lindman B, Olsson U, Söderman O. Surfactant solutions: Aggregation phenomena and microheterogeneity. In: Delpuech J-J, editor. Dynamics of Solutions and Fluid Mixtures by NMR. 1st ed. Chichester: Wiley; 1995. pp. 345-396. ISBN: 047195411X

[2] Janmey PA, Kinnunen PKJ. Biophysical properties of lipids and dynamic membranes. TRENDS in Cell Biology. 2006;**16**(10):538-546. DOI: 10.1016/j.tcb.2006.08.00

[3] van Meer G, Voelker DR, Feigenson GW. Membrane lipids: Where they are and how they behave. Nature Reviews. Molecular Cell Biology. 2008;**9**:112-124. DOI: 10.1038/nrm2330

[4] Cevc G, Marsh D. Phospholipid bilayers. Physical principles and models. In: Bittar EE, editor. Cell Biology: A Series of Monographs. Vol. 5. 2nd ed. New York: Wiley; 1987. pp. 1-442

[5] Zant-Przeworska E, Stasiuk M, Gubernator J, Kozubek A. Resorcinolic lipids improve the properties of sphingomyelin-cholesterol liposomes. Chemistry and Physics of Lipids. 2010;**163**(7):648-654. DOI: 10.1016/j.chemphyslip.2010.05.202

[6] Shillcock JC, Lipowsky R. Equilibrium structure and lateral stress distribution of amphiphilic bilayers from dissipative particle dynamics simulations. Journal of Chemical Physics. 2002;**117**(10):548-561. DOI: 10.1063/1.1498463

[7] Patel AJ, Varilly P, Jamadagni SN, Acharya H, Garde S, Chandler D. Extended surfaces modulate hydrophobic interactions of neighbouring solutes. PANS. 2011;**108**(43):17678-17683. DOI: 10.1073/pnas.1110703108/-/DCSupplemental

[8] Yeagle PL. The Structure of Biological Membranes. 3rd ed. New York: CRC Press; 2005. p. 398. ISBN: 9781439809570-CAT#K10479

[9] Kahya N, Schwille P. How phospholipid-cholesterol interactions modulate lipid lateral diffusion, as revealed by fluorescence correlation spectroscopy. Journal of Fluorescence. 2006;**16**:671-678. DOI: 10.1007/s10895-006-0108-6

[10] Marcotte I, Separovic F, Auger M, Gagné SM. A multidimensional ^{1}H NMR investigation of the conformation of methionine–enkephalin in fast-tumbling bicelles. Biophysical Journal. 2004;**86**:1587-1600. DOI: 10.1016/S0006-3495(04)74226-5

[11] Kaatze U, Goepel KD, Pottel R. Zwitterion motions of differently aggregated phospholipids in aqueous and methanolic solution. A dielectric relaxation study. Journal of Physical Chemistry. 1985;**89**(12):2265-2571. DOI: 10.1021/j100258a028

[12] Klösgen B, Reichle C, Kohlsmann S, Kramer KD. Dielectric spectroscopy as a sensor of membrane headgroup mobility and hydration. Biophysical Journal. 1996;**71**(6):3251-3260. DOI: 10.1016/S0006-3495(96)79518-8

[13] Smith G, Shekunov BY, Shen J, Duffy AP, Anwar J, Wakerly MG, Chakrabarti R. Dielectric analysis of phosphoryl choline headgroup mobility in egg lecithin liposomes. Pharmaceutical Research. 1996;**13**:1181-1185. DOI: 10.1023/A:1016004001678

[14] Sabín J, Prieto G, Ruso JM, Hidalgo-Álvarez R, Sarmiento F. Size and stability of liposome: A possible role of hydration and osmotic forces. European Physics Journal E. 2006;**20**:401-408. DOI: 10.1140/epje/i2006-10029-9

[15] Peng A, Pisal DS, Doty A, Balu-Iyer SV. Phosphatidylinositol induces fluid phase formation and packing detects in phosphatidylcholine model membranes. Chemistry and Physics of Lipids. 2012;**165**:15-22. DOI: 10.1016/j.chemphyslip.2011.10.002

[16] Timoszyk A, Latanowicz L. Interactions of sialic acid with phosphatidylcholine liposomes studied by 2D NMR spectroscopy. Acta Biochimica Polonica. 2013;**60**(4):539-546

[17] Phole W, Gauger DR, Bohl M, Mrazkova E, Hobza P. Lipid hydration: Headgroup CH moieties are involved in water binding. Biopolymers. 2004;**74**(1-2):27-31. DOI: 10.1002/bip.20037

[18] Shimanouchi T, Sasaki M, Hiroiwa A, Yashimoto N, Miyagawa K, Umakoshi H, Kuboi R. Relationship between the mobility of phosphocholine headgroups of liposomes and the hydrophobicity at membrane interface: A characterization with spectrophotometric measurements. Colloids and Surfaces B. 2011;**88**(1):221-230. DOI: 10.1016/j.colsurfb.2011.06.036

[19] Milhaud J. New insights into water-phospholipid model membrane interactions. Biochimica et Biophysica Acta. 2004;**1663**:19-51. DOI: 10.1016/j.bbamem.2004.02.003

[20] Walde P, Blöchliger E. Circular dichroic properties of phosphatidylcholine liposomes. Langmuir. 1997;**13**(6):1668-1671. DOI: 10.1021/la9610157

[21] Tenchov B, Koynova R, Rapp G. New ordered metastable phases between the gel and subgel phases in hydrated phospholipids. Biophysical Journal. 2001;**80**(4):1873-1890. DOI: 10.1016/S0006-3495(01)76157-7

[22] Boroske E, Trahms L. A ^{1}H and ^{13}C NMR study of motional changes of dipalmitoyl lecithin associated with pretransition. Biophysical Journal. 1983;**42**(3):275-283. DOI: 10.1016/S0006-3495(83)84395-1

[23] Satoh K. Determination of binding Ca^{2+}, Na^{+}, and Cl^{-} ions to liposomal membranes of dipalmitoylphosphatidylcholine at gel phase by particle electrophoresis. Biochimica et Biophysica Acta. 1995;**1239**(2):239-248. DOI: 10.1016/0005-2736(95)00154-U

[24] Stein AD, Peterson KA, Fayer MD. Dispersive excitation transport at elevated temperatures (50-298 K): Experiments and theory. Journal of Chemical Physics. 1990;**92**(9):5622-5635. DOI: 25 10.1063/1.458494

[25] Okubo T. Phase transition between liquid-like and crystal-like structures of deionized colloidal suspensions. Journal of Chemical Society, Faraday Transactions. 1990;**86**:2871-2876. DOI: 10.1039/FT9908602871

[26] Kaufmann K, Silman I. The induction by protons of ion channels through lipid bilayer membranes. Biophysical Chemistry. 1983;**18**(2):89-90. DOI: 10.1016/0301-4622(83)85002-9

[27] Santos S, Medronho B, dos Santos T, Antunes FE. Amphiphilic molecules in drug delivery. In: Coelho J, editor. Advanced Technologies Potentially Applicable in Personalised Treatment. Advances in Predictive, Preventive and Personalised Medicine. 4th ed. Dordrecht: Springer Science & Business Media Dordrecht; 2013. pp. 37-85. DOI: 10.1007/978-94-007-6010-3_2

[28] Lindman B, Olsson U. Structure of microemulsions studied by NMR. Berichte der Bunsengesellschaft für physikalische Chemie. 1996;**100**(3):344-363. DOI: 10.1002/bbpc.19961000321

[29] Berger N, Sachse A, Bender J, Schubert R, Brandl M. Filter extrusion of liposome using different devices: Comparison of liposome size, encapsulation efficiency, and process

characteristics. International Journal of Pharmaceutics. 2001;**223**:55-68. PII: S0378-5173 (01)00721-9

[30] Fa N, Ronkart S, Schanck A, Deleu M, Gaigneaux A, Goormaghtigh E, Mingeot-Leclercq M-P. Effect of the antibiotic azithromycin on thermotropic behavior of DOPC or DPPC bilayers. Chemistry and Physics of Lipids. 2006;**144**:108-116. DOI: 10.1016/j.chemphyslip.2006.08.002

[31] Leal C, Rögnvaldsson S, Fossheim S, Nilssen EA, Topgaard D. Dynamic and structural aspects of PEGylated liposomes monitored by NMR. Journal of Colloid and Interface Science. 2008;**325**:485-493. DOI: 10.1016/j.jcis.2008.05.051

[32] Stark B, Pabst G, Prassl R. Long-term stability of sterically stabilized liposomes by freezing and freeze-drying: Effects of cryoprotectants on structure. European Journal of Pharmaceutical Sciences. 2010;**41**(3-4):546-555. DOI: 10.1016/j.ejps.2010.08:010

[33] Huynh GH, Deen DF, Szoka Jr. FC. Barriers to carrier mediated drug and gene delivery to brain tumors. Journal of Controlled Release. 2006;**110**(2):236-259. DOI: 10.1016/j.conrel. 2005.09.053

[34] Cui Z, Edwords K, Orellana AN, Bastiat G, Benoit J-P, Lafleur M. Impact of interfacial cholesterol-anchored polyethylene glycol on sterol-rich non-phospholipid liposomes. Journal of Colloid and Interface Science. 2014;**428**:111-120. DOI: 10.1016/j.jcis.2014.04.031

[35] de Azambuja CRL, dos Santos LG, Rodrigues MR, Rodrigues RFM, de Silveira EF, Azambuja JH, Flores AFC, Horn AP, Dora CL, Muccillo-Baisch AL, Branganhol E, da Silva Pinto L, Parize AL, de Lima VR. Physico-chemical characterization of asolectin–genistein liposomal system: An approach to analyze its *in vitro* antioxidant potential and effect in glioma cells viability. Chemistry and Physics of Lipids. 2015;**193**:24-35. DOI: 10.1016/j.chemphyslip.2015.10.001

[36] Jeżowska I, Wolak A, Gruszecki WI, Strzałka K. Effect of β-carotene on structural and dynamic properties of model phosphatidylcholine membranes. II. A ^{31}P-NMR and ^{13}C-NMR study. Biochimica et Biophysica Acta. 1994;**1194**:143-148. DOI: 10.1016/005-2736(94) 90213-5

[37] Gabrielska J, Gagoś M, Gubernator J, Gruszecki WI. Binding of antibiotic amphotericin B to lipid membranes: A ^{1}H NMR study. FEBS Letters. 2006;**580**:2677-2685. DOI: 10.1016/j. febslet.2006.04.021

[38] Timoszyk A, Gdaniec Z, Latanowicz L. The effect of polysialic acid on molecular dynamics of model membranes studied by ^{31}P NMR spectroscopy. Solid State Nuclear Magnetic Resonance. 2004;**25**:142-145. DOI: 10.1016/j.ssnmr.2003.03.023

[39] Lindberg G. Resialylation of sialic acid deficit vascular endothelium circulating cells and macromolecules may counteract the development of atherosclerosis: A hypothesis. Atherosclerosis. 2007;**192**(2):243-245. DOI: 10.1016/j.atherosclerosis.2007.03.011

[40] Peuvot J, Schanck A, Deleers M, Brasseur R. Piracetam-induced changes to membrane physical properties. Biochemical Pharmacology. 1995;**50**(8):1129-1134

[41] Osajca A, Timoszyk A. Application of ^{1}H and ^{31}P NMR of topological description of a model of biological membrane fusion. Acta Biochimica Polonica. 2012;**59**(2):219-224

[42] Verkleij AJ, van Echteld CJA, Gerritsen WJ, Cullis PR, de Kruijff B. The lipidic particle as an intermediate structure in membrane fusion process and bilayer to hexagonal (H_{II}) transitions. Biochimica et Biophysica Acta. 1980;**600**(3):620-624. DOI: 10.1016/0005-2736(80)90465-4

[43] Siegel DP. Inverted micellar intermediates and the transition between lamellar, cubic, and inverted hexagonal lipid phases. I. Mechanism of $L_{\alpha} \rightarrow H_{II}$ phase transitions. Biophysical Journal. 1986;**49**(6):1155-1170. DOI: 10.1016/S0006-3495(86)83744-4

[44] Hermandez JM, Stein A, Behrmann E, Riedel D, Cypionka A, Farsi Z, Walla PJ, Raunser S, Jahn R. Membrane fusion intermediates via directional and full assembly of the SNARE complex. Science. 2012;**336**(6088):1581-1584. DOI: 10.1126/science.1221976

[45] Schultze KD, Sprinz H. Electrochemical and NMR spectroscopic investigations of the influence of the probe molecule $Eu(fod)_3$ on permeability of lipid membranes to ions. Biochimica et Biophysica Acta. 2000;**1467**(1):27-38. DOI: 10.1016/S0005-2736(00)00193-0

[46] Hunt GRA, Tipping LRH. A ^{1}H NMR study of the effects of metal ions, cholesterol and *n*-alkanes on phase transitions in the inner and outer monolayers of phospholipid vesicular membranes. Biochimica et Biophysica Acta. 1978;**507**(2):242-261. DOI: 10.1016/0005-2736(78)90420-0

[47] Amari L, Layden B, Rong Q, Geraldes FGC, de Freitas DM. Comparison of fluorescence, ^{31}P and ^{7}Li NMR spectroscopic methods for investigating Li^{+}/Mg^{2+} competition for molecules. Analytical Biochemistry. 1999;**272**:1-7. URL: http://hdl.handle.net/10316/3898

[48] Levin J, Bertach U, Kretzschmar H, Gise A. Single practical analysis of manganese-induced prion protein aggregates. Biochemical and Biophysical Research Communications. 2005;**329**:1200-1207

[49] Verkleij AJ, Bombers C, Gerritsen WJ, Leunissen-Bijvelt L, Cullis PR. Fusion of phospholipids vesicles in association with the appearance of lipidic particles as visualized by freeze fracturing. Biochimica et Biophysica Acta. 1979;**555**:358-361. DOI: 10.1016/0005-2736(79)90175-5

[50] Siegel DP. Energetics of intermediates in membrane fusion: Comparison of stalk and inverted micellar intermediate mechanisms. Biophysical Journal. 1993;**65**(5):2124-2140. DOI: 10.1016/S0006-3495(93)81256-6

[51] Timoszyk A, Latanowicz L. Physical stability of temperature-sensitive liposomes. In: Bryjak M, Majewska-Nowak K, Kobsch-Korbutowicz M, editors. The Impact of Membrane Technology to Human Life. 1st ed. Wrocław: Technical University of Wrocław Publishing; 2006. pp. 27-34. ISBN: 83-7085-922-4

[52] Li J, Wang X, Zhang T, Wang Ch, Huang Z, Luo X, Deng Y. A review on phospholipids and their main applications in drug delivery. Asian Journal of Pharmaceutical Sciences. 2014;**10**(2):81-98. DOI: 10.1016/j.ajps.2014.09.004

[53] Ghanbarzadeh S, Arami S, Pourmoazzen Z, Ghasemian-Yadegari J, Khorrami A. Plasma stable, pH-sensitive fusogenic polymer-modified liposomes: A promising carrier for mitoxantrone. Journal of Biomaterials Application. 2014;**29**(1):81-92. DOI: 10.1177/0885328213515288

[54] Weinstein JN, Margin RL, Yatvin MB, Zaharko DS. Liposomes and local hyperthermia: Selective delivery of methotrexate to heated tumors. Science. 1979;**204**(4389):188-191. DOI: 10.1126/science.432641

[55] Gaber MH, Hong K, Huang SK, Papahadjopoulos D. Thermosensitive sterically stabilized liposomes: Formulation and *in vitro* studies on mechanism of doxorubicin release by bovine serum and human plasma. Pharmaceutical Research. 1995;**12**(10):1407-1416. DOI: 10.1023/A:1016206631006

[56] Landon ChD, Park J-Y, Needham D, Dewhirst MW. Nanoscale drug delivery and hyperthermia: The material design and preclinical and clinical testing of low temperature-sensitive liposomes used in combination with mild hyperthermia in treatment of local cancer. Open Nanomedicine Journal. 2013;**3**:38-64. DOI: 10.2174/1875933501103010038

[57] Pentak D, Sułkowska A, Sułkowski WW. Application of NMR and UV spectroscopy in the study of interactions between anticancer drugs and their phospholipid carriers. Journal of Molecular Structure. 2008;**887**:187-193. DOI: 10.1016/j.molstruc.2008.02.043

[58] Peleg-Shulman T, Gibson D, Cohen R, Abra R, Berenholz Y. Characterization of sterically stabilized cisplatin liposomes by nuclear magnetic resonance. Biochimica et Biophysica Acta. 2001;**1510**(1-2):278-291. DOI: 10.1016/S005-2736(00)00359-X

Homo- and Hetero-Covariance NMR Spectroscopy and Applications to Process Analytical Technology

Martin Jaeger and Robin Legner

Additional information is available at the end of the chapter

http://dx.doi.org/10.5772/intechopen.68981

Abstract

Covariance processing of data and spectra has established itself among the computer-based NMR spectroscopy methodologies to increase sensitivity and resolution and to facilitate spectral analysis. While homo-correlations yield two-dimensional (2D) diagonally symmetric or antisymmetric spectra, hetero-covariance transformations allow to transfer NMR chemical shift information to other spectroscopic techniques, such as near infra-red or Raman. This is visualized as a 2D correlation map, provided a common indirect or perturbation domain, such as time, concentration change, and pressure. Covariance spectra can be generated as synchronous or asynchronous maps. The synchronous map relates the signals of species, e.g., educts and products. The asynchronous spectrum allows to derive the sequential order in which such species occur relative to each other. After a theoretical introduction into covariance NMR, its application in process analytical technology is discussed for wine fermentation, a radical polymerization reaction, a continuous process ethanol production using immobilized yeast, and a Knoevenagel condensation in a microreaction system. The covariance approach is extended toward two perturbation variables and quantitative relationships through PARAFAC kernel analysis and is illustrated for the preparation of polylactic acid nanocomposites. The advantages and added values of using synchronous and asynchronous spectra to gain process knowledge and control are demonstrated.

Keywords: homo- and hetero-correlation spectroscopy, covariance NMR, synchronous and asynchronous spectra, process analytical technology, Raman spectroscopy

1. Introduction

Striving for enhanced sensitivity, specificity, and resolution NMR spectroscopy traditionally turned to creating stronger magnets, thus higher magnetic field strengths. The implementation of pulsed-field gradients and the development of cryogenically cooled probes contributed further to increasing instrumental sensitivity. In recent years, vivid interest was paid to so-called fast NMR methods for taking another step in ameliorating the signal-to-noise ratio. Fast methods followed several approaches. These consisted of pulse-sequence-based methods, such as time-shared experiments, hardware oriented strategies, such as parallel acquisition and detection, and the combination of two or more NMR experiments into one pulse sequence. They all aimed at optimization to take advantage of a given experimental timeframe. Not only the long-time used spectral acquisition schemes were re-evaluated, the spectral processing procedure was also equally subjected to re-investigation. As a consequence, the so-far untouched Fourier Transformation (FT), being at the heart of multi-dimensional NMR spectroscopy, was challenged. Statistic data treatment turned out to transform experimentally acquired data into spectra evenly well. Covariance transformations were applied to raw data sets as well as pre-processed data. Covariance NMR and covariance processing methods have been recently reviewed in great detail [1–7]. Due to the purely mathematical nature, the computer power and the algorithms applied determine the speed with which covariance spectra can be obtained. The experimentally acquired data determine the sensitivity observed in the covariance spectrum [8].

Beyond NMR, covariance transformations have been known to be of a very general nature according to Eq. (1) [9, 10]. The potential of generalized covariance processing was soon recognized, thus allowing traditional one-dimensional (1D) spectroscopic techniques such as infra-red (IR) and Raman spectroscopy to yield two-dimensional (2D) spectra [11, 12]. To fully exploit Eq. (1), data matrices of two distinct spectroscopic techniques, such as NMR and IR, were transformed to yield a two-dimensional IR-NMR spectrum, and the technique was baptized hetero-spectroscopy [13, 14]. As a prerequisite for its application, the spectra need to possess a common dimension prior to transformation, e.g., reaction time or change in sample pressure, called the perturbation dimension. The technique proved not only suitable for the transformation of heterogeneous data arrays or spectra but also helpful to visualize valuable information via correlation signals and their phases [12, 15]. Correlation signals indicated spectral constituents that share a common fate. The phases reflect simultaneous or asynchronous increase or decrease of the spectral constituents.

In this report, covariance NMR spectroscopy, in particular correlation and hetero-covariance NMR, shall be described in theory and practice for the investigation of chemical reactions and batch characterization. Illustrative examples shall be given how NMR spectroscopy can help attribute and distinguish signals from different spectroscopic techniques that provide lower spectral resolution or ambiguity for the assignment. In this respect, contributions of correlation spectroscopy, homo- and hetero-covariance NMR spectroscopy to the field of Process Analytical Technologies (PAT), shall be reported.

2. The concept of homo- and hetero-covariance spectroscopy

Covariance stems from statistical mathematics. Variances represent the deviation from the mean of a series of data. The covariance **C** in matrix form according to Eq. (1) is understood as the difference between the correlated and the uncorrelated products of a series of data [9, 10, 16].

$$\mathbf{C}(x, y) = \langle (x - \langle x \rangle)(y - \langle y \rangle) \rangle = \langle xy \rangle - \langle x \rangle \langle y \rangle \tag{1}$$

where $\langle x \rangle$ and $\langle y \rangle$ are the mean values, and $\langle \ \rangle$ represents any type of correlation function.

Let x and y in Eq. (1) be spectroscopic data series and be arranged such that S represents a spectrum of N_1 data points, and the elements C_{ij} of the covariance matrix or covariance map $\mathbf{C}$ are calculated in Eq. (2) as follows:

$$C_{ij} = \frac{1}{N_1 - 1} \sum_{k=1}^{N_1} \left(S(k, i) - \langle S(i) \rangle \right) \left(S(k, j) - \langle S(j) \rangle \right) \tag{2}$$

$$\langle S(i) \rangle = \frac{1}{N_1} \sum_{k=1}^{N_1} S(k, i) \tag{3}$$

where the average spectrum is defined as $\langle S(i) \rangle$ in Eq. (3). Substitution of i by j defines $\langle S(j) \rangle$ analogously. In mathematical contexts, Eqs. (1) and (2) are common. For spectroscopy, the symbols for time and frequency, t and ν or ω, are more often used. Applying Parseval's theorem (4) to Eqs. (1) and (2), the covariance matrix can be expressed by Eq. (5).

$$\int_{-\infty}^{\infty} f(t) g^*(t) dt = \frac{1}{2\pi} \int_{-\infty}^{\infty} F(\omega) G^*(\omega) d\omega \tag{4}$$

The two data sets denoted either s, S, Φ, or Ψ in Eq. (5) consist of mixed time-frequency data before and frequency-frequency data after transformation. They also share a common indirect dimension. The latter can be interpreted in terms of a perturbation, and the dimension is hence called perturbation dimension [9].

$$\begin{aligned} C(\omega_{2,A}, \omega_{2,B}) &= \langle s(t_{inc}, \omega_{2,A}) \cdot s(t_{inc}, \omega_{2,B}) \rangle \\ &= \frac{1}{2\pi(T_{max} - T_{min})} \int_{-\infty}^{\infty} S(\omega_{inc}, \omega_{2,A}) \cdot S^*(\omega_{inc}, \omega_{2,B}) d\omega_{inc} \\ &= \Phi(\omega_{2,A}, \omega_{2,B}) + i\Psi(\omega_{2,A}, \omega_{2,B}) \end{aligned} \tag{5}$$

The index *inc* refers to the second or indirect spectral dimension. In a typical experiment, this dimension is recorded as discrete time intervals between a maximum limit T_{max} and a minimum limit T_{min}. The direct dimension may stem from two different data sets, $A \neq B$, or from the same data set, $A = B$. In the latter case, the data sets are transposed with respect to each other.

The spectra or maps $\mathbf{\Phi}$ and $\mathbf{\Psi}$ are defined according to Eqs. (6) and (7).

$$\Phi(\omega_{2,A}, \omega_{2,B}) = \frac{1}{T_{max} - T_{min}} \int_{T_{max}}^{T_{min}} s(t_{inc}, \omega_{2,A}) \cdot s(t_{inc}, \omega_{2,B}) dt_{inc} \tag{6}$$

$$\Psi(\omega_{2,A}, \omega_{2,B}) = \frac{1}{T_{max} - T_{min}} \int_{T_{max}}^{T_{min}} s(t_{inc}, \omega_{2,A}) \cdot h \cdot s(t_{inc}, \omega_{2,B}) dt_{inc} \tag{7}$$

where h is the Noda-Hilbert transform [15]. The reader is also referred to Eqs. (17) and (18) for definition and matrix notation. Integration of Eqs. (6) and (7) results in Eqs. (8) and (9).

$$\Phi(\omega_{2,A}, \omega_{2,B}) = p(\cos\varphi)^{A,B} Abs(\omega_{2,A}) Abs(\omega_{2,B}) \tag{8}$$

$$\Psi(\omega_{2,A}, \omega_{2,B}) = q(\sin\varphi)^{A,B} Abs(\omega_{2,A}) Abs(\omega_{2,B}) \tag{9}$$

Equations (8) and (9) are lengthy expressions when fully written for p and q. Yet, the phase φ is readily recognized. It may be considered as an internal reference according to Eqs. (10) and (11), which present the important parts of the complete definition for p and q.

$$p(\cos\varphi)^{A,B} \sim \cos(\omega_{2,\alpha} t_{inc} + \varphi),\ \alpha = A, B \tag{10}$$

$$q(\sin\varphi)^{A,B} \sim \sin(\omega_{2,\alpha} t_{inc} + \varphi),\ \alpha = A, B \tag{11}$$

The comparison of Eqs. (11) and (12), the latter being an analogous expression but obtained after Fourier Transformation, readily reveals that an internal reference φ is absent in Eq. (12), i.e., manual phase correction after FT is required in contrast to the covariance transformed version of the spectral representation.

$$\begin{aligned} S(\omega_{inc}, \omega_2) &= \int s(\omega_{inc}, \omega_2) \exp(-i\omega_2 t_{inc}) dt_{inc} \\ &= \int s(t_{inc}, \omega_2) \cos(\omega_{inc}, t_{inc}) dt_{inc} \\ &+ i\int s(t_{inc}, \omega_2) \sin(\omega_{inc}, t_{inc}) dt_{inc} \end{aligned} \tag{12}$$

A spectrum after FT often consists of the real part data, with the imaginary part discarded. Yet, the phase still needs to be adjusted. The interested reader is referred to NMR textbooks and to the recent works in the context of covariance NMR [17, 18].

Equation (13) is the general form of Eq. (3). Here, f and ω denote spectral variables, such as frequencies, that may be recorded using any type of spectroscopy. A common perturbation such as a time domain t relates them to each other. Nevertheless, the perturbation could also be a series of samples, pressure, crystallization, etc.

$$C(f, \omega) = \langle s_1(f, t) \cdot s_2(\omega, t) \rangle \tag{13}$$

Spectra generated using Eq. (13) represent hetero-spectral correlation maps [14]. For pure NMR spectroscopy, unsymmetrical indirect covariance NMR was the first type of hetero-correlation spectroscopy, relating, e.g., ^{15}N and ^{13}C signals via the proton dimension, to each other [19–21]. Taken a step further, NMR and IR or NMR and mass spectrometry data were correlated to each other [22].

The covariance matrix contains as its elements the covariance C_{ij}, i.e., the amplitudes of positions i and j of the 1D spectra. Rewriting Eq. (2) in matrix form yields the relationship between $\boldsymbol{C}$ and the spectroscopic data set $\boldsymbol{S}$. The matrix multiplication of $\boldsymbol{S}$ with its transpose $\boldsymbol{S}^T$ is equal to $\boldsymbol{C}^2$, cf. Eq. (14).

$$C^2 = \boldsymbol{S}^T \cdot \boldsymbol{S} \tag{14}$$

The complete mathematical derivation and proofs have been accomplished by Brüschweiler et al. and Noda et al. [16, 23, 24].

Defining $\boldsymbol{S}$ as the mixed time-frequency matrix, $\boldsymbol{S}(t_1,\omega_2)$, the product $\boldsymbol{S}^T\boldsymbol{S}$ is the symmetric matrix $\boldsymbol{C}(\omega_1,\omega_2)$. A common two-dimensional NMR spectroscopic data set $\boldsymbol{S}$ often has $N_1 = 2$ k and $N_2 = 256$ k data points. Hence, the resulting covariance map will be of dimensions N_1 x N_2 = 2 k x 2 k. This has been assumed as a projection of the direct or acquisition dimension onto the indirect or incremented dimension. It is readily recognized that the indirect dimension is thus substantially enlarged. Two data matrices $\boldsymbol{F}^T$ and $\boldsymbol{F}$ that have been the results of two-dimensional Fourier transformation may also be multiplied to form the covariance spectrum according to Eq. (15).

$$C^2 = \boldsymbol{S}^T \cdot \boldsymbol{S} = \boldsymbol{F}^T \cdot \boldsymbol{F} \tag{15}$$

The equality of transformations of the mixed time-frequency data and the completely Fourier transformed data is a consequence of Parseval's theorem (4) and ensures that the transformations of the mixed time-frequency data and the Fourier transformed data are equal [16, 24]. From another perspective, the spectral reconstruction can be considered as relating two direct dimensions through an indirect dimension or perturbation, which is discarded. The physical meaning of the indirect dimension is therefore of little importance. Thus, it relates Noda's model two IR wavenumber dimensions via a common perturbation, which may be time, pressure, temperature, sample space, or many more [13]. The matrix representation form reveals that Noda's synchronous matrix $\boldsymbol{\Phi}$, in Eqs. (6) and (16), corresponds to the covariance map according to Eq. (15), if mean centered spectra are the elements of the data matrices giving $\boldsymbol{\Phi}$. The asynchronous map $\boldsymbol{\Psi}$ of Eqs. (7) and (17) corresponds to the indirect covariance correlation spectrum. Equations (15) and (13) further extend covariance spectroscopy to hetero-correlation spectroscopy.

Eqs. (16) and (17) finally represent the matrix notation of equations as the synchronous map or spectrum and as the asynchronous map.

$$\Phi = \overline{\boldsymbol{X}}^T \cdot \overline{\boldsymbol{X}} \tag{16}$$

$$\Psi = \overline{\boldsymbol{X}}^T \cdot \boldsymbol{N} \cdot \overline{\boldsymbol{X}} \tag{17}$$

where $\overline{\boldsymbol{X}}$ is the matrix of mean-centered spectra and $\boldsymbol{N}$ the Noda-Hilbert orthogonalization matrix with $N_{ik} = 0$ if $i = k$ and $1/(\pi(k - i))$ otherwise.

Synchronous and asynchronous maps or spectra have some particular features [11]. Since synchronous homo-correlation spectra are computed from a data matrix and its transposed matrix, they are symmetric. They exhibit diagonal peaks, also called autopeaks, that are the autocorrelation functions of spectral intensity variations. They hence reflect the amount of change the corresponding signal experiences along the perturbation dimension. Off-diagonal signals correlate two signals changing simultaneously or coincidently under the influence of the perturbation. When both signals increase or decrease, the sign of the crosspeak is equal to that of the diagonal peaks. If they behave adversely, the sign is opposite. It is readily recognized that the resolution of spectra can be enhanced by the spread into two dimensions. Furthermore, the occurrence of two or more components, such as educt and product, can be readily seen and facilitate signal assignments. An example for a synchronous spectrum is given in **Figure 1(a)**. As will be shown below, synchronous spectra are useful in homo- and hetero-covariance NMR spectroscopy.

The asynchronous spectrum in general is less easily interpreted. As a consequence of the Noda-Hilbert orthogonalization, cf. Eq. (17), no diagonal peaks are observed. The spectrum visualizes successive or sequential changes of signal intensities, which forbids the occurrence of autopeaks. The asynchronous map is antisymmetric with respect to the diagonal. Noda has shown that $\Psi(\omega_i, \omega_j) = -\Psi(\omega_j, \omega_i)$. [11]. The sign of a crosspeak is positive—positive is defined as in phase with the diagonal peak in the corresponding synchronous spectrum—if the intensity in dimension 1 changes predominantly before that in dimension 2 in the sequence of the perturbation. This is valid for crosspeaks above the diagonal, i.e., $\omega_1 > \omega_2$. A negative crosspeak is obtained when the order is reversed. An illustration is given in **Figure 1(b)**. Specie B of the example hence occurs before A, and C before D. Thorough derivations and discussions have been accomplished previously [11].

Despite its ability to correlate non-simultaneous occurrence of signals, the asynchronous map does not allow the analysis of population dynamics as in a two-step chemical reaction. As a

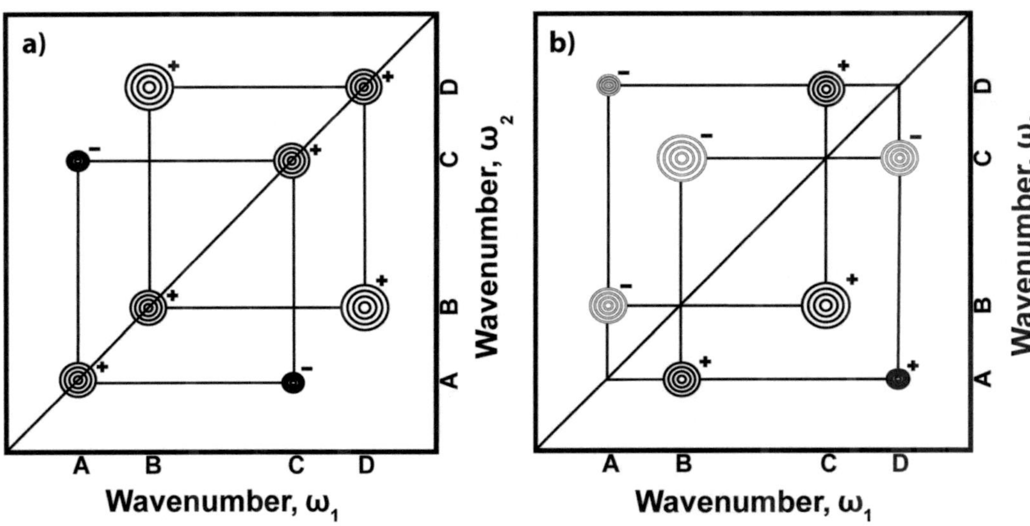

Figure 1. Schematic contour map of synchronous (a) and asynchronous (b) 2D correlation spectra. Peaks located at the diagonal are autopeaks. The signs of the correlation peaks are indicated. The intensity changes and signs are interpreted according to Noda's rules [11].

remedy to this problem, Noda devised two-dimensional codistribution spectroscopy [25]. Here, a moment analysis of spectral intensity distribution over the perturbation dimension was included, which accounted for the sequential attribution of species within a model chemical reaction A→B→C.

Out of the manifold of variations to combine raw and Fourier transformed data, a variety of covariance-transformed spectral representations have been introduced and their applications have been demonstrated: Among those used in NMR spectroscopy, the most often used or described were direct covariance, indirect covariance, doubly indirect covariance, unsymmetrical indirect covariance, generalized indirect covariance, which replaced the previous one, multidimensional covariance in form of Triple-Rank Covariance and 4D Covariance [2, 26–29]. Furthermore, the family of Statistical Correlation Spectroscopy (STOCSY) has been introduced, and its usefulness is demonstrated in many applications [22, 30–32].

For other spectroscopic techniques or combinations thereof, covariance spectroscopy is often referred to as 2D correlation spectroscopy, and hetero-covariance spectroscopy is called hetero-spectral, hetero-perturbation, and hetero-sample correlation spectroscopy [33]. Noda has further coined the term multiple perturbation 2D correlation, where the use of the parallel factor (PARAFAC) kernel analysis may play a key role in future spectral analysis [34–36]. As another variant, orthogonal sample design (OSD) was introduced and applied [37–39].

Multiple perturbation 2D correlation spectroscopy has been introduced recently by Shinzawa et al. [40, 41]. It is based on the extension of Eq. (3) yielding Eqs. (18) and (19) as follows:

$$\langle S_p(\omega, q)\rangle = \frac{1}{P}\sum_{p=1}^{P} S(\omega, p, q) \tag{18}$$

$$\tilde{S}(\omega, p, q) = S(\omega, p, q) - \langle S_p(\omega, q)\rangle \tag{19}$$

where S is a set of spectra depending on frequency ω exposed to multiple perturbations p = 1,2,…P and q = 1,2,…Q, such as time, temperature, concentration, etc. ⟨ ⟩ denote the average spectrum. Partial synchronous and asynchronous correlation spectra are computed according to Eqs. (20) and (21) in analogy to Eqs. (6) and (7).

$$\Phi_p(\omega_1, \omega_2, q) = \frac{1}{P-1}\sum_{p=1}^{p} \tilde{S}_p(\omega_1, p, q)\cdot \tilde{S}_p(\omega_2, p, q) \tag{20}$$

$$\Psi_p(\omega_1, \omega_2, q) = \frac{1}{P-1}\sum_{p=1}^{P} \tilde{S}_p(\omega_1, p, q)\cdot \tilde{S}_p^{\neq}(\omega_2, p, q) \tag{21}$$

where $\tilde{S}_p^{\neq}$ denotes the Hilbert-Noda transformation in this case given by Eq. (22).

$$\tilde{S}_p^{\neq}(\omega_2, p, q) = \sum_{r=1}^{P} N_{pr}\tilde{S}_p(\omega_2, r, q) \tag{22}$$

with $N_{pr} = 0$ for $p = r$ and $N_{pr} = 1/((r-p)\pi)$ otherwise.

The PARAFAC kernel decomposes the data into scores and loading vectors. The original three-way data array is rearranged into a two-way data array by means of the so-called Kathri-Rao

$(|\otimes|)$ product, which implies the use of the Kronecker product $\otimes$. The matrix decomposition is usually achieved through solving an alternating least-squares problem iteratively. Disregarding the matrix of the residuals for the minimization problem, Eq. (23) is the fundamental matrix representation of the multiple perturbation correlation analysis.

$$\boldsymbol{X} = \boldsymbol{A}(\boldsymbol{C}| \otimes |\boldsymbol{B})^{T} \tag{23}$$

where $\boldsymbol{X}$ contains spectral data, $\boldsymbol{A}$ and $\boldsymbol{C}$ refer to perturbations 1 and 2, and $\boldsymbol{B}$ contains the spectral variable. The p-synchronous and p-asynchronous kernel matrices are similar to their analogs in Eqs. (16) and (17) but formed mean-centered and normalized score-vector matrix $\boldsymbol{A}$. The ij-element of the p-synchronous kernel matrix as well as of the asynchronous one assumes values between −1 and +1, giving a similarity measure in the synchronous case and a dissimilarity measure in the asynchronous case between the score vectors of the ith and jth components. Evenly comparable, the sequential order of signal changes can be derived from the signs of the kernel matrix elements. The signal of the ith species changes before that of the jth when the signs of the ij-elements of the synchronous and asynchronous kernel matrix are the same. The order is reverted if the elements possess opposite signs. Spectral analysis can be carried out as well by performing the computation with the score matrix $\boldsymbol{C}$ instead of $\boldsymbol{A}$. Complete mathematical descriptions have been published by Shinzawa et al. [34, 41].

Software suitable for covariance processing has recently been reviewed as well [3, 42]. With respect to some of the work performed in this report, we would like to direct the reader's attention to 2DShige. While this program is not especially dedicated to NMR spectroscopy, it is capable of performing hetero-spectroscopic covariance transformations. The program was devised by Morita and may be accessed for download via https://sites.google.com/site/shigemorita/home/2dshige. Covariance transformations applied therein follow the work by Noda. Synchronous and asynchronous maps are computed from data in CSV format.

The following section will focus on Process Analytical Technology (PAT) such that the stage will be set for the applications of covariance processing and NMR spectroscopy to process monitoring or process understanding.

3. A brief outline of process analytical technologies and microreaction processes

Process analytical technologies (PAT) have grown into an integral part of industrial manufacturing processes. The development of a process on a laboratory scale, the collection of data as well as monitoring of the production process in place are directed toward a well-understood process to ensure final product quality [43, 44].

This knowledge first enables process control and then process improvement. The envisaged process optimization is aimed at cost reduction, sustainability, and safety. Generally, production processes proceed on a large scale. The analytical instruments used close to the process are robust, relatively easy to operate instruments. Only for the development or validation of the analytical method are the dimensions of such large-scale processes reduced to laboratory scales. The analytical instruments yet may be of the same size but more complex and of higher sensitivity and resolution.

Process analytical technologies often make use of spectroscopic and chromatographic as well as of integral methods. Today, Raman spectroscopy and near IR (NIR) spectroscopy play major roles, whereas pH, pressure, and refractivity techniques are typical non-specific methods, inexpensive still ubiquitous, and powerful within well-controlled processes [45]. The conditions of the production process often demand for greater robustness, stability, and performance of the analytical instruments, because of the close proximity to the manufacturing line. Process monitoring and control require prompt or real-time data recording, processing, and feeding the data back to the process control unit. These constraints necessitate in-line, on-line, or at least at-line analytical methods [46].

Microprocesses or microreactions are conducted in very small-scale reactors and mixing devices equipped with tubing, pumps, and valves. The reaction set-up is composed in a Lego-like manner, cf. **Figure 2**. Microdevices allow for a highly efficient heat transfer as compared to

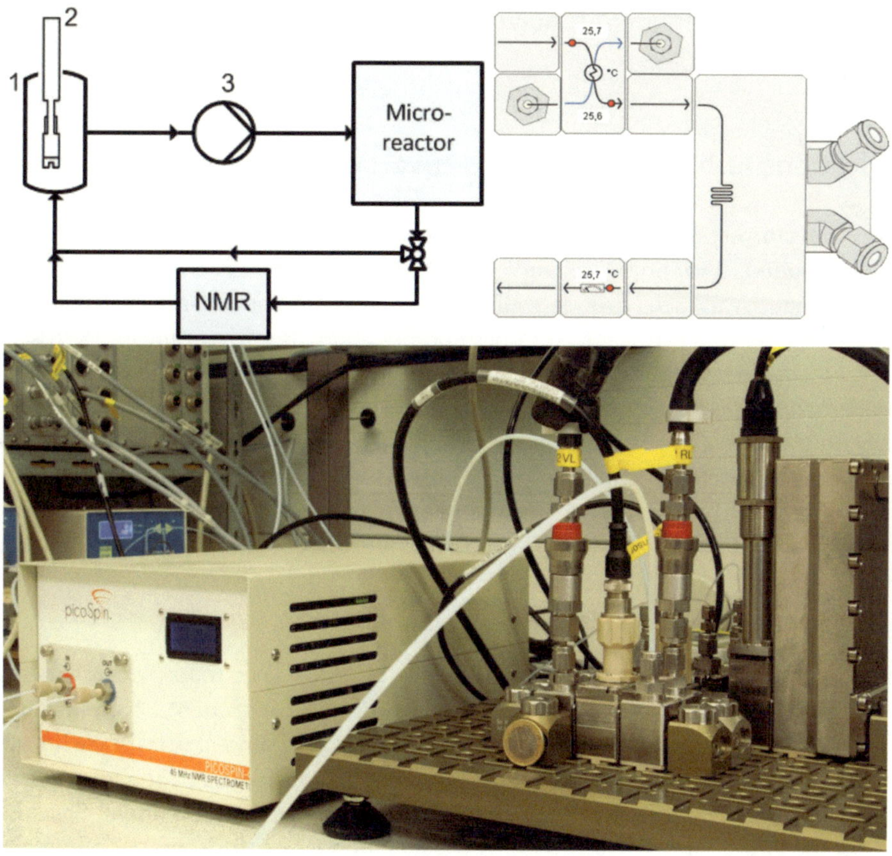

Figure 2. Microreaction assembly with on-line low-field ^{1}H-NMR spectrometer (bottom); process flow chart of the set-up of the microprocess analytics (top left): (1) storage vessel, (2) transflectance NIR immersion probe, (3) pump; zoom of the microreactor assembly (top right).

large-scale vessels. While they sometimes enable a superior mass transfer, they sometimes cause an inferior mixing of the reactants due to microfluidic effects. Typical yields may range from milligrams to a few grams per day depending on the reaction conducted in batch or flow mode [47–49].

Microprocesses with respect to scale, volumetric flow, and yield demand for microanalytics if implemented in-line or on-line. At-line installations merely require a sample cell of suitable size and sensitivity. Two different ways have been described to monitor microprocesses in-line or on-line with spectroscopic methods. The probes or sample cells were located either in the reaction vessel or a by-pass similar to large-scale facilities. Alternatively, the reaction was conducted within the sample cell of a spectrometer, e.g., UV/Vis or Nuclear Magnetic Resonance. Miniaturized analytical devices are preferable in case of microreaction vessels, whereas standard laboratory instruments may be used for the second case. So-called bench-top instruments are particularly interesting for microprocess analytical technology. Bench-top instruments may be found as the size of a microwave oven [50–53].

In the following sections, illustrative examples for the application of NMR spectroscopy, covariance, homo- and hetero-correlation spectroscopy to process monitoring, and process understanding will be given.

4. Applications of homo- and hetero-covariance spectroscopy

Covariance transformations of NMR data with or without prior Fourier transformation today are widely applied. Prominent examples comprise generalized indirect covariance and multidimensional covariance NMR as well as the combination of covariance and non-uniform sampling of data [54–57]. While the concept of homo- and hetero-covariance spectroscopy was developed nearly three decades ago, there are relatively few reports on the use of synchronous and asynchronous spectra involving NMR spectroscopy [3, 14, 55, 58, 59]. In contrast, an abundant number of investigations have applied so-called statistical hetero-spectroscopy (STOCSY) that has delivered important contributions to the field of metabolomics and whose variants have recently been depicted like a phylogenetic tree [22, 32, 60]. In the current report, the focus is however laid on examples from chemical processes rather than metabolomics.

4.1. Reaction monitoring of a wine fermentation

Kirwan et al. monitored a wine fermentation by ^{1}H NMR spectroscopy, drawing samples daily [61]. After careful preprocessing by segmentation, alignment, normalization, and smoothing, the data were covariance transformed, yielding homo-spectral synchronous and asynchronous matrices. While the synchronous map was found less prone to small chemical shift and linewidth variations, the asynchronous matrix was very sensitive. Sasic had also reported on the effects of linewidth [62]. In his metabonomics study on vasculitis analyzing rat urine samples, butterfly-like signal shapes were observed as a result of shifting peak positions. The lack of uniform pre-processing led to numerous artifacts and problems that severely hampered spectral interpretation in contrast to the wine study. The spectra recorded in the wine fermentation study were hence ameliorated in a successive approach by imposing a fixed linewidth prior to covariance transformation such that the effects of linewidths changing during the fermentation were compensated for [63]. Extracted regions of both spectra are shown in **Figure 3**.

The spectra contained strong signals from sugars, fructose, and glucose, in the early period. In the later phase, ethanol signals became predominant. Other molecular species were emerging and vanishing as well. Their temporal relationship was said difficult to assess, which can be seen from inspection of **Figure 3**. A manifold of correlations are present in the covariance maps. Most clearly, the interdependence of the sugar and ethanol signals is recognized. Kirwan et al. already pointed out the difficulty of interpreting the spectra due to the high resolution of the initial ^{1}H NMR spectra leading to the large number of signals and correlations [61]. The authors suggested the use of slices through the synchronous map allowing the signal attribution and further extraction of the sequential information out of the corresponding slices of the asynchronous map. Careful analysis revealed that glucose was consumed and transformed at a higher rate than fructose, which was interpreted in terms of the different diffusion rates of the two sugars across the fermenting yeast cell membrane. The authors thus demonstrated the usefulness of correlation NMR spectroscopy for monitoring concentrations and sequential relationships in a biochemical process.

4.2. IR-NMR hetero-covariance spectroscopy applied to radical polymerization

Ryu et al. used 2D IR-NMR hetero-spectroscopy to characterize a chain transfer reaction during the radical polymerization of N-vinylpyrrolidone (NVP) [64]. Polyvinylpyrrolidone (PVP) was polymerized to form nanoparticles through a chain transfer reaction initiated by silver nitrate. Upon reduction via electron transfer, PVP polymer silver nanoparticles were

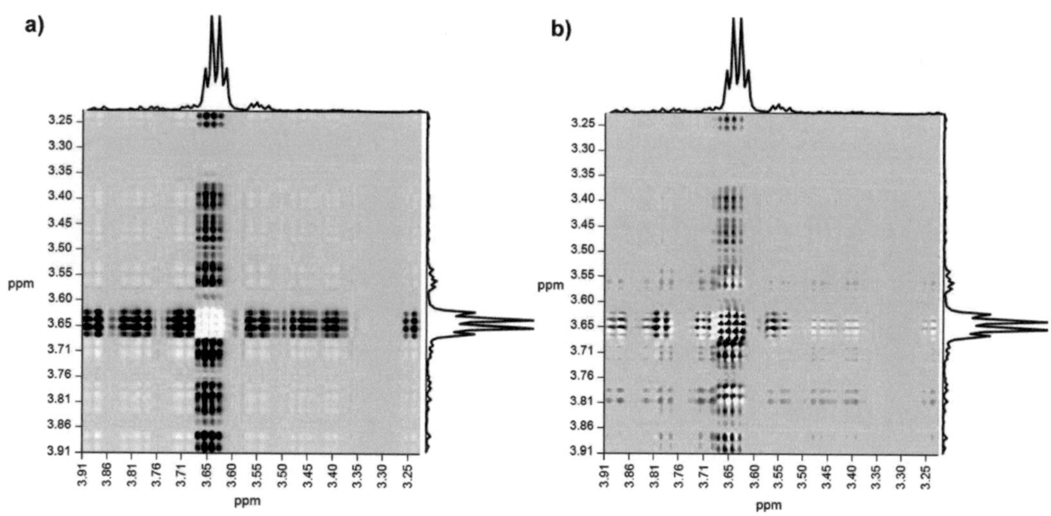

Figure 3. The synchronous (a) and asynchronous (b) maps of a section of the mean-centered 1D ^{1}H NMR spectra at 500 MHz of a series of wine fermentation samples. Reprinted from Kirwan et al. [63]. Copyright 2008, with permission from Elsevier.

formed. The resulting nanoparticles possessed a carbon-carbon double bond at the end of the PVP chain after chain transfer termination. Radical formation was initiated through azobisisobutyronitrile. The reaction was monitored using IR and 1D ^{1}H NMR spectroscopy. That is, a series of IR and NMR spectra depending on reaction time as perturbation domain were obtained. In the IR synchronous homo-correlation spectrum, bands at 1660 and 1676 cm^{-1} were revealed that could be attributed to the stretching vibration of the carbon-carbon double bond and of the carbonyl group, respectively. The asynchronous map was interpreted in terms of an intensity decrease of the band at 1660 cm^{-1} preceding the increase of the carbonyl band at 1676 cm^{-1}, cf. **Figure 4**.

Following Noda's rules on analyzing the synchronous and asynchronous spectral matrices, one might also come to a reversed conclusion concerning the sequential order [11, 12]. Both educt and product after chain transfer termination do exhibit carbon-carbon double bonds, where the NMR signals of the monomeric educts should lead to more intense signals due to less relaxation broadening. Yet, IR-NMR hetero-spectral correlation maps were used to unequivocally attribute the less-resolved IR bands in the product to the carbon-carbon double

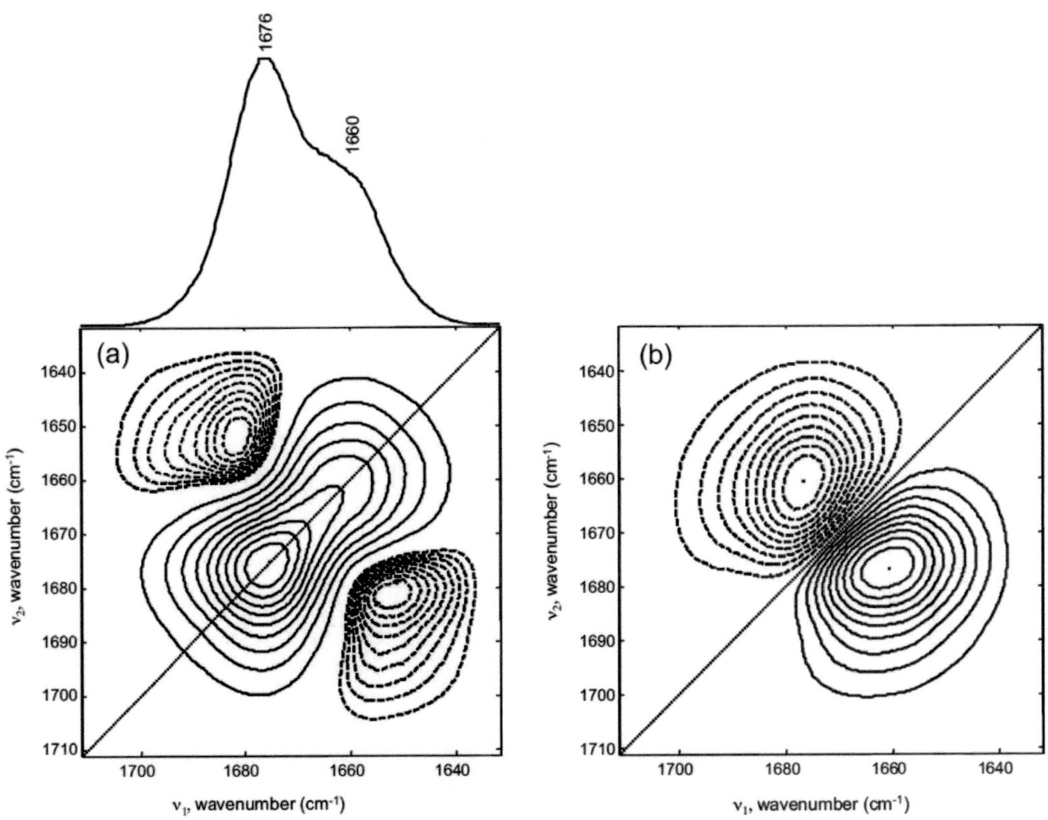

Figure 4. Synchronous (a) and asynchronous (b) 2D FTIR correlation spectra of PVP during polymerization with 400 ppm silver nitrate. The autopower spectrum extracted along the diagonal line in the synchronous 2D correlation spectrum is given on the top of (a). The solid and dashed lines in the spectra represent the positive and negative crosspeaks, respectively. Reprinted from Ryu et al. [64]. Copyright 2012, with permission from Elsevier.

bond and to the carbonyl group. Thus, both homo- and hetero-spectral correlations are of considerable value to increase spectral resolution and cross-fertilize the spectral analysis or assignment of one spectroscopic technique by taking advantage of another technique. This is especially helpful in process analytical technologies when signal crowding or strong overlap due to conditions unfavorable for a certain spectroscopic method occurs frequently.

4.3. Study of polylactic acid nanocomposites at varied temperatures and compositions using PARAFAC kernel analysis

Shinzawa et al. investigated polylactic acid nanocomposites using solid-state cross-polarization magic angle spinning (CP-MAS) ^{13}C NMR experiments [34]. They prepared four samples with varying clay content through a melt-blend process to obtain pellets. The properties of the sample exposed to temperature variation were studied by thermomechanical analysis. The elongation of the sample measured under imposture of a load occurred most notably at the glass transition temperature of the samples around 60°C. After a certain increase, a plateau was reached. The finding was interpreted that the plastic deformation observed was related to the glass-to-rubber transition of the amorphous polylactic acid component. When the elongation did no longer increase, a network structure due to physical crosslinkage induced by the crystalline domain was assumed. The dependence on the clay content suggested that with increasing clay inclusion, the tendency to elongate with temperature decreases. Thus, inclusion of clay led to enhanced stiffness. By applying NMR spectroscopy, Shinzawa et al. strove to probe the macroscopic properties on a molecular scale. To this purpose, they inspected the ^{13}C NMR resonance around 170 ppm, which originates from two peaks at 169 ppm and 170 ppm attributed to the crystalline and amorphous phases, respectively. Since NMR spectra depended on two separate perturbations, i.e., clay content and temperature, the PARAFAC kernel analysis according to Eq. (23) was employed for a detailed analysis. As described above, two sets of synchronous and asynchronous correlation spectra were obtained after the covariance transformations and matrix decompositions: the temperature-dependent and the clay-dependent homo-correlations. The partial correlations from composition-dependent NMR spectra at fixed temperature are exemplarily presented in **Figure 5**. Whereas the partial temperature-dependent correlation spectra revealed

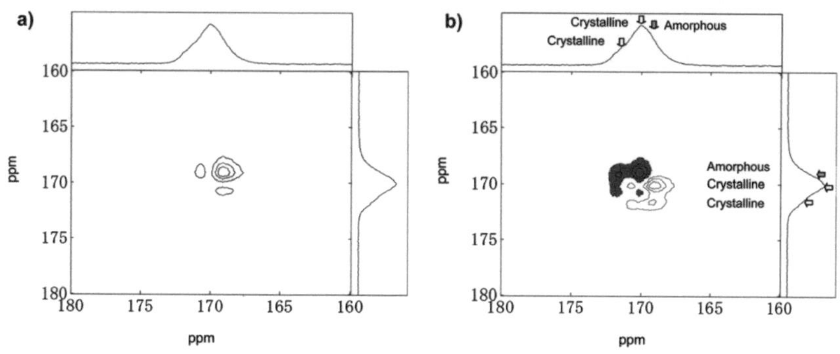

Figure 5. Partial synchronous correlation (a) and partial asynchronous correlation (b) spectra calculated from clay weight-dependent 1D CP-MAS ^{13}C NMR spectra recorded at 100.56 MHz. Reproduced from Shinzawa et al. [34] with permission of The Royal Society of Chemistry (RSC).

that the amorphous preceded the crystalline component upon temperature increase, the spectra in **Figure 5** showed that the amorphous content occurred predominantly before the crystalline content on increasing clay content. This was assumed due to the clay acting as a nucleating agent. It would foster additional crystallization of the polylactic acid. Upon decrease of the amorphous phase, the phase transitioning from glass to rubber should be reduced. These results were supported by the thermomechanical analysis.

The added value of the PARAFAC kernel analysis is that it furnishes quantitative data. The score, $\boldsymbol{A}$ and $\boldsymbol{C}$, and loading, $\boldsymbol{B}$, matrices reflect the change in signal intensity separated into composition and temperature dependence. They also provide abstract information on the dynamic behaviors of the crystalline and amorphous phases. The synchronous and asynchronous pair of the kernel matrix is exemplarily presented in **Figure 6** for the spectral intensity change of the nanocomposite samples due to clay content variation. The so-called *q*-synchronous correlation intensity, cf. above, $\boldsymbol{H}_{q\ \text{amorphous, crystalline}} = -0.98$ and *q*-asynchronous correlation intensity $\boldsymbol{K}_{q\ \text{amorphous, crystalline}} = 0.06$ were interpreted in terms of similarity of changes in the amorphous and crystalline components due to the presence of clay. Yet, the negative sign indicated opposite direction, i.e., increase in clay content augmented the crystalline and decreased the amorphous phase, which agreed well with the homo-spectral correlation results. In practice, the application of PARAFAC kernel analysis was envisioned to provide opportunities to gain detailed information on sequences of species occurring under multiple perturbations.

4.4. Monitoring of ethanol production from immobilized yeast using homo- and hetero-covariance spectroscopy

As an example for process monitoring of biochemical processes, the conversion of glucose into alcohol by *Saccharomyces cerevisiae*, baker's or brewer's yeast, was monitored using low-field 1D ^{1}H NMR and Raman spectroscopy [65]. Monitoring of fermentation processes was described

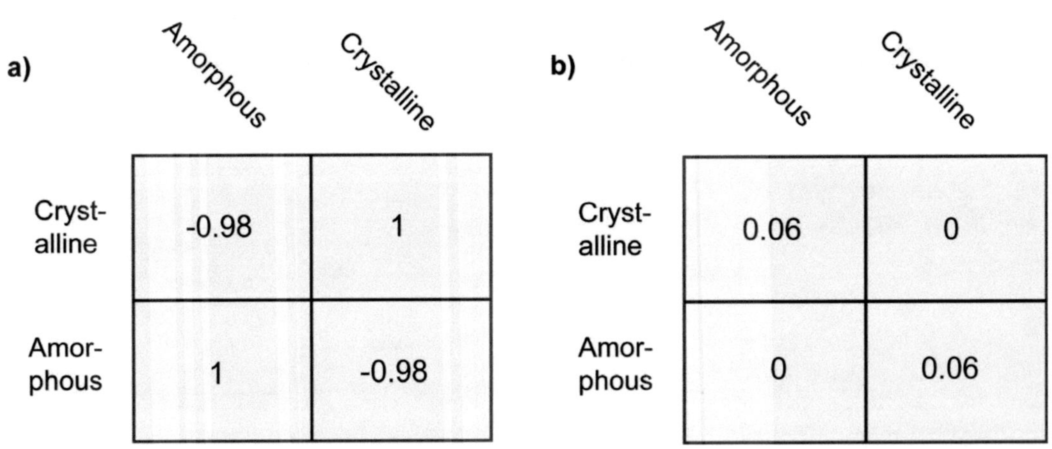

Figure 6. Representations of the *q*-synchronous kernel (a) and *q*-asynchronous kernel (b) matrices computed from the score matrix $\boldsymbol{C}$ of the clay weight-dependent 1D CP-MAS ^{13}C NMR spectra recorded at 100.56 MHz as used in the PARAFAC kernel analysis [34].

earlier, and NIR became the standard methodology [66, 67]. Later, attempts were made to use Raman spectroscopy [68]. Recently, hetero-spectral correlation NIR-IR spectroscopy was applied [69]. The fermentation described in the current report was conducted as a continuous process feeding glucose solution at a constant flow into a 2 L fermenter. Yeast immobilized within an alginate hydrocolloid converted the sugar to ethanol. The aqueous ethanolic solution was diverted at a constant flow. The flow rate was optimized such that during the residence time of a given volume, the glucose was fully converted into ethanol. On-line monitoring, i.e., through analysis of the ethanol signals and potential remainders of the glucose signals, was applied to control the efficiency of the process from the initial induction phase to the final stable production. After optimization, a sugar concentration of about 17% could be successfully transformed into ethanol.

Since no deuterated solvents were used, the series of 1D ^{1}H NMR spectra exhibited a dominant water signal and the typical ethanol resonances, cf. projections in **Figure 7**.

Homo- and hetero-covariance transformations were computed after spectral alignment and normalization to the water resonance at 4.8 ppm. The synchronous NMR spectrum displayed the expected positive intra-ethanol correlation at (3.8 ppm, 1.2 ppm). It exhibited further positive correlations between the signal at 4.8 ppm and the ethanol resonances at 1.2 and 3.8 ppm, suggesting that both signals increased or decreased in phase. As the spectra were normalized to the water resonance, the tentative change was traced back to changes in the linewidth of the water signal and should therefore not further be considered. Inspection of the asynchronous spectrum, cf. **Figure 7**, showed no intramolecular correlations at 3.8 and 1.2 ppm as would be expected, since the ethanol signals would change in phase. Yet, correlations between the signal at 4.8 ppm and the ethanol resonances were observed. The sign of the correlation suggested that water or an underlying component would grow before ethanol

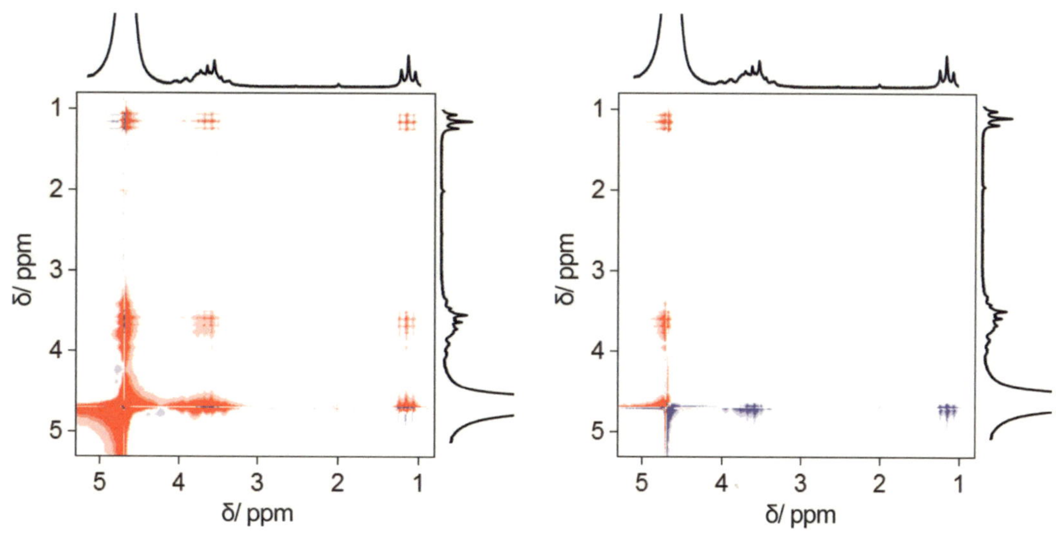

Figure 7. Synchronous (left) and asynchronous (right) 2D NMR homo-correlation spectra of ethanol production by immobilized yeast using on-line 1D ^{1}H NMR spectroscopy (82 MHz, T = 36°C).

increased. Since change on the water resonance was traced back to linewidth variations, no conclusion with respect to the sequence of change should be drawn. Taking into account the earlier observation by Kirwan et al. and Sasic with respect to changes in linewidth and spectral alignment, the described preprocessing procedures were found difficult to apply to low-field spectra with relatively poor resolution and signals with strongly differing linewidths.

In contrast, the hetero-covariance NMR-Raman spectrum proved very useful for the quick analysis and assignment of the signals in the Raman spectrum, cf. **Figure 8**.

Only the band at 1360 cm^{-1} showed a negative correlation with the NMR resonances of ethanol, thus identifying this band as educt related. All other Raman bands were found in phase with the ethanol NMR signals and could thus serve for product monitoring. The hetero-correlation spectrum was hence able to readily visualize that nearly all Raman bands at least predominantly originated from ethanol, but in contrast to low-field NMR signals provided an educt signal, which appeared only as a shoulder in the corresponding 1D Raman spectrum.

4.5. Reaction monitoring of a Knoevenagel condensation in a microreaction system

A Knoevenagel condensation reaction was conducted in a microreaction system, cf. **Figure 2** [70]. Neat malonic acid diethylester and 2-propanal were flowed through the microreactor at a

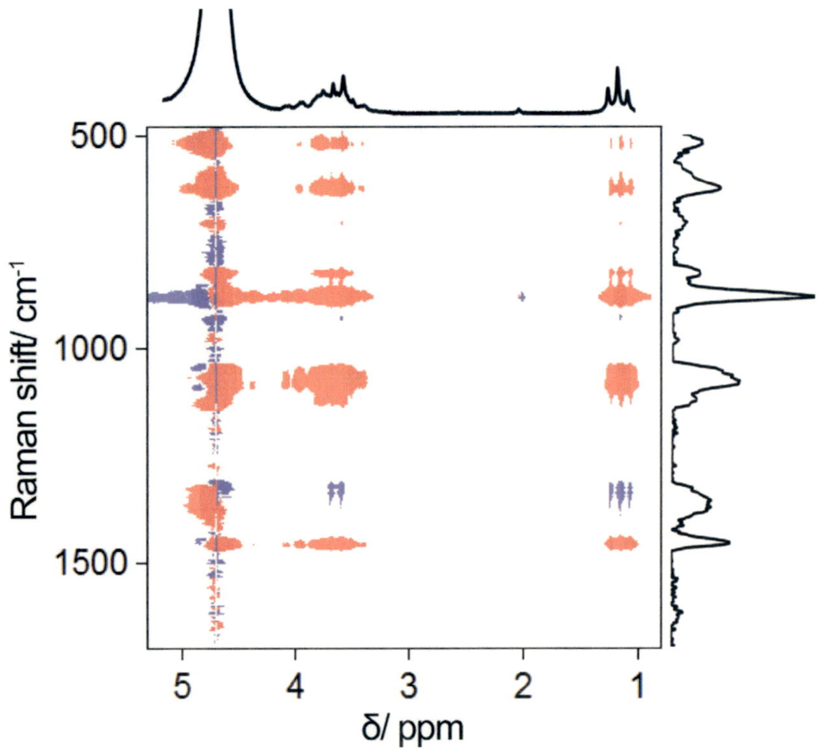

Figure 8. 2D NMR-Raman hetero-correlation spectra of ethanol production by immobilized yeast using on-line 1D ^{1}H NMR spectroscopy (82 MHz, T = 36°C) and in-line Raman spectroscopy (laser wavelength 785 nm).

http://dx.doi.org/10.5772/intechopen.68981

temperature of 82°C. Butylidene malonic acid diethylester and water were obtained as products. The solution was re-circulated for 1200 min and monitored using on-line low-field benchtop 1D ^{1}H NMR (82 MHz), in-line NIR, and in-line Raman spectroscopy (laser excitation wavelength 785 nm). Despite the relatively poor resolution of the low-field instrument, the series of 1D ^{1}H NMR spectra showed well-resolved signals for each educt and product, cf. **Figure 9**. Therefore, signals could be integrated and concentration-time plots were established.

The covariance transformation to synchronous and asynchronous homo-spectral correlation maps helped quickly visualize the interdependence of the signals, cf. **Figure 10**. The assignment of educt and product signals was in agreement with correlation crosspeaks and their signs. The signals were attributed as 3.2 ppm for malonic acid diethylester, 6.8 ppm for the product butylidene malonic diethylester, and 9.5 ppm for 2-propanal. Since the intensity of the autopeaks in the homo-covariance map reflects the amount of change, peaks appeared at strongly different intensity levels such that representations of the spectra should be prepared using different thresholds with emphasis on either strong or weak signals. Nevertheless, the sign of the crosspeaks was found in accordance with the expectancy values. With respect to the

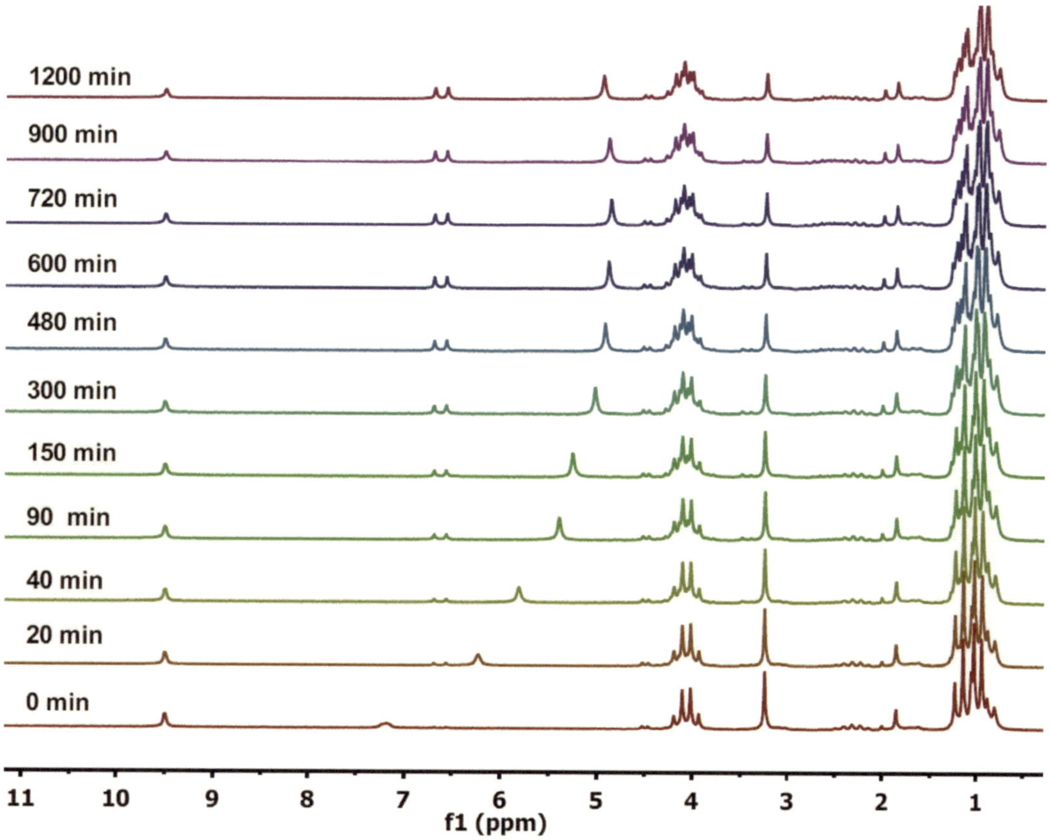

Figure 9. 1D ^{1}H NMR spectra (82 MHz, T = 36°C) recorded for 1200 min during on-line monitoring of a Knoevenagel condensation of neat malonic acid diethylester and 2-propanal yielding butylidene malonic acid diethylester conducted in a microreaction system presented in Figure 2.

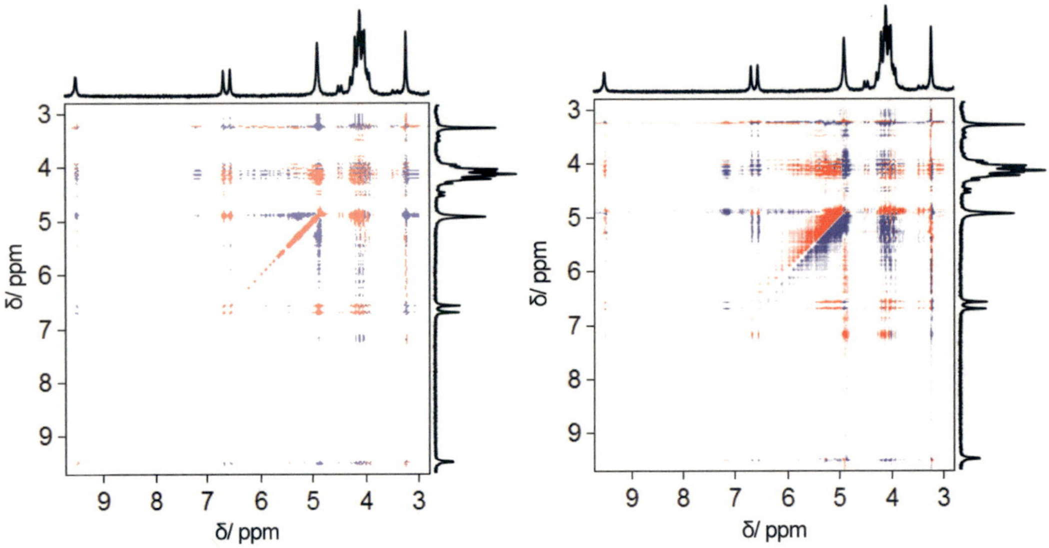

Figure 10. Synchronous (left) and asynchronous (right) 2D NMR correlation spectra of a Knoevenagel condensation of neat malonic acid diethylester and 2-propanal yielding butylidene malonic acid diethylester conducted in a microreaction system and monitored during 1200 min at a reaction temperature of 82°C using on-line 1D ^{1}H NMR (82 MHz, T = 36°C).

asynchronous map, the positive crosspeak at 9.5 and 6.8 ppm was seen indicative for the aldehyde reaction preceding the final product formation, i.e., the formation of the double bond.

Although the interpretation of the 1D ^{1}H NMR spectra was readily achieved, the use of in-line applicable techniques such as Raman and NIR spectroscopy was considered preferable from a process analytical perspective. This required the interpretation of the vibrational spectra. The increasing intensity of the Raman band at 1600 cm^{-1}, which originated from the carbon-carbon double bond vibration, was to some extent obvious in the series of 1D Raman spectra recorded. Further attribution of bands useful for reaction component monitoring was not readily achieved. To this purpose, hetero-correlation maps were computed from NMR and Raman spectra as well as from NMR and NIR, shown in **Figure 11**. Preprocessing of all spectra with respect to baseline correction, alignment, normalization, and data reduction or binning was found of utmost importance. The NMR signal assignment was readily transferred to the bands at 1600 and 800 cm^{-1} through correlations. The band at 800 cm^{-1} exhibited negative correlations to the product chemical shift and was hence found due to one of the educts. On inspection of the aldehyde resonance at 9.5 ppm, positive signs were found, which would be expected for a correlation between educts. The band at 1450 cm^{-1}, which was assigned to a methylene group bending vibration, showed only weak correlations. One of them correlated that band to the resonance at 3.2 ppm, indicating an educt-educt relationship. Analyzing the NIR-NMR hetero-covariance spectra, the enhancing power of NMR spectroscopy becomes even more evident. While NIR provides very broad bands that are due to either C-H or O-H overtone or combination frequencies and thus seemingly non-specific, the well-resolved signals of NMR spectroscopy assist in finding regions that can be attributed to educts or products and thus used for

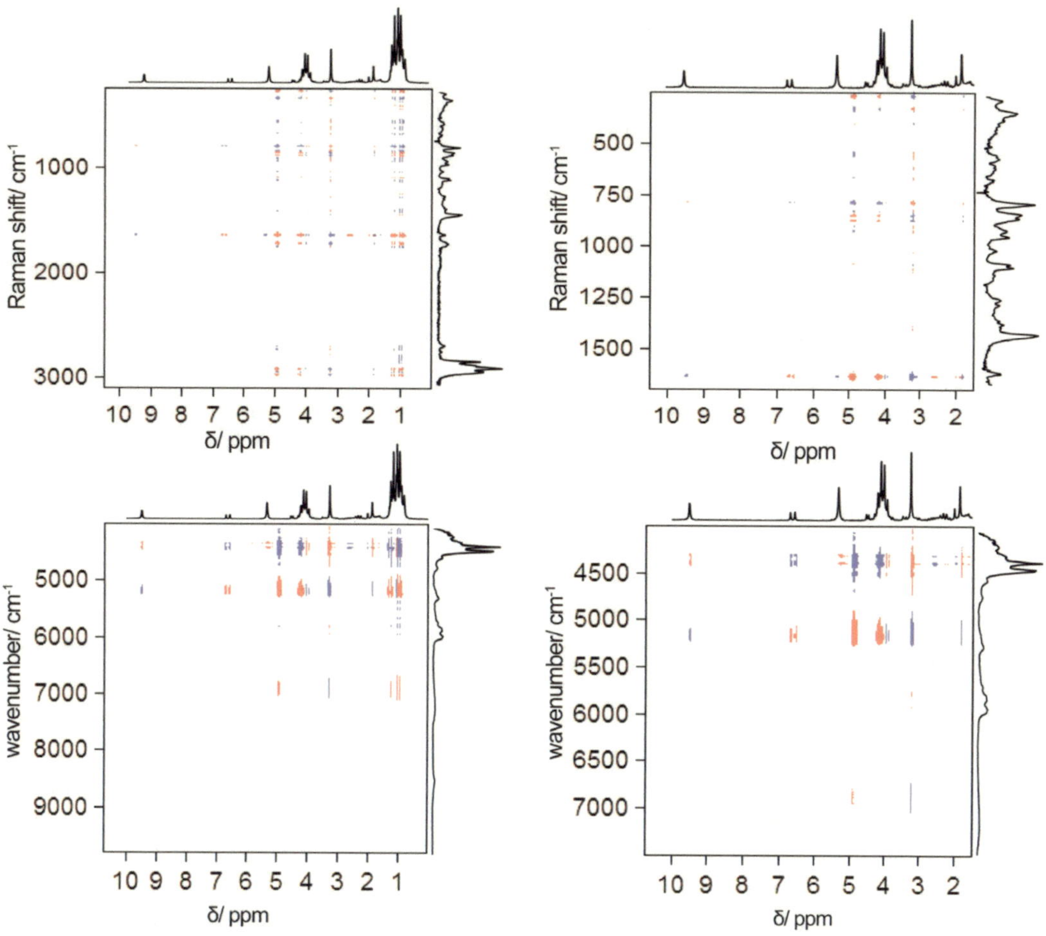

Figure 11. Synchronous Raman-NMR (top row) and NIR-NMR (bottom row) hetero-correlation spectra of a Knoevenagel condensation of neat malonic acid diethylester and 2-propanal yielding butylidene malonic acid diethylester conducted in a microreaction system and monitored during 1200 min at a reaction temperature of 82°C using on-line 1D ^{1}H NMR (82 MHz, T = 36°C), in-line NIR and in-line Raman spectroscopy (laser wavelength of 785 nm); full spectrum (left) and enlarged region (right), 1D spectra recorded at 1200 min are shown top and right of the correlation map.

reaction monitoring. The NIR-NMR correlation signals in **Figure 11**, bottom row, indicate that the O-H resonance around 7000 cm^{-1} stemmed from product water, while that around 5200 cm^{-1} was due to an educt C-H combination frequency. Thus, two potential monitoring frequency ranges could be identified.

Based on the thus identified and attributed signals, intensity-time plots and hence concentration-time curves could be extracted from the series of one-dimensional NIR spectra. This allowed the comparison of reaction monitoring by three different spectroscopic techniques, NIR, Raman, and NMR. The results were found in rather good agreement with each other. The concentration-time curves could be computed using chemical kinetic models from which reaction rate constants and half-lives were obtained. The reaction was found to follow first- or pseudo first-order reaction kinetics. It was expected that the knowledge of reaction parameters could later be transformed into automatic process control [70].

5. Conclusion

Covariance NMR has become a valuable tool in the ensemble of NMR methodologies. Generalized covariance was often performed with techniques other than NMR to profit from synchronous and asynchronous correlation maps. The synchronous map as a substitute to or along with the traditional Fourier transformed spectrum was nevertheless employed quite frequently in NMR. Hetero-spectroscopic covariance was used to concatenate NMR and mass spectrometry, NIR, and Raman data allowing combining the information of two techniques. Resolution improvement was reported an advantage of both the homo- and hetero-covariance processing, since information could be transferred from the well-resolved NMR domain into the less obvious to interpret vibrational domains. Here, the synchronous spectrum helped to increase resolution and assign signals to either the same or different species. As had been reported for vibrational spectroscopy, homo-covariance transformation also gives rise to two-dimensional data when only series of 1D NMR spectra are available, e.g., due to the application of low-field NMR instruments. Although the asynchronous map provides information on the sequential occurrence of signals, it has been relatively rarely exploited for NMR purposes. In a more recent study, Noda showed that more sophisticated mathematical processing was needed to derive the order of three or more species within a chemical reaction. When the asynchronous spectra were computed and analyzed, the sequential attribution feature proved very useful for PAT applications, such as in fermentation or reaction monitoring. Examples of wine fermentation, ethanol production using immobilized yeast, and monitoring of a radical polymerization and a Knoevenagel condensation in a microreaction system with a low-field NMR instrument were discussed. For quantitation of signal intensity changes and conclusions therefrom, the PARAFAC kernel analysis applied to polylactic acid nanocomposites with various clay content and at varied temperatures was summarized.

The opportunities of homo- and hetero-covariance spectroscopy in the field of NMR combined with other spectroscopic and spectrometric techniques are numerous. Still, new mathematical extensions continue to be devised. The authors therefore expect that with commercial software becoming more available for non-developing users, the reports on applications of homo- and hetero-covariance spectroscopy yielding synchronous and asynchronous spectra to chemical problems will steadily grow.

Acknowledgements

R. Legner is very grateful for a grant from the German Academic Scholarship Foundation. The authors thank the Niederrhein University of Applied Sciences for financial support. They are indebted to Professor Dr. Anna Nickisch-Hartfiel and Professor Dr. Peter Naderwitz for biochemical and technical expertise.

Author details

Martin Jaeger* and Robin Legner

*Address all correspondence to: martin.jaeger@hs-niederrhein.de

Department of Chemistry, Instrumental Analytical Chemistry and ILOC Institute for Coatings and Surface Chemistry, Niederrhein University, Krefeld, Germany

References

[1] Snyder DA, Bruschweiler R, Williams T, Martin G, Rovnyak D. Multi-dimensional spin correlations by covariance NMR. In: Modern NMR Approaches to the Structure Elucidation of Natural Products: Volume 1: Instrumentation and Software. The Royal Society of Chemistry. Cambridge, UK, 2016; pp. 244–258. DOI: 10.1039/9781849735186-00244

[2] Snyder DA, Bruschweiler R, Brüschweiler R, et al, Morris GE, Emsley JW. Multidimensional correlation spectroscopy by covariance NMR. In: Multidimensional NMR Methods for the Solution State. Chichester, UK: John Wiley & Sons Ltd, 2009; pp. 97–105. DOI: 10.1002/9780470034590.emrstm1098

[3] Jaeger M, Aspers RLEG. Covariance NMR and small molecule applications. Annual Reports on NMR Spectroscopy. 2014;**83**:271–349. DOI: 10.1016/B978-0-12-800183-7.00005-8

[4] Jaeger M, Aspers RLEG, Voigt M. Covariance NMR. In: Lindon JC, Tranter GE, Koppenaal DW (eds.) The Encyclopedia of Spectroscopy and Spectrometry, 3rd edition, 2017;**1**:396–400. Oxford: Academic Press. DOI: 10.1016/B978-0-12-409547-2.12106-7

[5] Park Y, Noda I, Jung YM. Novel developments and applications of two-dimensional correlation spectroscopy. Journal of Molecular Structure. 2016;**1124**:11–28. DOI: http://dx.doi.org/10.1016/j.molstruc.2016.01.028

[6] Noda I. Frontiers of two-dimensional correlation spectroscopy. Part 1. New concepts and noteworthy developments. Journal of Molecular Structure. 2014;**1069**:3–22. DOI: http://dx.doi.org/10.1016/j.molstruc.2014.01.025

[7] Noda I. Frontiers of two-dimensional correlation spectroscopy. Part 2. Perturbation methods, fields of applications, and types of analytical probes. Journal of Molecular Structure. 2014;**1069**:23–49. DOI: 10.1016/j.molstruc.2014.01.016

[8] Blinov KA, Larin NI, Williams AJ, et al. Unsymmetrical covariance processing of COSY or TOCSY and HSQC NMR data to obtain the equivalent of HSQC-COSY or HSQC-TOCSY spectra. Journal of Heterocyclic Chemistry. 2006;**43**:163–166. DOI: 10.1002/jhet.5570430124

[9] Noda I. Two-dimensional infrared spectroscopy. Journal of the American Chemical Society. 1989;**111**:8116–8118. DOI: 10.1021/ja00203a008

[10] Frasinski LJ, Codling K, Hatherly PA. Covariance mapping: A correlation method applied to multiphoton multiple ionization. Science. 1989;**246**:1029–1031. DOI: 10.1126/science.246.4933.1029

[11] Noda I. Two-dimensional infrared (2D IR) spectroscopy: Theory and applications. Applied Spectroscopy. 1990;**44**:550–561. DOI: 10.1366/0003702904087398

[12] Noda I. Generalized two-dimensional correlation method applicable to infrared, raman, and other types of spectroscopy. Applied Spectroscopy. 1993;**47**:1329–1336. DOI: 10.1366/0003702934067694

[13] Noda I, Dowrey AE, Marcott C, et al. Generalized two-dimensional correlation spectroscopy. Applied Spectroscopy. 2000;**54**:236A-248A. DOI: 10.1366/0003702001950454

[14] Eads CD, Noda I. Generalized correlation NMR spectroscopy. Journal of the American Chemical Society. 2002;**124**:1111–1118. DOI: 10.1021/ja011819v

[15] Noda I. Determination of two-dimensional correlation spectra using the Hilbert transform. Applied Spectroscopy. 2000;**54**:994–999. DOI: 10.1366/0003702001950472

[16] Bruschweiler R. Theory of covariance nuclear magnetic resonance spectroscopy. The Journal of Chemical Physics. 2004;**121**:409–414. DOI: 10.1063/1.1755652

[17] Ernst RR, Bodenhausen G, Wokaun A. Principles of Nuclear Magnetic Resonance in One and Two Dimensions. Oxford: Clarendon Press; 1990. Available from: http://books.google.de/books?id=4bohtceju9kC

[18] Bax A. Two-Dimensional Nuclear Magnetic Resonance in Liquids. Dordrecht, Holland: D. Reidel Publishing Company; 1982

[19] Martin GE, Irish PA, Hilton BD, et al. Utilizing unsymmetrical indirect covariance processing to define 15N-13C connectivity networks. Magnetic Resonance in Chemistry. 2007;**45**:624–627. DOI: 10.1002/mrc.2029

[20] Martin GE, Hilton BD, Irish PA, et al. Application of unsymmetrical indirect covariance NMR methods to the computation of the 13C.tautm.15N HMBC-IMPEACH correlation spectra. Magnetic Resonance in Chemistry. 2007;**45**:883–888. DOI: 10.1002/mrc.2064

[21] Martin GE, Hilton BD, Blinov KA, et al. Using indirect covariance spectra to identify artifact responses in unsymmetrical indirect covariance calculated spectra. Magnetic Resonance in Chemistry. 2008;**46**:138–143. DOI: 10.1002/mrc.2141

[22] Crockford DJ, Holmes E, Lindon JC, et al. Statistical heterospectroscopy, an approach to the integrated analysis of NMR and UPLC-MS data sets: Application in metabonomic toxicology studies. Analytical Chemistry. 2006;**78**:363–371. DOI: 10.1021/ac051444m

[23] Bruschweiler R, Zhang F. Covariance nuclear magnetic resonance spectroscopy. The Journal of Chemical Physics. 2004;**120**:5253–5260. DOI: 10.1063/1.1647054

[24] Hu B-W, Zhou P, Noda I, et al. An NMR approach applicable to biomolecular structure characterization. Analytical Chemistry. 2005;**77**:7534–7538. DOI: 10.1021/ac051061o

[25] Noda I. Two-dimensional codistribution spectroscopy to determine the sequential order of distributed presence of species. Journal of Molecular Structure. 2014;**1069**:50–59. DOI: 10.1016/j.molstruc.2014.01.024

[26] Martin GE. Posaconazole: Application of HSQC-ADEQUATE from general indirect covariance processing. Journal of Heterocyclic Chemistry. 2012;**49**:716–720. DOI: 10.1002/jhet.892

[27] Aspers RLEG, Geutjes PETJ, Honing M, et al. Using indirect covariance processing for structure elucidation of small molecules in cases of spectral crowding. Magnetic Resonance in Chemistry. 2011;**49**:425–436. DOI 10.1002/mrc.2766

[28] Zhang F-L, Bruschweiler-Li L, Bruschweiler R. Simultaneous de novo identification of molecules in chemical mixtures by doubly indirect covariance NMR spectroscopy. Journal of the American Chemical Society. 2010;**132**:16922–16927. DOI: 10.1021/ja106781r

[29] Bingol K, Salinas RK, Brueschweiler R. Higher-rank correlation NMR spectra with spectral moment filtering. The Journal of Physical Chemistry Letters. 2010;**1**:1086–1089. DOI: 10.1021/jz100264g

[30] Lindon JC, Nicholson JK. Spectroscopic and statistical techniques for information recovery in metabonomics and metabolomics. Annual Review of Analytical Chemistry. 2008;**1**:45–69. DOI: 10.1146/annurev.anchem.1.031207.113026

[31] Robinette SL, Zhang F, Bruschweiler-Li L, et al. Web server based complex mixture analysis by NMR. Analytical Chemistry (Washington, DC, USA). 2008;**80**:3606–3611. DOI: 10.1021/ac702530t

[32] Robinette SL, Lindon JC, Nicholson JK. Statistical spectroscopic tools for biomarker discovery and systems medicine. Analytical Chemistry. 2013;**85**:5297–5303. DOI: 10.1021/ac4007254

[33] Noda I. Recent developments in two-dimensional (2D) correlation spectroscopy. Chinese Chemical Letters. 2015;**26**:167–172. DOI: 10.1016/j.cclet.2014.10.006

[34] Shinzawa H, Nishida M, Kanematsu W, et al. Parallel factor (PARAFAC) kernel analysis of temperature- and composition-dependent NMR spectra of poly(lactic acid) nanocomposites. Analyst. 2012;**137**:1913–1921. DOI: 10.1039/C2AN16019F

[35] Tomasi G, Bro R. A comparison of algorithms for fitting the PARAFAC model. Computational Statistics & Data Analysis. 2006;**50**:1700–1734. DOI: http://dx.doi.org/10.1016/j.csda.2004.11.013

[36] Smilde A, Bro R, Geladi P. Three-Way component and regression models. In: Multi-Way Analysis with Applications in the Chemical Sciences. John Wiley & Sons, Ltd, Chichester, UK, 2004; pp. 57–87. DOI: 10.1002/0470012110.ch4

[37] Qi J, Li H, Huang K, et al. Orthogonal sample design scheme for two-dimensional synchronous spectroscopy and its application in probing intermolecular interactions. Applied Spectroscopy. 2007;**61**:1359–1365. DOI: 10.1366/000370207783291993

[38] Qi J, Huang K, Gao X, et al. Orthogonal sample design scheme for two-dimensional synchronous spectroscopy: Application in probing lanthanide ions interactions with organic ligands in solution mixtures. Journal of Molecular Structure. 2008;**883-884**:116–123. DOI: 10.1016/j.molstruc.2008.01.036

[39] Liu Y, Zhang C, Liu S, et al. Modified orthogonal sample design scheme to probe intermolecular interactions. Journal of Molecular Structure. 2008;**883-884**:124–128. DOI: 10.1016/j.molstruc.2007.12.025

[40] Shinzawa H, Morita S-I, Awa K, et al. Multiple perturbation two-dimensional correlation analysis of cellulose by attenuated total reflection infrared spectroscopy. Applied Spectroscopy. 2009;**63**:501–506. DOI: 10.1366/000370209788346977

[41] Shinzawa H, Hashimoto K, Sato H, et al. Multiple-perturbation two-dimensional (2D) correlation analysis for spectroscopic imaging data. Journal of Molecular Structure. 2014;**1069**:176–182. DOI: 10.1016/j.molstruc.2014.02.013

[42] Noda I. Two-dimensional correlation spectroscopy—Biannual survey 2007-2009. Journal of Molecular Structure. 2010;**974**:3–24. DOI: 10.1016/j.molstruc.2010.01.069

[43] FDA. Guidance for industry PAT: A framework for innovative pharmaceutical development, manufacuring, and quality assurance. FDA Off Doc. 2004;**16**. DOI: http://www.fda.gov/CDER/guidance/6419fnl.pdf

[44] (a) EMEA PAT Team, Reflection Paper, Chemical, pharmaceutical and biological information to be included in dossiers when Process Analytical Technology (PAT) is employed. EMEA/INS/277260/2005, March 2006. (b) EMEA Note for guidance on analytical development, EMEA/CHMP/ICH/167068/2004

[45] Kessler RW, editor. Prozessanalalytik: Strategien und Fallbeispiele aus der industriellen Praxis. Weinheim: Wiley-VCH; 2006

[46] Bakeev KA, editor. Process Analytical Technology. Spectroscopic Tools and Implementation Strategies for the Chemical and Pharmaceutical Industries. 2nd ed. Chichester: John Wiley & Sons, Ltd; 2010

[47] Jähnisch K, Hessel V, Löwe H, et al. Chemistry in microstructured reactors. Angewandte Chemie International Edition. 2004;**43**:406–446. DOI: 10.1002/anie.200300577

[48] Bogdan AR, Poe SL, Kubis DC, et al. The continuous-flow synthesis of ibuprofen. Angewandte Chemie. 2009;**121**:8699–8702. DOI: 10.1002/ange.200903055

[49] Ehrfeld W, Hessel V, Löwe H. Microreactors—New technology for modern chemistry. Organic Process Research & Development. 2001;**5**:89–89. DOI: 10.1021/op000071i

[50] Zalesskiy SS, Danieli E, Blümich B, et al. Miniaturization of NMR systems: Desktop spectrometers, microcoil spectroscopy, and 'NMR on a Chip' for chemistry, biochemistry, and industry. Chemical Reviews. 2014;**114**:5641–5694. DOI: 10.1021/cr400063g

[51] Danieli E, Perlo J, Duchateau ALL, et al. On-line monitoring of chemical reactions by using Bench-Top nuclear magnetic resonance spectroscopy. ChemPhysChem. 2014;**15**:3060–3066. DOI: 10.1002/cphc.201402049

[52] McGill CA, Nordon A, Littlejohn D. Comparison of in-line NIR, Raman and UV-visible spectrometries, and at-line NMR spectrometry for the monitoring of an esterification reaction. Analyst. 2002;**127**:287–292. DOI: 10.1039/B106889J

[53] Maiwald M, Fischer HH, Kim YK, et al. Quantitative high-resolution on-line NMR spectroscopy in reaction and process monitoring. Journal of Magnetic Resonance. 2004;**166**:135–146. DOI: 10.1016/j.jmr.2003.09.003

[54] Li Y, Hu B, Chen Q, et al. Comparison of various sampling schemes and accumulation profiles in covariance spectroscopy with exponentially decaying 2D signals. Analyst. 2013;**138**:2411–2419. DOI: 10.1039/C3AN36375A

[55] Lafon O, Hu B, Amoureux J-P, et al. Fast and high-resolution stereochemical analysis by nonuniform sampling and vovariance processing of anisotropic natural abundance 2D 2H NMR datasets. Chemistry—A European Journal. 2011;**17**:6716–6724. S6716/1–S6716/6. DOI: 10.1002/chem.201100461

[56] Chen Y, Zhang F, Bermel W, et al. Enhanced covariance spectroscopy from minimal datasets. Journal of the American Chemical Society. 2006;**128**:15564–15565. DOI: 10.1021/ja065522e

[57] Li Y, Wang Q, Zhang Z, et al. Covariance spectroscopy with a non-uniform and consecutive acquisition scheme for signal enhancement of the NMR experiments. Journal of Magnetic Resonance. 2012;**217**:106–111. DOI: 10.1016/j.jmr.2012.02.016

[58] Takeda K, Kusakabe Y, Noda Y, et al. Homo- and heteronuclear two-dimensional covariance solid-state NMR spectroscopy with a dual-receiver system. Physical Chemistry Chemical Physics. 2012;**14**:9715–9721. DOI: 10.1039/c2cp41191a

[59] Garcia H, Barros AS, Gonçalves C, et al. Characterization of dextrin hydrogels by FTIR spectroscopy and solid state NMR spectroscopy. European Polymer Journal. 2008;**44**:2318–2329. DOI: 10.1016/j.eurpolymj.2008.05.013

[60] Robinette SL, Bruschweiler R, Schroeder FC, et al. NMR in metabolomics and natural products research: Two sides of the same coin. Accounts of Chemical Research. 2012;**45**:288–297. DOI: 10.1021/ar2001606

[61] Kirwan GM, Clark S, Barnett NW, et al. Generalised 2D-correlation NMR analysis of a wine fermentation. Analytica Chimica Acta. 2008;**629**:128–135. DOI: 10.1016/j.aca.2008.09.046

[62] Sasic S. Two-dimensional correlation analysis of nuclear magnetic resonance metabonomics data. Applied Spectroscopy. 2008;**62**:840–846. DOI: 10.1366/000370208785284439

[63] Kirwan GM, Adams MJ. Peak width issues with generalised 2D correlation NMR spectroscopy. Journal of Molecular Structure. 2008;**892**:225–230. DOI: 10.1016/j.molstruc.2008.05.031

[64] Ryu SR, Bae WM, Hong WJ, et al. Characterization of chain transfer reaction during radical polymerization of silver nanocomposite polyvinylpyrrolidone by using 2D hetero-spectral IR/NMR correlation spectroscopy. Vibrational Spectroscopy. 2012;**60**:168–172. DOI: 10.1016/j.vibspec.2011.12.009

[65] Legner R, Wirtz A, Tiedt M, et al. On-line monitoring of ethanol fermentation using homo and hetero covariance low-field NMR, Raman and NIR spectroscopy. Prep

[66] Cavinato AG, Mayes DM, Ge Z, et al. Noninvasive method for monitoring ethanol in fermentation processes using fiber-optic near-infrared spectroscopy. Analytical Chemistry. 1990;**62**:1977–1982. DOI: 10.1021/ac00217a015

[67] Blanco M, Peinado AC, Mas J. Analytical monitoring of alcoholic fermentation using NIR spectroscopy. Biotechnology and Bioengineering. 2004;**88**:536–542. DOI: 10.1002/bit.20214

[68] Burikov S, Dolenko T, Patsaeva S, et al. Raman and IR spectroscopy research on hydrogen bonding in water–ethanol systems. Molecular Physics. 2010;**108**:2427–2436. DOI: 10.1080/00268976.2010.516277

[69] Nishii T, Genkawa T, Watari M, et al. Selection of the NIR region for a regression model of the ethanol concentration in fermentation process by an online NIR and Mid-IR Dual-Region spectrometer and 2D heterospectral correlation spectroscopy. Analytical Sciences. 2012;**28**:1165–1170. DOI: 10.2116/analsci.28.1165

[70] Legner R, Haefner S, Voigt M, et al. Micro process analytical technology. GIT Lab J. 2016;**3-4**:38–40. Available from: http://www.laboratory-journal.com/science/chemistry-physics/microprocess-analytical-technology

Chapter 5

1H and 13C NMR for the Profiling of Natural Product Extracts: Theory and Applications

Fabian M. Dayrit and Angel C. de Dios

Additional information is available at the end of the chapter

http://dx.doi.org/10.5772/intechopen.71040

Abstract

Nuclear magnetic resonance (NMR) spectroscopy and mass spectrometry (MS) are the principal methods of metabolomics, the branch of '-omics' that deals with small molecules. Although MS is gaining popularity in metabolomics, NMR enjoys a number of key advantages because it is nondestructive, unbiased, quantitative, does not require separation or derivatization, and is amenable to compounds that are difficult to analyze by gas chromatography-mass spectrometry (GC-MS) and liquid chromatography-mass spectrometry (LC-MS). There are two general approaches to the use of NMR for profiling studies: an untargeted approach, which uses chemometric analysis; and a targeted approach, which aims to quantify known compounds in the extract. These approaches, however, are not mutually exclusive and will likely converge in the future. This paper will describe the basic theoretical principles that should be considered to develop NMR into a standard quantitative method. Although 1H NMR is more sensitive, 13C NMR spectra are simpler with less overlapping signals and are less affected by different magnetic field strengths. Various applications of 1H and 13C NMR for the profiling of natural products are described. The use of two-dimensional 1H NMR has been used to overcome problems of spectral overlap. The standardization of the NMR protocol will make it a more useful tool for the profiling of natural products extracts.

Keywords: nuclear magnetic resonance, 1H NMR, 13C NMR, natural products profiling, metabolomics, chemometrics

1. Introduction

The objective of this paper is to review the applications of 1H and 13C nuclear magnetic resonance (NMR) for the quantitative profiling of plant natural products extracts and the theoretical parameters that should be considered, if it is to become a more useful tool.

NMR and mass spectrometry (MS) are the principal methods of metabolomics, the branch of '-omics' that deals with small molecules. The Metabolomics Society describes metabolomics as: "the comprehensive characterization of the small molecule metabolites in biological systems" [1]. NMR has a number of characteristics that meet the requirements of metabolomics: it is accurate, quantitative, comprehensive, unbiased, and is able to provide information that can be used to determine molecular structure. The review will discuss these aspects in detail.

1.1. NMR and MS

Although MS techniques, such as gas chromatography-mass spectrometry (GC-MS) and liquid chromatography-mass spectrometry (LC-MS), are most commonly used in metabolomics, NMR still enjoys a number of key advantages. In particular, NMR is nondestructive, unbiased, quantitative, does not require separation or derivatization, and is amenable to compounds that are difficult to analyze by GC-MS and LC-MS. For example, GC-MS often requires derivatization of compounds, such as sugars and amines. LC-MS, on the other hand, generally requires sample preparation, chromatographic separation, specific experimental and ionization conditions, instrumentation and operator skill [2]. These make it difficult to standardize MS analysis. In contrast, NMR does not require elaborate sample preparation and fractionation, is highly reproducible, and is able to provide both qualitative and quantitative information on chemically diverse compounds [3, 4]. The standardization of the NMR protocol will further improve the usefulness of NMR as a tool for the profiling of natural products extracts. Because NMR is able to detect compounds only down to 0.1% level, it is not suitable for the detection of trace components. NMR is less sensitive than MS, which can detect compounds down to parts per million (ppm) levels. Because of the distinct advantages of each method, NMR and MS are considered as complementary techniques.

NMR is a quantitative spectroscopic tool because the intensity of the peaks is directly proportional to the number of nuclei. With improvements in electronics and the use of higher magnetic field strengths, the sensitivity and resolving power of NMR has improved. However, the lack of standardized protocols has limited its quantitative application and many consider NMR mainly as a qualitative method, mainly for chemical structure determination and molecular dynamics [5].

The use of NMR as a quantitative method has been expanding, giving rise to the term "quantitative NMR" (qNMR). The pharmaceutical industry, which has stringent requirements of analysis, has been turning to the use of qNMR in early drug development to address the need for rapid, selective, and accurate analysis without requiring expensive and tedious chromatographic methods. It is also worth noting that qNMR meets the stringent regulatory standards of the pharmaceutical industry, including the International Conference on Harmonization. qNMR has been applied mainly to 1H nuclei although 19F and 31P NMR have also been used where appropriate because of their 100% isotopic abundance [6]. The main advantages of qNMR are its accuracy, reproducibility, and flexibility with respect to the nature of the analyte, the only requirement being the presence of protons and carbon, and its ability to simultaneously quantify multiple analytes, especially when validated using external calibration. Quantitative 1H NMR (qHNMR) has been shown to have an accuracy and precision of ±1%

and an uncertainty of measurement of less than 0.1%. This makes it suitable as a metrological technique for the certification of purity of organic compounds [7].

There are two general approaches to the use of NMR for profiling or metabolomics studies. In the first approach, only the spectral patterns (chemical shifts and intensities) are recorded and are used to compare and group samples. In this approach, compounds are not initially identified. Because statistical tools, such as principal components analysis (PCA) are used, this is sometimes called a chemometric approach. In the second approach, particular compounds which are known to be present in the extract are identified and quantified using a reference spectral library. This approach is referred to as quantitative or targeted metabolomics [8]. These approaches, however, are not mutually exclusive and will likely converge in the future with improved statistical tools and bigger NMR spectral databases.

Because of the large amount of data that are produced, statistical methods, known as chemometrics, are applied to reduce the number of variables. Chemometrics is a family of statistical techniques that are applied to large sets of chemical data, such as NMR chemical shift peaks, with the objective of gaining insights into the characteristics of the samples through the use of graphical representation [9]. Because chemometrics is able to process large amounts of data, it is an ideal tool for NMR which produces a lot of data (chemical shifts). This can be used to find patterns of groupings and correlations among natural product samples which can be used for quality control and standardization [10]. Since chemometrics started to be applied to NMR around the year 2000, progress has been very rapid. Chemometrics has been used to classify whole plant samples based on their NMR profiles according to species, origin, processing treatment, age, and various quality parameters [11].

2. 1H and 13C NMR as profiling methods

In a talk given during the William Draper Harkins Lecture, University of Chicago in 1991, Alexander Pines mentioned that his organic chemistry colleagues at Berkeley consider two vital instruments in a research laboratory: a balance and an NMR spectrometer. His view is not surprising as decades of improvement in both instrumentation and techniques had rendered the NMR spectrometer as a tool of choice in characterizing molecules, from the structures of natural products and synthetic organic compounds to biomolecules and organometallic complexes. NMR spectroscopy takes advantage of the interaction between nuclei that are acting as tiny magnets and an external magnetic field and this provides a powerful means of probing the chemical bonding and environment of the nucleus. These phenomena are key to the applicability of 1H and 13C NMR to natural products.

2.1. 1H NMR spectroscopy

Hydrogen is present in almost every organic molecule, and its major isotope, 1H, has an abundance of 98.985%. The 1H nucleus reports a frequency specific to its immediate vicinity in an NMR spectrum. This frequency is extremely sensitive to the electronic environment thus giving each 1H nucleus in an organic compound a type of identification number, called the

NMR chemical shift. Magnetic nuclei, such as 1H, also interact with each other. In solution or liquid-state NMR spectroscopy, these interactions, called couplings, are observed as "splitting" of lines in an NMR spectrum. The magnitude of these couplings not only depends on the number and type of bonds separating the interacting pair of 1H nuclei but also on the spatial orientation between the nuclei. Both NMR chemical shifts and coupling constants provide immense information regarding structure and environment. Hence, NMR spectroscopy has become a powerful tool for the determination of organic structure.

These NMR interactions (chemical shifts and coupling constants), although very sensitive, are quite weak such that improvements in their detection have been one of the primary goals of developments in NMR instrumentation. Such limitations are no longer severe. The use of pulses and data processing by Fourier transformation, first introduced by Ernst and Anderson [12] and the availability of high-field superconducting magnets have allowed for efficient signal averaging such that nowadays, with an 11 T magnet (500 MHz), a 1H NMR spectrum can be obtained even from very dilute solutions (micromolar concentration).

Pulse Fourier transform NMR spectroscopy, as in other spectroscopic methods, involves transitions between energy levels. However, unlike other spectroscopic methods, the transition probability in an NMR excitation is the same regardless of chemical environment. NMR spectroscopy does not need to consider oscillator strengths or extinction coefficients, which are important for infrared and UV-visible spectroscopy, respectively. The intensity of an NMR signal is determined solely by the excitation pulse, strength of the external magnetic field, and temperature. The magnetic field strength and temperature determine the Boltzmann population difference between the two energy levels while the excitation pulse dictates the extent of the transition. Since only one pulse is often used to excite all of the 1H nuclei in a sample, the extent of transitions is the same for all. Furthermore, the NMR chemical shift, which reflects the differences in resonance frequencies of inequivalent 1H nuclei, is very small: the differences are in parts per million (ppm). Hence, the Boltzmann distribution for the two spin states is essentially the same for every 1H in a molecule. Indeed, as early as 1963, the area under each peak in a 1H NMR spectrum has been shown to correspond proportionally to the number of hydrogen atoms sharing the same environment in a given compound [13]. This quantitative aspect applies not only to pure substances but also to mixtures. In fact, during the same year, a successful quantitative analysis by 1H NMR spectroscopy of a mixture of aspirin, phenacetin, and caffeine was demonstrated [14].

2.2. 13C NMR spectroscopy

13C also has a spin of ½ and is therefore likewise NMR active. However, because the 13C isotope occurs at only 1.108%, it is difficult to observe. (The major carbon isotope, 12C, is not NMR-active.) David Grant and coworkers published a series of papers on 13C NMR spectroscopy that spanned two decades [15, 16]. In the first paper of this series, inherent difficulties in observing 13C NMR spectra were addressed by proton decoupling and sample spinning. Since carbon atoms are frequently attached to hydrogen atoms in organic compounds, 13C-1H coupling is present and leads to splitting of 13C resonances. Proton decoupling removes this interaction, consolidating multiple 13C peaks into a single taller peak.

Moreover, additional enhancement of 13C signals is observed when the 1H spin populations are perturbed, similar to the effect observed by Overhauser with electron spins [17]. Taking advantage of both the nuclear Overhauser effect (NOE) and the increased signal due to the collapse of multiple peaks, measurement of 13C NMR spectra became routine and easy to interpret. Being in the proximity of more than one pair of electrons, 13C nuclei offer a much wider range of chemical shifts than 1H (200 ppm for 13C versus 10 ppm for 1H). In addition, since the probability that a 13C nucleus is attached to another 13C nucleus is very small (about 0.0001), 13C-13C couplings are usually not observed thereby providing a much simpler 13C NMR spectrum.

Using 13C NMR spectroscopy as a powerful analytical tool can be easily appreciated by considering the three isomers of a simple hydrocarbon C_5H_{12} (see **Figure 1**). n-Pentane ($CH_3CH_2CH_2CH_2CH_3$), produces three peaks with a 1:2:2 intensity ratio, 2-methylbutane ($(CH_3)_2CHCH_2CH_3$) displays four peaks with a 1:1:2:1 intensity ratio, and neopentane ($(CH_3)_4C$) gives two peaks of 4:1 intensity ratio. For the above reasons, a qualitative and quantitative analysis that is nondestructive and requires no separation is possible with 13C NMR spectroscopy [18]. All that one needs is a library of 13C NMR spectra of all possible components, a good spectral prediction software, and an efficient algorithm that can do the search and construct a simulated spectrum that matches the observed spectrum. All of these requirements are already available today. A similar treatment has been shown to be feasible in determining the acyl profile in various vegetable oils [19] and in characterizing the various sesquiterpenes in essential oils from juniper, rosemary, cedarwood, and ginger [20].

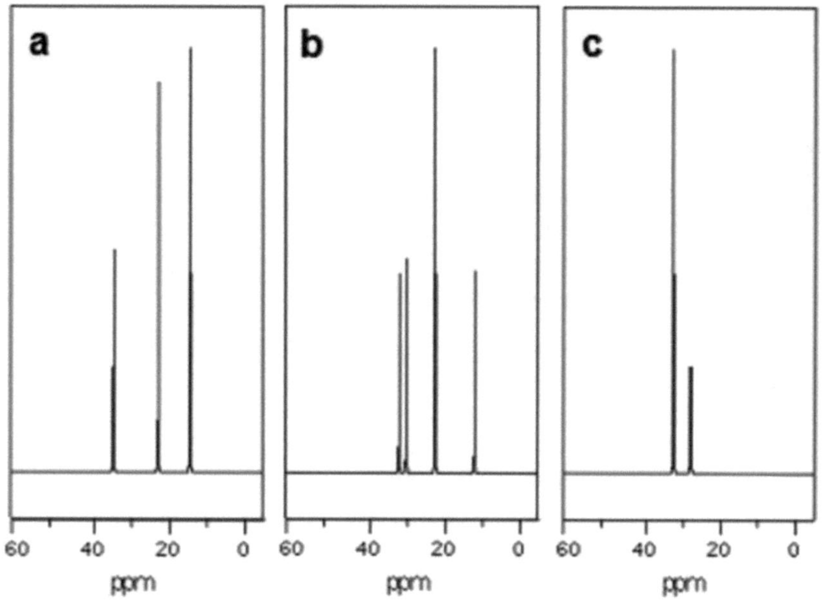

Figure 1. 13C NMR spectra of (a) n-pentane, (b) 2-methylbutane, and (c) neopentane.

The promise of a wealth of information that NMR spectroscopy offers, however, comes also with challenges. Since the frequencies observed depend on the magnetic field strength, the peaks' shapes and widths are sensitive to the homogeneity of the magnetic field throughout the sample. Experimentally, corrections to field homogeneity are done through a process called shimming, which involves adding small magnetic field gradients. Shimming used to be an art and both symmetry and narrowness of an NMR peak depended on the expertise of the NMR operator. Fortunately, with new superconducting magnets and automated shimming, 13C NMR spectra can now be made reproducible and comparable regardless of who operates the spectrometer. However, there are still numerous factors that are independent of the NMR operator which can affect the appearance of an NMR spectrum.

2.3. NMR chemical shifts and coupling constants

Since the frequencies (in hertz) observed for each NMR-active nucleus are dependent on the field strength, chemical shifts are reported in dimensionless units of parts per million (ppm), which then becomes independent of the magnetic field strength. Interactions between nuclei, on the other hand, are independent of field strength, so these are recorded in units of frequency, hertz. Since the ppm equivalent of a hertz is determined by the strength of the magnetic field, splittings will appear narrower in a high-field magnet than in a low-field magnet. When the coupling interactions are of the same magnitude as the chemical shift differences, the coupling pattern is complex [21]. A hypothetical example for 1H NMR is shown in **Figure 2**, where the coupling constant is equal to the chemical shift difference in a spectrometer operating with a 1H frequency of 100 MHz. As the strength of the magnetic field increases, chemical shift differences (in hertz) also increase, which can dramatically change the appearance of the spectrum. In this particular example, the spectrum only begins to appear simpler with a spectrometer operating at 1 GHz, in which the chemical shift difference is now 10 times bigger than the coupling constant; this is called a first-order spectrum. Thus, 1H spectra taken at different magnetic field strengths appear different. On the other hand, 13C spectra appear similar at different magnetic field strengths. This is because 13C-13C coupling is not observed due to low natural abundance, 13C-1H couplings, although present, are always several orders of magnitude lower than the frequency difference between these two nuclei, and proton-decoupled 13C NMR spectra are singlets. Therefore, although 13C presents detection challenges due to its lower frequency and low natural abundance, 13C has the advantage over 1H with regard to simplicity of NMR spectra.

Absolute frequencies for NMR transitions are seldom used since these numbers are dependent on the strength of the external magnetic field. Chemical shift differences are instead reported in ppm, which is the ratio of the absolute frequency with respect to the frequency of a reference compound, such as tetramethylsilane (TMS). Alternatively, the solvent can be utilized as internal reference. Due to the sensitivity of the NMR chemical shift, intermolecular effects are also frequently observed [22]. Since solvents are known to induce shifts, it is important that when comparing different spectra, the same solvent should be used. Since 13C has a much wider chemical shift range, the effect of solvent on chemical shift is smaller for 13C (2/200) than that of 1H (0.7/10). Furthermore, since carbon atoms, unlike hydrogen atoms, reside on the interior of the molecule, 13C is generally shielded from environmental effects, such as intermolecular interactions and solvent effects. This is one reason why 13C NMR chemical shifts are nearly exclusively dependent only on its covalent bonding interactions [23]. The greater susceptibility of 1H NMR chemical shifts to solvent effects makes 13C NMR spectroscopy a better alternative in profiling natural products. Solvent effects on both 1H and 13C NMR chemical shifts are expected to be dominated by van der Waals interactions with the solvent. These interactions are largely nonspecific thus an approximation that the solvent simply causes a constant offset on all resonances may be valid. Using an internal reference can therefore easily remove effects of the medium on the observed chemical shifts. Attention, however, is still required for sites that can participate in electrostatic interactions and hydrogen bonding. 13C in carbonyl groups is one example [24].

Temperature can also affect observed chemical shifts through changes in the density of the sample as well as changes in the internal motions of the molecule [25]. For a fair comparison of library and sample spectra, it is important that spectra are taken at the same temperature.

Lastly, a quantitative 13C NMR spectrum requires uniform excitation of all 13C nuclei. The wider chemical shift range and lower frequency for 13C necessitate excitation pulses with much higher power with a pulse that is less than 15 μs long so that the entire chemical shift

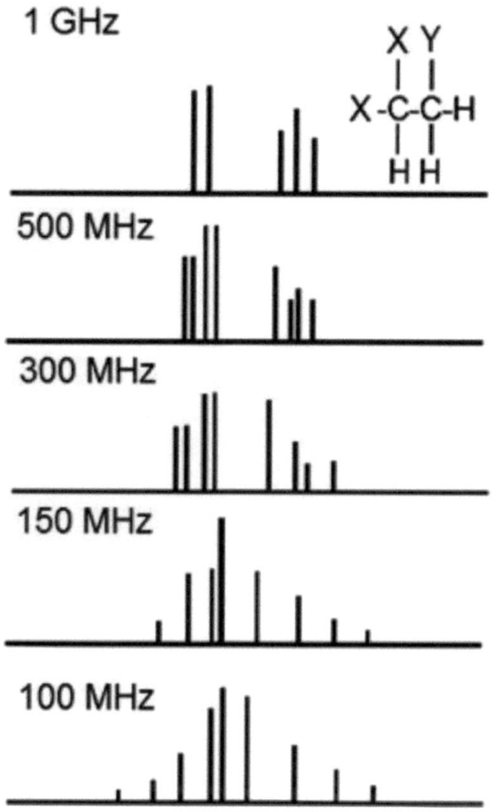

Figure 2. 1H NMR spectra of strongly coupled nuclei at various magnetic field strengths.

range is uniformly irradiated [26]. Furthermore, proton decoupling is also regularly used to collapse multiple 13C peaks, but this can lead to NOEs which enhance 13C nuclei that are directly bound to protons, making 13C NMR no longer uniform for carbon nuclei with different numbers of attached protons. The inverse-gated 13C NMR experiment can be used to overcome these problems. This involves turning the proton decoupler on only during acquisition and providing adequate time for all the 13C nuclei to relax [27]. 13C nuclei are most often relaxed by a nearby 1H nucleus. Thus, the needed relaxation time (equal to $5 \times T_1$) can be quite long for compounds that contain quaternary carbons. These quaternary 13C nuclei may require minutes to relax and this dramatically increases the time required for NMR experiments. Because running such a lengthy 13C NMR experiment is not practical, it is normal practice to run proton-decoupled 13C NMR using standard conditions and to use the resulting spectra for pattern recognition but not for quantitation.

2.4. Reproducibility of NMR spectra

The use of a library of NMR spectra in the analysis of a mixture of natural products requires reproducibility of both chemical shift and peak intensity. Since samples of natural products are often dissolved in either dimethyl sulfoxide or methanol, confining both library and sample data to these two solvents can easily ameliorate the confounding effects of the solvent on the observed chemical shifts. Modern NMR spectrometers are normally equipped with temperature control so the measurements can be made at a given temperature, which also avoids the temperature dependence of NMR chemical shifts, thus eliminating this problem. The reproducibility of peak intensities, however, requires additional considerations.

The robustness of current NMR instrumentation is evident in successful indirect detection methods during which resonances from 1H nuclei bound to 12C are separated from those attached to 13C [28]. Indirect detection is possible only if the scans or transients are highly reproducible such that these can be added to extract the desired resonances and remove completely the unwanted signals. However, this robustness only entails the reproducibility of an NMR experiment from one transient to the next. It does not address the reproducibility of NMR experiments among different laboratories. Thus, there is a need to standardize both NMR acquisition conditions and processing parameters.

The intensity of an NMR peak depends on the duration of the pulse. Equalizing the $m = +½$ and $m = -½$ spin populations requires what is called in NMR spectroscopy as a 90° pulse. Peak intensities are at a maximum with this pulse. Since all nuclei in a sample are subject to the same pulse, it is not necessary that a 90° pulse is always employed. For an NMR spectrum to be quantitative, the relative, not the absolute, peak intensities are sufficient. However, the extent of the pulse determines how much time is required for relaxation between transients. For signal averaging to be effective, one still needs to make sure that the spins have reached equilibrium before applying the next pulse so as to avoid saturation, which leads to loss of signal [29]. When a 90° pulse is employed, the time between transients must be at least five times as long as the relaxation time. The time required between pulses can be reduced by using a smaller flip angle. For example, a 30° pulse requires a delay that is three times shorter. This reduces the peak intensity for each scan but reduces the delay time required between scans enabling the acquisition of more scans for the same amount of time. Another parameter that can affect the appearance of an NMR spectrum is acquisition time, which determines spectral resolution. What is directly acquired from an NMR experiment is a free induction decay (FID), which still needs to be processed to produce the frequency spectrum. During processing, apodization, zero-filling, and baseline and phase corrections are normally applied. All of these can significantly alter the integrated areas under the peaks of an NMR spectrum. Thus, a list of universal parameters for quantitative NMR has been established by national and international round robin tests [30] which includes temperature (300 K), pulse angle (30°), preacquisition delay (5 s), acquisition time (3.4 s), relaxation delay (7/3 of relaxation time), and line broadening (0.3 Hz). For processing, careful manual phase and baseline corrections are recommended since automated features of popular NMR processing software packages are unreliable. This validation has been performed with 1H NMR, but these can be applied to 13C. With 13C, relaxation times are appreciably longer so relaxation agents such as paramagnetic compounds have been used as in the earlier work on petroleum distillates [31].

2.5. Sensitivity and dynamic range

For the unbiased profiling of natural products extracts, one needs to consider the problems of sensitivity and dynamic range. A natural product extract typically contains major and minor components. Oftentimes, in order to detect minor components, it is necessary to employ separation techniques, such as successive fractionation and chromatography which have the effect of increasing sensitivity to minor constituents and improving dynamic range. However, this introduces bias.

Limits of detection and quantification are often given in terms of signal to noise ratios. The International Conference on Harmonization of Technical Requirements recommends a signal to noise ratio of 3 and 10 for the detection limit and quantification limit, respectively (ICH Expert Working Group, 1994). In practice, for error values less than 1%, a signal to noise ratio of 150 is recommended [30]. The signal-to-noise ratio (S/N) in NMR spectroscopy however depends not only on concentration but also on other factors [32]:

$$\frac{S}{N} \approx \frac{N \gamma_n^{\frac{5}{2}} B_0^{\frac{3}{2}} T_2 \sqrt{ns}}{T} \tag{1}$$

In this equation, N is concentration, γ_n is the magnetogyric ratio of the nucleus, B_0 is the strength of the external field, T_2 is the transverse relaxation time, ns is the number of transients, and T is temperature. Considering both magnetogyric ratio and natural abundance, one can therefore estimate that the detection limit for 13C will be orders of magnitude higher than that of 1H. Since the number of transients depends on how much time is available for data acquisition, one can improve S/N by simply taking more scans for 13C measurements. Since nuclei with longer relaxation times give sharper lines, these likewise yield higher S/N, making the detection limit dependent on the size of the molecule and the solvent. The above equation does not include factors dependent on the spectrometer's probe, receiver, and filters. In an analysis of diesel fuel, detection limits of 0.01 and 0.5 mol% are cited for 1H and 13C, respectively [33].

Another consideration is dynamic range. For 1H NMR, signals arising from the solvent, in particular water, can easily use up most of the higher bits in a spectrometer's digitizer thereby decreasing the precision of signals coming from the natural product constituents. This can be alleviated by suppressing solvent resonances, but this introduces the problem of reproducibility between runs and remains a problem for components which have signals near the solvent.

A quantitative comparison using three magnetic field strengths—300, 400, and 500 MHz—showed that there was no difference in the sensitivity and that the standard protocol could differentiate plant samples which were spiked with 0.2 mg/mL of rutin (MW 610.5; 328 μM). This is due to the mild dependence of S/N on the field strength.

For the application of 1H NMR for pattern recognition, the use of the magnitude spectrum has been suggested [34]. The standard 1H NMR spectrum utilizes the phase-corrected real component of the Fourier transform of the free induction decay (FID), discarding the imaginary component. This yields the absorption spectrum which is useful for normal qualitative analysis due to its good peak resolution. However, this procedure sacrifices reproducibility. The use of the magnitude spectrum, which utilizes the absolute value of both the real and imaginary components of the FID improves the reproducibility of the spectra thereby improving its accuracy for pattern recognition. This method is applicable to one-dimensional 1H NMR.

Peak integrals in an NMR spectrum unfortunately are also sensitive to data processing. Apodization, zero-filling, phase and baseline corrections, and the integration itself can affect the signal-to-noise ratio of an NMR spectrum. Thus, the current limit in the sensitivity of NMR-based metabolomics is not due to magnetic field strength, but is due to the current data processing methodology which uses spectral binning (alternatively called bucketing) and PCA. The usual bin size for 1H NMR is 0.04 ppm. This divides a 10 ppm 1H spectrum into 250 bins, which effectively becomes the resolution of the method. A smaller bin size can be used if the variability in the chemical shift can be minimized. Another problem observed is the effect of different solvent (see below) to move the position of chemical shifts, which will make identification using database comparisons difficult [35].

2.6. Effect of solvent

Because plant samples contain a wide variety of compounds with corresponding differences in polarity, the solvent used for extraction and the NMR analysis is very important. The solvent system must balance the ability to perform a comprehensive extraction with solvent complexity and reproducibility. In particular, multi-component solvent systems are prone to variation, and if there is a wide difference in vapor pressures (boiling points), the solvent composition may change if care is not taken. Acetone and acetonitrile are effective solvents but their use is limited by their low boiling points. The use of methanol-D_4 in combination with deuterated water (1:1) have been reported. By using these deuterated solvents, the extracts can be measured directly after extraction without need for evaporation and reconstitution. However, use of water will introduce a strong water peak in the 1H NMR spectrum that must be irradiated. This becomes a source of variability around the water peak across different operators and instruments. To avoid shifts due to differences of pH in 1H NMR measurements a buffer, such as KH_2PO_4, is used [36].

3. Recent applications

NMR is capable of providing simultaneous access to both qualitative (chemical structure) and quantitative information. Unfortunately, NMR has been more generally associated with multidimensional qualitative NMR used in structural analysis and qNMR has been living under this shadow. Fan (1996) pointed out that comprehensive metabolite profiling of complex food products can be done using one- and two-dimensional NMR analysis [37]. However, it is in the use of NMR combined with chemometric methods that the extraordinary potential of both the qualitative and quantitative applications have been realized [38].

In view of its ability to be used as an exhaustive molecular fingerprinting technique, 1H NMR has been found to be a suitable method for the identification, quality control, and fraud detection of essential oils, a function normally reserved for GC-MS [39]. NMR fingerprinting involves obtaining 1H or 13C spectra of whole solvent extracts under standardized conditions and ignoring, at least initially, the assignment of peaks. Multivariate statistical methods, such as PCA, are used to compare spectra from the samples to identify clusters so that inferences can be drawn about the classification of individual plant samples. The identities of metabolites responsible for differences between groups can be investigated from loadings plots generated by PCA [40]. The following section will cover applications of 1H NMR in one- and two-dimensions and 13C NMR together with the statistical tools.

3.1. Metabolomic profiling using 1H NMR

One-dimensional 1H NMR (1D HNMR) can be used in the untargeted and targeted mode. The earliest use of 1D HNMR for the profiling of complex extracts had the objective of monitoring the major components of exudates of plants, such as its root system. The relative increase or decrease of primary metabolites, such as lactate, ethanol, and certain amino acids, could be observed [41]. However, its application to natural product compounds is more challenging due to their more complex structures and lower concentrations. Because of its simplicity and speed, 1D HNMR in the untargeted mode can be used by itself or as a first-pass screening to obtain cluster and profile information using HCA and PCA [42]. The majority of HNMR studies combine 1D HNMR for PCA analysis with two-dimensional homonuclear (1H-1H) or heteronuclear (1H-13C) NMR methods for identification of natural product metabolites.

3.1.1. One-dimensional 1H NMR

This section discusses applications that make use of 1D HNMR alone. The number of such studies is limited because of the presence of overlapping signals and the need for high magnetic fields. 1D HNMR at 500 MHz was used to authenticate grapes for wine making by analyzing their skin and pulp at maturity. Spectral data were reduced by binning using 0.04 ppm bin size and normalized to generate 183 variables to describe each spectrum. Chemometric methods, in particular PCA and partial least squares (PLS), enabled the identification of compounds that contributed to differences between berries, due to the sugars (glucose, fructose, and sucrose), organic acids (tartaric, malic, citric, and succinic acids), and amino acids (proline, arginine, gamma-aminobutyric acid, valine, alanine, leucine, and isoleucine) [43].

A set of green teas selected from a Japanese tea contest were analyzed by 1D HNMR at 750 MHz to classify tea quality with respect to that judged by tea tasters and to propose a quality prediction model. PCA metabolomics profiling revealed a separation between the high- and the low-quality green teas. The taste marker compounds contributing to the discrimination of tea quality were identified from 1D HNMR as caffeine, theanine, epigallocatechin-3-gallate, epigallocatechin, epicatechin-3-gallate, and epicatechin [44].

The use of the magnitude spectrum showed good reproducibility in the analysis of 4 diverse natural product samples (12 tea extracts, 8 liquor samples, 9 hops extracts, and 25 cannabis extracts) using 1D HNMR at 500-MHz and various statistical tools [45].

3.1.2. Two-dimensional 1H NMR

Because of problems of signal overlaps in 1D HNMR spectra, two-dimensional NMR techniques are usually used to overcome these limitations. The 2D methods include 1H *J*-resolved NMR (2D JNMR), 1H-1H correlation spectroscopy (2D COSY) and total correlation spectroscopy (2D TOCSY), 1H-13C heteronuclear single quantum coherence (2D HSQC), and 1H-13C heteronuclear multiple bond coherence (2D HMBC).

1D and 2D NMR at 600 MHz together with chemometric analysis were used to differentiate the origin, purity, and processing methods of chamomile flowers which were obtained from three different countries. The extracts were dissolved in D_2O phosphate buffer adjusted to pH 7.4. 1D NMR data were analyzed by PCA analysis to determine the groupings by pattern recognition and 2D COSY and 2D TOCSY pulse sequences were used to assign the resonances and identify constituents [46].

Several NMR-based metabolomic studies have been done on green tea (*Camellia sinensis*, L.). In one study, 191 green tea samples from different countries were analyzed using 1D HNMR and 2D NMR at 400 MHz to determine origin, quality, effects of climate and season, growth conditions, and even plucking position. The highest quality Chinese tea showed higher levels of theanine, gallic acid, caffeine, epigallocatechin gallate, and epicatechin gallate and lower levels of epigallocatechin when compared with other teas. These new markers were suggested to be useful for the authentication of tea [47]. In another study, the effects of climatic conditions (temperature, sun exposure, and precipitation) and plucking positions on the tea plant were investigated using 1D HNMR profiling combined with multivariate pattern recognition methods. Assignment of NMR signals was done using 2D TOCSY, 2D HMBC, and 2D HSQC. The variations in the composition of specific tea compounds were obtained [48, 49]. The sensitivity of the NMR method at 400 MHz was demonstrated in a study on three different varieties of green tea. 1D HNMR, 2D JNMR, and 2D COSY spectra were run and identification of constituents was done using MestRenova version 11.0.0. The following compounds were identified: theanine, alanine, threonine, succinic acid, aspartic acid, lactic acid, caffeine, and derivatives of epigallocatechin [50].

The same strategy was used for chemotaxonomic classification of 11 South American *Ilex* species. Data from 1D HNMR at 600 MHz were combined with PCA, partial least square-discriminant analysis (PLS-DA), and hierarchical cluster analysis (HCA) to reveal four distinct groups. 1H signal overlaps were addressed using 2D JNMR and 2D HSQC. The combined use of 1D- and 2D-NMR and chemometric analysis enabled unambiguous chemotaxonomic discrimination of the *Ilex* species and varieties [51].

1D HNMR fingerprinting followed by 2D TOCSY and 2D HSQC methods were used to distinguish four Asian and four Korean ginseng products, as well as their commercial products. In this way, the major metabolites—glutamine, arginine, sucrose, malate, and myo-inositol—were identified as chemical markers for quality assurance [52]. In a study on Indian ginseng, *Withania somnifera* (L.) Dun., 1D HNMR profiling was performed on the leaves, stems, and roots to obtain a profile of this plant. PCA and hierarchical cluster analysis (HCA) were performed to group samples which were collected from six different regions of India. 2D JNMR, 2D COSY, 2D HSQC, and 2D HMBC, were then used to identify specific metabolites. The ratio of two withanolides was found to be a key discriminating feature of *W. somnifera* leaf samples from different regions [53].

This NMR-based metabolomic strategy was applied to analyze seven spices used in traditional Mediterranean cuisine and to detect metabolic changes over different seasons. Both primary and secondary metabolites were identified and quantified. The major secondary metabolites identified were polyphenols, including flavonoids (apigenin, quercetin, and kaempferol derivatives) and phenylpropanoid derivatives (chlorogenic and rosmarinic acid). This study was performed using a 300 MHz NMR instrument [54].

The application of NMR-based metabolomics method in plant breeding has been reported. Using a 500 MHz instrument, the NMR-based metabolomics was applied to the identification of sugar beet (*Beta vulgaris* L.) genotypes which were susceptible to the *Cercospora* leaf diseases of sugar beet plants worldwide. This approach was able to successfully profile foliar metabolites without inoculation tests which would have required a significant amount of time and effort. In this study, field-grown leaves which had different levels of resistance were collected from 12 sugar beet genotypes at 4 growth time points. The aqueous extracts were studied using 1D HNMR, 2D COSY, 2D TOCSY, and 2D HSQC. Thirty metabolites were identified and annotated using the SpinAssign program from the PRIMe web service. PCA of the NMR data revealed clear differences among the growth stages, in terms of the content of sugar, glycine betaine, and choline [55].

3.2. Metabolomic profiling using 13C NMR

Because of its lower sensitivity and longer acquisition time, 13C NMR is used less often than 1H NMR. However, 13C NMR spectra are simpler, have less severe problems with overlapping peaks, are more comparable across different magnetic field strengths, and are less susceptible to solvent effects. In addition, the singlet nature of 13C NMR signals makes it easier to determine the identity of individual compounds in a mixture.

13C NMR methodology was used to study the triacylglycerols of the oil extracted from the seeds of *Moringa oleifera*, Lam. It was able to simultaneously detect specific unsaturated acyl chains according to their positions on the glycerol backbone through carboxylic, olefinic, and methylene carbons [56]. However, at this time, its use was not specifically identified as a profiling method. Later, 13C NMR was applied to the fingerprinting of lipids for the authentication of marine and fish oils. In this work, 13C NMR was combined with chemometrics and database information and compared with relevant authentic samples [57]. 13C NMR in combination with multivariate data analysis have been used in the analysis of lipids from various fishes. In one application, this method was used to discriminate between farmed and wild Atlantic salmon (*Salmo salar*, L.), between samples from different geographical origins [58], and to detect mislabeling and adulteration [59].

13C NMR was used in a dereplication strategy for the identification of natural product compounds directly from plant extracts. The whole extract was first separated into fractions of simpler composition, which were then analyzed by 13C NMR. The 13C spectra of all the fractions were aligned and subjected to pattern recognition by HCA. This yielded correlations among 13C signals within each fraction which were visualized as chemical shift clusters, which were assigned to specific compounds in a 13C database. This strategy was applied to the analysis of 5 g of a bark extract from the African tree *Anogeissus leiocarpus* which resulted in the unambiguous identification of seven major compounds [60].

Chemical profiling and standardization of the methanol extract from the leaves of *Vitex negundo*, L. were carried out using 13C NMR followed by chemometric analysis. Because PCA analysis gave an explained variability of only 41% for PC1 and PC2, an alternative method, called k-means clustering, was employed. This was able to successfully differentiate samples that were deliberately allowed to degrade. The multivariate control chart, which is analogous to the analytical control chart method, classified samples whose quality exceeded the upper control limit (UCL). The plant samples were also analyzed by quantitative thin layer chromatography (qTLC) using agnuside as marker compound. Comparison of the univariate qTLC results with the multivariate control chart showed poor correspondence: some samples that gave high agnuside values exceeded the UCL while others that had low agnuside values were below the UCL. This means that a univariate analysis of a plant sample using a marker compound does not adequately represent the overall plant profile [61].

13C NMR is being used more often for dereplication of natural product extracts without fractionation. This approach is being enhanced by availability of 13C NMR databases and predictive software which list compounds that are most likely to be present in the extract. These results have been found to be comparable to those obtained using LC-MS and GC-MS, which require fractionation and sample preparation [62].

The combined use of high-resolution 1H and 13C NMR analysis has the potential to reveal more details that are not available using only one technique. This combined approach was employed to detect and quantify a wide range of triacylglycerols and their component fatty acids in marine cod liver oil supplements. The combination of 1H and 13C spectra permitted the detailed analysis of components, including sn-1 monoacylglycerols, sn-1,2- and sn-1,3-diacylglycerol adducts, and other minor components, such as trans-fatty acids, free glycerol and cholesterol, and added vitamins A and E and synthetic compounds, such as ethyl docosahexaenoate or eicosopentaenoate. The identity of each compound was confirmed using 2D COSY [63].

4. Future prospects

The use of 1H and 13C NMR for the profiling of natural products extracts is a rapidly growing branch of metabolomics. It will further accelerate with the increasing use of NMR in quality management, the growth of NMR databases, the development of portable and benchtop NMR instrumentation, and advances in the use of statistical analysis. Despite its considerable potential, the routine application of this method is limited by the lack of expertise to run sophisticated NMR experiments and the lack of computational tools for NMR spectral deconvolution, in particular of 1H spectra [64].

4.1. NMR in quality management

NMR has been used for the monitoring and quality management of foods, beverages, cosmetics, and pharmaceuticals. The same can be done for the profiling of natural products. In order to ensure reproducibility and reliability and to minimize experimental artifacts, the entire process—from sample collection and storage, extraction, NMR measurement and data processing, and statistical analysis—should be optimized and standardized [65, 66]. The NMR solvent is of particular importance because of its influence on the chemical shift positions of protons in phenolic compounds [67] and other solvent effects. This problem is more severe for 1H as compared with 13C NMR.

It has been claimed that periodic calibration can deliver accuracy as high as 99.9% and precision as good as 0.59%, and if calibration is performed with each study, the accuracy and precision can reach 100 and 0.35%, respectively [68]. The various experimental parameters are listed below:

- Sample preparation: homogeneity of sample, extraction solvent, extraction method, and NMR solvent.
- Acquisition parameters: temperature, acquisition time, pulse angle, number of data points, time delay (relaxation time), and electronic amplification.
- NMR data processing: smoothing, phase correction, baseline correction, and signal integration.

4.2. NMR databases in natural products

The usefulness of NMR databases is premised on the reproducibility of the NMR experiment—starting with sample preparation, NMR acquisition, and processing—across different laboratories. It is important to avoid conditions that alter the position of chemical shifts, which will make identification using database comparisons difficult. Open-access and user-contributed 1H and 13C NMR spectral databases have a high potential as a useful tool for natural products researchers provided that sample preparation, instrumentation, and acquisition parameters are standardized. For sample preparation, only selected NMR solvents should be used. Magnetic field strength is more critical for 1H than 13C NMR. Acquisition and processing parameters should be standardized. As of 2015, 1829 1H NMR and 1383 13C NMR spectra have been available in open-access chemical databases. To further promote participation by researchers, the entire process, from data acquisition, conversion of vendor-specific raw data files, and data deposition have to be simplified and standardized [69].

4.3. Portable and benchtop NMR instrumentation

NMR is usually considered to be an expensive analytical technique which is used for research purposes only. However, for NMR to become more useful for the natural products industry where many of the companies are small to medium in size, more affordable instrumentation is needed. There have been numerous announcements regarding the development of portable and benchtop NMR instruments with full spectrum 1H and 13C NMR capability using microcoils with small portable magnets of up to 2 T (approximately 85 MHz 1H) [70]. Although these are limited in capability and reproducibility compared with a full laboratory NMR instrument, they can be used in the field or production site where cryogenic liquids and stable power are not available. Because there is a demand for such instrumentation for other purposes, such as forensic investigation, detection of explosives, and medical diagnostics, their development is certain to accelerate. This will expand the use of NMR for the profiling of natural products.

4.4. Advances in the use of statistical analysis

Although the use of NMR in the analysis of biological extracts was already being done in the 1980s, it was the application of statistical methods that enabled researchers to make use the large amount of NMR data to find patterns and correlations. The first step usually involves the simplification of large NMR data sets to find relationships, groupings, or dependencies using PCA. Second, the groups can be classified with or without a training set which has known information or characteristics against which other sample sets are compared. Linear discriminant analysis (LDA) and soft independent modeling of class analogy (SIMCA) are used for this purpose. For quantitative analysis of constituents, in particular for strongly overlapping peaks, principal component regression (PCR) or PLS regression can be used [71]. Although these statistical techniques are now commonly used, new ones continue to be developed and reported.

One of the most exciting areas of development is the use of statistical methods to correlate NMR signals with biological activity. Since the NMR signals can be related to specific compounds, this in effect allows one to correlate specific compounds with biological activity. Although it has to be emphasized that correlation is not proof of biological activity, this strategy nevertheless allows one to shortcut the process of discovering bioactive compounds in a complex natural product mixture. This also allows one to detect multiple active compounds.

5. Conclusions

1H and 13C NMR are rapidly expanding it role from its traditional use mainly as a qualitative spectroscopic technique for the determination of chemical structure to a quantitative tool for the metabolomic study of natural product extracts, whether for quality control of phytomedicine products, analysis of the metabolome for plant profiling, identification of constituents as plant markers, or for plant biotechnology. A major enabler for the use of NMR for metabolomic studies is the application of various statistical techniques which are able to find patterns and correlations in the large NMR data sets. The continued expansion of the use of NMR for the metabolomic profiling of natural product extracts will likely depend on the further development of statistical methods and the availability of NMR databases for both 1H and 13C nuclei. It is likely that more compounds will be identified as techniques are improved.

An NMR spectrum is quantitative. An understanding of the physical principles of NMR provides the theoretical basis for its use as a quantitative tool. NMR spectroscopy does not require a standard for each component since the intensity of each signal is directly proportional to the number of nuclei being observed regardless of environment. NMR spectroscopy also offers detailed information regarding molecular structure. Using NMR spectroscopy as a tool in the profiling of natural product extracts therefore not only provides accurate and precise composition, but also structural evidence for each of the components. Since the NMR signal dependence on various factors is already well known, resonance positions and intensities are highly reproducible. These are important characteristics which give NMR a unique advantage over other analytical methods.

Abbreviations

1D HNMR	One-dimensional 1H NMR
2D COSY	Two-dimensional 1H-1H correlation spectroscopy
2D JNMR	Two-dimensional 1H J-resolved spectroscopy
2D HMBC	Two-dimensional 1H-13C heteronuclear multibond coherence
2D HSQC	Two-dimensional, 1H-13C single quantum coherence
2D TOCSY	Two-dimensional total correlation spectroscopy
GC-MS	Gas chromatography-mass spectrometry
HCA	Hierarchical cluster analysis
LC-MS	Liquid chromatography-mass spectrometry
MS	Mass spectrometry
NMR	Nuclear magnetic resonance

PCA	Principal components analysis
PLS	Partial least squares
PLS-DA	Partial least squares-discriminant analysis
qNMR	Quantitative NMR
qHNMR	Quantitative proton (1H) NMR
qTLC	Quantitative thin layer chromatography

Author details

Fabian M. Dayrit[1]* and Angel C. de Dios[2]

*Address all correspondence to: fdayrit@ateneo.edu

1 Department of Chemistry, Ateneo de Manila University, Quezon City, Philippines

2 Department of Chemistry, Georgetown University, Washington, DC, USA

References

[1] Metabolomics Society. Metabolomics. [Cited 14-09-2014]. Available from: http://www.metabolomicssociety.org/

[2] Commisso M, Strazzer P, Toffal K, Stocchero M, Guzzo F. Untargeted metabolomics: An emerging approach to determine the composition of herbal products. Computational and Structural Biotechnology Journal. 2013;**4**(5):e201301007. Available from: http://dx.doi.org/10.5936/csbj.201301007

[3] Wishart DS. Quantitative metabolomics using NMR. Trends in Analytical Chemistry. 2008;**27**(3):228-237

[4] Markley JL, Bruschweiler R, Edison AS, Eghbalnia HR, Powers R, Raftery D, Wishart DS. The future of NMR-based metabolomics. Current Opinion in Biotechnology. 2017;**43**:34-40

[5] Malz F, Jancke H. Validation of quantitative NMR. Journal of Pharmaceutical and Biomedical Analysis. 2005;**38**:813-823

[6] Webster GK, Kumar S. Expanding the analytical toolbox: Pharmaceutical application of quantitative NMR. Analytical Chemistry. 2014;**86**:11474-11480

[7] Simmler C, Kulakowski D, Lankin DC, McAlpine JB, Chen SN, Pauli GF. Holistic analysis enhances the description of metabolic complexity in dietary natural products. Advances in Nutrition. 2016;**7**:179-189

[8] Wishart DS. Quantitative metabolomics using NMR. Trends in Analytical Chemistry. 2008;**27**(3):228-237

[9] Wold S. Chemometrics; what do we mean with it, and what do we want from it? Chemometrics and Intelligent Laboratory Systems. 1995;**30**:109-115

[10] Bansal A, Chhabra V, Rawal RK, Sharma S. Chemometrics: A new scenario in herbal drug standardization. Journal of Pharmaceutical Analysis. 2014;**4**(4):223-233

[11] Kim HK, Saifullah KS, Wilson EG, Prat Kricun SD, Meissner A, Goraler S, Deelder AM, Choi YH, Verpoorte R. Metabolic classification of South American Ilex species by NMR-based metabolomics. Phytochemistry. 2010;**71**:773-784

[12] Ernst RR, Anderson WA. Application of Fourier transform spectroscopy to magnetic resonance. The Review of Scientific Instruments. 1966;**37**:93-102

[13] Jungnickel JL, Forbes JW. Quantitative measurement of hydrogen types by integrated nuclear magnetic resonance intensities. Analytical Chemistry. 1963;**35**:938-942

[14] Hollis DP. Quantitative analysis of aspirin, phenacetin, and caffeine mixtures by nuclear magnetic resonance spectrometry. Analytical Chemistry. 1963;**35**:1682-1684

[15] Paul EG, Grant DM. Carbon-13 magnetic resonance. I. Improved carbon-13 magnetic resonance spectra obtained by proton decoupling and rapid sample spinning. Journal of the American Chemical Society. 1964;**86**:2977-2983

[16] Dalling DK, Ladner KH, Grant DM, Woolfenden WR. Carbon-13 magnetic resonance. 27. The dependence of chemical shifts on methyl rotational conformations and dynamics in the methylated benzenes and naphthalenes. Journal of the American Chemical Society. 1977;**99**:7142-7150

[17] Overhauser AW. Polarization of nuclei in metals. Physics Review. 1953;**92**:411-415

[18] Laude DA, Wilkins CL. Identification of organic mixture components without separation: Quantitative and edited carbon-13 nuclear magnetic resonance spectrometry data for analysis of petroleum distillates. Analytical Chemistry. 1986;**58**:2820-2824

[19] Wollenberg KF, Kurt F. Quantitative high resolution 13C nuclear magnetic resonance of the olefinic and carbonyl carbons of edible vegetable oils. Journal of the American Oil Chemists' Society. 1990;**67**:487-494

[20] Tomi F, Bradesi P, Bighelli A, Casanova J. Computer-aided identification of individual components of essential oils using carbon-13 NMR spectroscopy. Journal of Magnetic Resonance Analysis. 1995;**1**:25-34

[21] Swalen JD, Reilly CA. Analysis of complex NMR spectra. An iterative method. The Journal of Chemical Physics. 1962;**1**:21-29

[22] Jameson CJ. Gas-phase NMR spectroscopy. Chemical Reviews. 1991;**91**:1375-1395

[23] De Dios AC, Pearson JG, Oldfield E. Secondary and tertiary structural effects on protein NMR chemical shifts: An ab initio approach. Science. 1993;**260**:1491-1496

[24] De Dios AC, Oldfield E. Chemical shifts of carbonyl carbons in peptides and proteins. Journal of the American Chemical Society. 1994;**116**:11485-11488

[25] De Dios AC, Jameson CJ. The NMR chemical shift: Insight into structure and environment. Annual Reports on NMR Spectroscopy. 1994;**29**:1-69

[26] Skinner TE, Reiss TO, Luy B, Khaneja N, Glaser SJ. Application of optimal control theory to the design of broadband excitation pulses for high-resolution NMR. Journal of Magnetic Resonance. 2003;**163**:8-15

[27] Freeman RH, Hill DW, Kaptein R. Proton-decoupled NMR. Spectra of carbon-13 with the nuclear Overhauser effect suppressed. Journal of Magnetic Resonance. 1972;**7**:327-329

[28] Bax A, Subramanian S. Sensitivity-enhanced two-dimensional heteronuclear shift correlation NMR spectroscopy. Journal of Magnetic Resonance. 1986;**67**:565-569

[29] Ernst RR, Anderson WA. Application of Fourier transform spectroscopy to magnetic resonance. The Review of Scientific Instruments. 1966;**37**:93-102

[30] Malz F, Jancke H. Validation of quantitative NMR. Journal of Pharmaceutical and Biomedical Analysis. 2005;**38**:813-823

[31] Laude DA, Wilkins CL. Identification of organic mixture components without separation: Quantitative and edited carbon-13 nuclear magnetic resonance spectrometry data for analysis of petroleum distillates. Analytical Chemistry. 1986;**58**:2820-2824

[32] Bharti SK, Roy R. Quantitative 1H NMR spectroscopy. Trends in Analytical Chemistry. 2012;**35**:5-26

[33] Hsieh PY, Widegren JA, Slifka AJ, Hagen AJ, Rorrer RA. Direct measurement of trace polycyclic aromatic hydrocarbons in diesel fuel with 1H and 13C NMR spectroscopy: Effect of PAH content on fuel lubricity. Energy and Fuels. 2015;**29**(7):4289-4297

[34] Harrington PB, Wang X. Spectral Representation of proton NMR spectroscopy for the pattern recognition of complex materials, Published online: 24 February 2017. Journal of Analysis and Testing. 2017. DOI: 10.1007/s41664-017-0003-y

[35] Van der Kooy F, Venkataya B, Pearson JL, Torres A, Li CG, Chang D. Sensitivity of NMR-based metabolomics in drug discovery from medicinal plants. European Journal of Medicinal Plants. 2015;**5**(2):191-203

[36] Verpoorte R, Choi YH, Kim HK. NMR-based metabolomics at work in phytochemistry. Phytochemistry Reviews. 2007;**6**:3-14

[37] Fan TWM. Metabolite profiling by one- and two-dimensional NMR analysis of complex mixtures. Progress in Nuclear Magnetic Resonance Spectroscopy. 1996;**28**:161-219

[38] Pauli GF, Jaki BU, Lankin DC. Quantitative 1H NMR: Development and potential of a method for natural products analysis. Journal of Natural Products. 2005;**68**:133-149

[39] Guerrini A, Sacchetti G, Muzzoli M, Rueda GM, Medici A, Besco E, Bruni R. Composition of the volatile fraction of *Ocotea bofo* Kunth (Lauraceae) calyces by GC-MS and NMR fingerprinting and its antimicrobial and antioxidant activity. Journal of Agricultural and Food Chemistry. 2006;**54**:7778-7788

[40] Ward JL, Baker JM, Beale MH. Recent applications of NMR spectroscopy in plant metabolomics. The FEBS Journal. 2007;**274**:1126-1131

[41] Fan TWM, Higashi RM, Lane AN. Monitoring of hypoxic metabolism in superfused plant tissues by in vivo 1H NMR. Archives of Biochemistry and Biophysics. 1986;**251**(2):674-687

[42] PSimmler C, Napolitano JG, McAlpine JB, Chen SN, Pauli GF. Universal quantitative NMR analysis of complex natural samples. Current Opinion in Biotechnology. 2014. DOI: 10.1016/j.copbio.2013.08.004

[43] Pereira GE, Gaudillere JP, van Leeuwen C, Hilbert G, Maucourt M, Deborde C, Moing A, Rolin D. 1H NMR metabolite fingerprints of grape berry: Comparison of vintage and soil effects in Bordeaux grapevine growing areas. Analytica Chimica Acta. 2006;**563**:346-352

[44] Tarachiwin L, Ute K, Kobayashi A, Fukusaki E. 1H NMR based metabolic profiling in the evaluation of Japanese green tea quality. Journal of Agricultural and Food Chemistry. 2007;**55**:9330-9336

[45] Harrington PB, Wang X. Spectral Representation of proton NMR spectroscopy for the pattern recognition of complex materials, Published online: 24 February 2017. Journal of Analysis and Testing. 2017. DOI: 10.1007/s41664-017-0003-y

[46] Wang Y, Tang H, Nicholson JK, Hylands PJ, Sampson J, Whitcombe I, Stewart CG, Caiger S, Oru I, Holmes E. Metabolomic strategy for the classification and quality control of phytomedicine: A case study of chamomile flower (*Matricaria recutita* L.). Planta Medica. 2004;**70**:250-255

[47] Le Gall G, Colquhoun IJ, Defernez M. Metabolite profiling using 1H NMR spectroscopy for quality assessment of green tea, *Camellia sinensis* (L.). Journal of Agricultural and Food Chemistry. 2004;**52**:692-700

[48] Lee JE, Lee BJ, Chung JO, Hwang JA, Lee SJ, Lee CH, Hong YS. Geographical and climatic dependencies of green tea (*Camellia sinensis*) metabolites: A 1H NMR-based metabolomics study. Journal of Agricultural and Food Chemistry. 2010;**58**:10582-10589

[49] Lee JE, Lee BJ, Hwang JA, Ko KS, Chung JO, Kim EH, Lee SJ, Hong YS. Metabolic dependence of green tea on plucking positions revisited: A Metabolomic study. Journal of Agricultural and Food Chemistry. 2011;**59**:10579-10585

[50] Wahyuni DSC, Kristanti MW, Putri RK, Rinanto T. NMR metabolic profiling of green tea (*Camellia sinensis* L.) leaves grown at Kemuning, Indonesia. IOP Conference Series: Journal of Physics: Conference Series. 2017;**795**:012013

[51] Kim HK, Saifullah KS, Wilson EG, Prat Kricun SD, Meissner A, Goraler S, Deelder AM, Choi YH, Verpoorte R. Metabolic classification of South American *Ilex* species by NMR-based metabolomics. Phytochemistry. 2010;**71**:773-784

[52] Lee EJ, Shaykhutdinov R, Weljie AM, Vogel HJ, Facchini PJ, Park SU, Kim YK, Yang TJ. Quality assessment of ginseng by 1H NMR metabolite fingerprinting and profiling analysis. Journal of Agricultural and Food Chemistry. 2009;**57**:7513-7522

[53] Namdeo AG, Sharma A, Yadav KN, Gawande R, Mahadik KR, Lopez-Gresa MP, Kim HK, Choi YH, Verpoorte R. Metabolic characterization of *Withania somnifera* from different regions of India using NMR spectroscopy. Planta Medica. 2011;**77**(17):1958-1964

[54] Scognamiglio M, D'Abrosca B, Esposito A, Fiorentino A. Chemical composition and seasonality of aromatic mediterranean plant species by NMR-based metabolomics. Journal of Analytical Methods in Chemistry. 2015 Article ID 258570, 9 pages. http://dx.doi.org/10.1155/2015/258570

[55] Sekiyama Y, Okazaki K, Kikuchi J, Ikeda S. NMR-based metabolic profiling of fieldgrown leaves from sugar beet plants Harbouring different levels of resistance to *Cercospora* leaf spot disease. Metabolites. 2017;**7**:(4). 13 pages. DOI: 10.3390/metabo7010004

[56] Vlahov G, Chepkwony PK, Ndalut PK. 13C NMR characterization of triacylglycerols of moringa oleifera seed oil: An "oleic-Vaccenic acid" oil. Journal of Agricultural and Food Chemistry. 2002;**50**:970-975

[57] Standal IB, Axelson DE, Aursand M. Authentication of marine oils using 13C NMR spectroscopy. Lipid Technology. 2011;**23**(7):152-154

[58] Aursand M, Standal IB, Prael A, McEvoy L, Irvine J, Axelson DE. 13C NMR pattern recognition techniques for the classification of Atlantic Salmon (*Salmo salar* L.) according to their wild, farmed, and geographical origin. Journal of Agricultural and Food Chemistry. 2009;**57**:3444-3451

[59] Aursand M, Standal IB, Axelson DE. High resolution 13C nuclear magnetic resonance spectroscopy pattern recognition of fish oil capsules. Journal of Agricultural and Food Chemistry. 2007;**55**:38-47

[60] Hubert J, Nuzillard JM, Purson S, Hamzaoui M, Borie N, Reynaud R, Renault JH. Identification of natural metabolites in mixture: A pattern recognition strategy based on 13C NMR. Analytical Chemistry. 2014;**86**:2955-2962

[61] Lagurin LG, Magsalin JDJ, Zosa AR, Dayrit FM. Chemical profiling and chemical standardization of *Vitex negundo* using 13C NMR. Journal of Medicinal Plants Research. 2017;**11**(1):11-21

[62] Bakiri A, Hubert J, Reynaud R, Lanthony S, Harakat D, Renault JH, Nuzillard JM. Computer-aided 13C NMR chemical profiling of crude natural extracts without fractionation. Journal of Natural Products 2017;**80**:1387-96

[63] Siddiqui N, Sim J, Silwood CJL, Toms H, Iles RA, Grootveld M. Multicomponent analysis of encapsulated marine oil supplements using high-resolution 1H and 13C NMR techniques. Journal of Lipid Research. 2003;**44**:2406-2427

[64] Mahrous EA, Farag MA. Two dimensional NMR spectroscopic approaches for exploring plant metabolome: A review. Journal of Advanced Research. 2015;**6**:3-15

[65] Rivas-Ubach A, Perez-Trujillo M, Sardans J, Gargallo-Garriga A, Parella T, Penuelas J. Ecometabolomics: Optimized NMR-based method. Methods in Ecology and Evolution. 2013;**4**:464-473

[66] Schönberger T, Monakhova YB, Lachenmeier DW, Kuballa T, the Non-Profit Expert Team (NEXT)-NMR working group Germany. Guide to NMR method development and validation – Part 1: Identification and quantification. In: EUROLAB Technical Report 1/2014

[67] Kim HK, Choi YH, Verpoorte R. NMR-based metabolomic analysis of plants. Nature Protocols. 2010;**5**:536-549

[68] Crouch R, Russell D. Easy, Precise and Accurate Quantitative NMR. Agilent Technologies Application Note. 2011. [Cited 14-09-2014] http://www.agilent.com

[69] Johnson SR, Lange BM. Open-access metabolomics databases for natural product research: Present capabilities and future potential. Frontiers in Bioengineering and Biotechnology. 2015;**3**(22):10

[70] Herberg JL, Demas V, Malba V, Bernhardt A, Evans L, Harvey C, Chinn S, Maxwell R, Reimer J, Pines A. Portable, low-cost NMR with laser-lathe lithography produced microcoils. Journal of Magnetic Resonance. 2007;**189**:121-129

[71] Monakhova YB, Kuballa T, Lachenmeier DW. Chemometric methods in NMR spectroscopic analysis of food products. Journal of Analytical Chemistry. 2013;**68**(9):755-766

Chapter 6

Application of Mass Spectroscopy in Pharmaceutical and Biomedical Analysis

Uttam Singh Baghel, Atamjit Singh,
Deeksha Singh and Manish Sinha

Additional information is available at the end of the chapter

http://dx.doi.org/10.5772/intechopen.70655

Abstract

Mass spectrometry (MS) is a powerful analytical tool with many applications in pharmaceutical and biomedical field. The increase in sensitivity and resolution of the instrument has opened new dimensions in analysis of pharmaceuticals and complex metabolites of biological systems. Compared with other techniques, mass spectroscopy is only the technique for molecular weight determination, through which we can predict the molecular formula. It is based on the conversion of the sample into ionized state, with or without fragmentation which are then identified by their mass-to-charge ratios (m/e). Mass spectroscopy provides rich elemental information, which is an important asset to interpret complex mixture components. Thus, it is an important tool for structure elucidation of unknown compounds. Mass spectroscopy also helps in quantitative elemental analysis, that is, the intensity of a mass spectra signal is directly proportional to the percentage of corresponding element. It is also a noninvasive tool that permits *in vivo* studies in humans. Recent research has looked into the possible applications of mass spectrometers in biomedical field. It is also used as a sensitive detector for chromatographic techniques like LC–MS, GC–MS and LC/MS/MS. These recent hyphenated technological developments of the technique have significantly improved its applicability in pharmaceutical and biomedical analyses.

Keywords: mass spectrometry, pharmaceutical, biomedical, phytochemical, structure elucidation

1. Introduction

Mass spectrometry (MS) is an advanced technique for determining the molecular weight of a compound. The first mass spectrometer was developed in 1912 by J.J. Thompson. The instrument has now a wide range of applications in pharmaceutical (drug discovery, pharmacokinetics, drug metabolism), clinical (neonatal screening, hemoglobin analysis, drug testing), environmental (water quality, food contamination, pollutant determination), geological (oil composition, hydrocarbon fraction determination in petroleum industry), metallurgy (determination of rare earth metals and metals at ppq (parts per quadrillion)), sports (dope test of drugs in athletes), forensic (poison and drug metabolite determination) and biotechnology (proteins, peptide analysis) like fields.

2. Principle

The mass spectroscopy is based on the positive ion generation. For its most popular model, the electron impact ionization with magnetic sector analyzer, the sample under investigation is converted into vapor phase and bombarded with electrons having energy sufficient to knock out one electron from it (>10 eV) to produce a positively charged ion called molecular ion or parent ion which is denoted by M^+.

Positively charged molecule M^+ is often unstable, and with increase in energy (10–70 eV) according to bond strength, they break into fragments called fragment or daughter ion which is denoted by M^{+1}. Ions formed are separated in analyzer under the influence of electric and magnetic field and are recorded by the detector to give rise a mass spectrum (**Figure 1**).

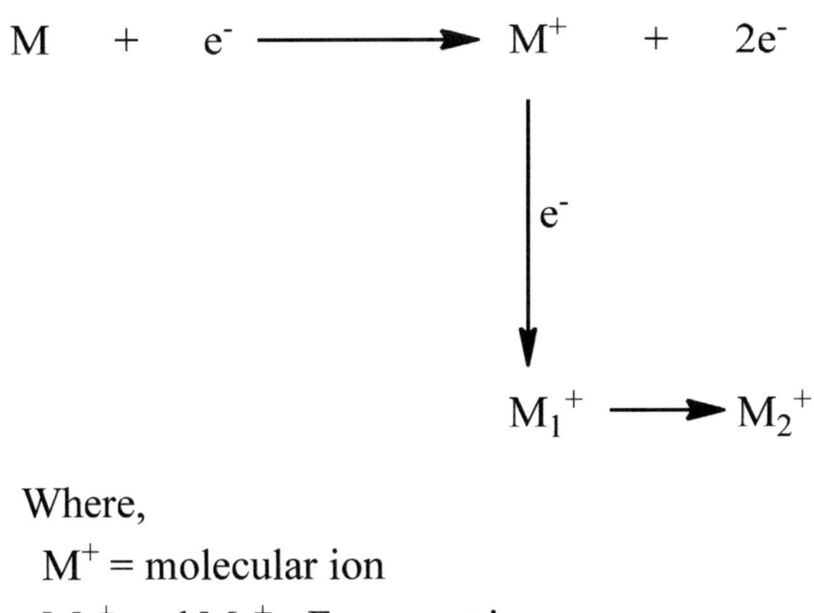

Figure 1. Ionization of molecule by electron bombardment.

3. Components of mass spectrometer

Mass spectrometer mainly consists of following components:

1. Inlet system
2. Ion generation chamber
3. Analyzer tube
4. Ion collector
5. Data collection system

The inlet system transfers the gaseous form of sample into the vacuum of the ion generation chamber of mass spectrometer. In the ion generation chamber, neutral sample molecules are ionized and then accelerated into the mass analyzer tube. The mass analyzer tube is the most important part on which a range of the mass spectrometer depends. This segment separates generated ions, either in space or in time, according to their mass-to-charge ratio (m/e). Once the ions are separated, they are collected and detected in ion collector chamber. Then, the signal is transferred to a data collection system for data investigation. The high vacuum is applied between the ion generation chamber, analyzer tube and ion collector. The vacuum system is maintaining the low pressure which minimizes the chances of ion-molecule reaction, scattering and neutralization of the ions (**Figure 2**).

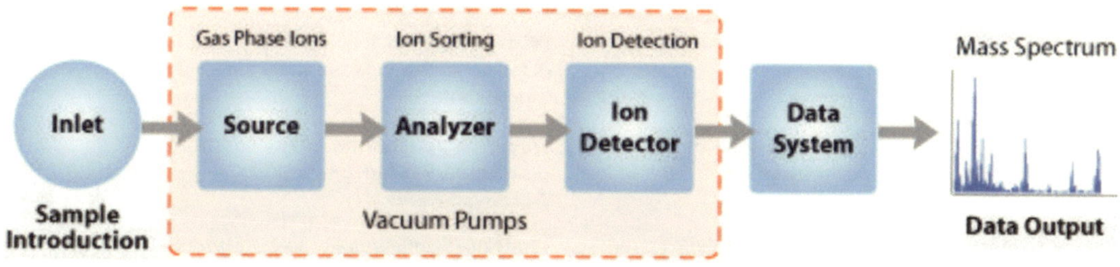

Figure 2. Components of mass spectrometer.

4. Applications

4.1. Phytochemical analysis

Mass spectroscopy is widely employed in phytochemical analysis due to its capability to identify and measure metabolites having very low molecular weight at very low concentration ranges below nanogram per milliliter (ng/mL). Therefore, it is considered as trace analysis methodology. A variety of analyte separation techniques like capillary electrophoresis, gas chromatography and high-performance liquid chromatography are united with mass spectroscopy for simultaneous separation and determination of analytes called CE-MS, GC–MS and HPLC-MS, respectively. Mass spectrometers like quadrupole or quadrupole-time-of-flight (Q-TOF) are frequently employed in combination along with gas chromatographic system. Several phytoconstituents are volatile and thermolabile, and they can be analyzed by electrospray ionization (ESI) and matrix-assisted laser desorption ionization (MALDI). ESI is commonly employed in HPLC-MS and CE-MS. Fourier transform ion cyclotron resonance (FT-ICR), orbitrap and TOF are emerged as high-performance mass analyzers that are able to screen metabolites with fraction of seconds due to their high resolution. Combination of TOF with one (Q-TOF) or two quadrupoles (Qq-TOF) is emerged as hybrid mass spectrometers that are able to cover unlimited mass range with high scan rates up to 10^6 u/s and high resolving power. Analytes having high molecular weight and temperature sensitive can be efficiently analyzed by HPLC coupled with atmospheric pressure ionization-mass spectrometer (API-MS) [1]. Some of the recent research articles depicting the application of mass spectrometry for the phytochemical analysis are listed in **Table 1**.

S. no.	Analytical technique	Sample source	Analytes	Reference
1	HPLC-ESI-MS	*Leontopodium* species (Asteraceae)	Fatty acids, sucrose, diterpenes, sesquiterpene	[2]
2	GC × GC–MS	Essential oil of *Pelargonium graveolens*	α-Pinene, myrcene, limonene, citrinellal, geraniol	[3]
3	GC–MS	Methanolic fruit extract of *Momordica charantia*	Vitamin E, gentisic acid, 1-pentadecyne	[4]
4	GC–MS	Extracts of *Aerva lanata*	(R)-(+)-ç-Valerolactone, 5,14-di (*N*-butyl)-octadecane, 9-octadecenoic acid, 2-propynoic acid	[5]
5	GC–MS	Ethanolic extract of *Azolla microphylla*	(R)-(+)-ç-Valerolactone, 5,14-di (*N*-butyl)-octadecane, 9-octadecenoic acid, 2-propynoic acid	[6]
6	UHPLC-ESI-MS	*Rhizopus microsporus* var. *oryzae* challenged soya bean seedlings	Prenylated isoflavonoids and isoflavonoids like daidzein, genistein, glycitein	[7]
7	HPLC-ESI-MS/MS	*Radix astragali*	Calycosin, calycosin-7-O-β-D-glycoside, formononetin, formononetin-7-O-glycoside	[8]
8	HPLC–MS/MS	*Glycyrrhiza uralensis* Fisch. extract	Glycyrrhizic acid, liquorice saponin G2, liquiritin, licuraside, ononin, glycycoumarin	[9]
9	LC/MS/MS	Dried plums	Hydroxycinnamics, including acids, esters and glycosides; hydroxybenzoic acids and one flavonoid	[10]
10	UHPLC–MS	*Licorice* root extract in 70% ethanol, ethanol and ethyl acetate	Prenylated flavonoids	[11]

Table 1. Applications of mass spectroscopy in phytochemical analysis.

http://dx.doi.org/10.5772/intechopen.70655

4.2. Structure elucidation

Mass spectroscopy has major use in structure elucidation of compounds. Mass spectrum is produced in the form of bar graph which is interpreted by using the following peaks.

1. *Base peak*: It is the most intense peak of the mass spectrum. It has 100% abundance.
2. *Molecular ion peak*: It has highest mass-to-charge ratio (m/e) in mass spectrum created due to loss of one electron from molecule, and its m/e value roughly represents the molecular weight of compound.
3. *Isotopic peaks:* Isotopic peaks are often seen in the mass spectrum due to the presence of isotopes. These are usually one or two mass units higher or lower than the ion peak. It can be represented by m + 1 and m + 2 for one and two molecular weight higher isotopes and *vice versa* for lower weight isotopes. Intensity of the peak depends on the natural abundance of isotopes. These are often used as characteristic of a compound for identification.
4. *Fragment ion peak*: Peak of fragment ion in mass spectrum is called fragment ion peak. Fragmentation pattern is characteristic for a particular organic compound. So we can confirm the compound by comparing with the library of fragmentation pattern of reference compounds.
5. *Metastable ion peak*: This ion is formed from the disintegration of fragment ion in the analyzer tube of mass spectrometer. It is generated due to loss of kinetic energy of ion during acceleration from ionization chamber to analyzer tube. This ion appears in the spectrum at m/e ratio, which depends upon the mass of original ion from which it is formed (m_1^+) and fragment ion mass (m_2^+). Metastable ion peaks usually appear at nonintegral values of m/e ratio and are often seen as broad peaks. This is used for the confirmation of proposed fragmentation pattern of a molecule.

$$m^* = \frac{(m_2^+)^2}{(m_1^+)}$$

Here, m^* is the mass of metastable ion observed in mass spectrum.

There are some rules which are employed in interpretation of mass spectra in structure elucidation process. These are:

1. *Nitrogen rule*: Nitrogen rule gives very useful clue about the presence or absence and number of nitrogen atoms. It is divided into two parts:

   i. If *m/e* value of molecular ion peak is odd number, then it may contain odd number of nitrogen atoms.

   ii. If *m/e* value of molecular ion peak is even number, then it may or may not contain even number of nitrogen atoms.

2. *Hydrogen deficiency index* **(HDI)**: Number of pairs of hydrogen required to saturate the compound is called hydrogen deficiency index (HDI). Hydrogen deficiency index is also called unsaturation index which gives the information about the number of π-bonds and/or rings present in a molecular structure.

Steps for determination of hydrogen deficiency index:

1. Make the correction in the predicted base formula with respect to the elements obtained from other spectral data, and the correction requires addition or removal of hydrogen atoms, which depends upon the type of element added.
   - **A.** For group V elements (N, P, As, Sb, Bi): Addition of one hydrogen atom is required with each added element.
   - **B.** For group VI elements (O, S, Se, Te): There is no need of addition of any hydrogen atom in the formula.
   - **C.** For group VII elements (F, Cl, Br, I): The removal of one hydrogen atom is required with each added element.
2. After specific needed corrections, compare molecular formula of unknown compound with the formula of saturated hydrocarbon.
3. Calculate the difference in hydrogen atoms between two formulas and corresponding pairs of hydrogen atom.

Interpretation of hydrogen deficiency index:

1. If hydrogen deficiency index is one, there must be one double bond or one ring present in the structure.
2. If hydrogen deficiency index is two, there must be triple bond or two double bonds, two rings, or one double bond and one ring present in the structure.
3. Similarly, benzene ring has hydrogen deficiency index four because it contains three double bonds and one ring in it.
4. Any substance with hydrogen deficiency index four or more possibly contains a benzenoid ring in it, and the compounds with hydrogen deficiency index less than four cannot contain such type of ring.

We can understand this application by the following examples:

Example 1: Calculate the HDI of C_2H_4.

C_2H_4 has two carbon atoms. Therefore, base molecular formula will be:

$=C_nH_{2n+2}$.

$=C_2H_{(2 \times 2)+2}$.

$=C_2H_6$.

Hydrogen atoms in predicted base molecular formula = 6.

Hydrogen atoms in actual molecular formula of compound = 4.

Difference in hydrogen atoms between two formulas (6–4) = 2 (which corresponds to one pair of hydrogen atom).

Hydrogen deficiency index is 1, because one pair of hydrogen atom is required to saturate the compound.

Example 2: Calculate the HDI of C_6H_6.

C_6H_6 has six carbon atoms. Therefore, base molecular formula will be:

$=C_nH_{2n+2}$.

$=C_6H_{(2 \times 6)+2}$.

$=C_6H_{14}$.

Hydrogen atoms in predicted base molecular formula = 14.

Hydrogen atoms in actual molecular formula of compound = 6.

Difference in hydrogen atoms between two formulas (14–6) = 8 (which corresponds to four pairs of hydrogen atom).

Hydrogen deficiency index is 4, because four pairs of hydrogen atoms are required to saturate the compound.

Fragmentation pattern is also an important component of mass spectra from which qualitative analysis of compounds can be done, and it is also useful in elucidation of structural arrangement of compound. From Beynon table, one can predict the possible elemental arrangement or composition of particular mass and determine the molecular formula of compound. The following examples from the literature support the present application of mass spectroscopy.

The structure of flavonoid monoglycosides like genistein-7-O-glucoside, genistein-4′-O-glucoside, 2′-hydroxygenistein-7-O-glucoside and apigenin, isolated from shoot of lupin (*Lupinus luteus* L.), was elucidated by using LSI-MS and EI-MS with double-focusing reversed geometry between mass spectrometer [12].

The analysis of sulfated heparin-like glycosaminoglycan oligosaccharides was done with the help of tandem mass spectroscopy (MS/MS) using quadrupole ion trap mass spectrometer and quadrupole orthogonal acceleration time-of-flight mass spectrometer, and their fragmentation pattern was also studied. This study suggested the use of tandem mass instruments like Q-TOF and metal cations in mass spectroscopy of heparin-like glycosaminoglycan oligosaccharides [13].

A one-step complete analysis method was developed for galacto-oligosaccharide mixtures obtained in lactose transgalactosylation using β-galactosidase from *Aspergillus oryzae* based on ion-mobility spectrometry-tandem mass spectrometry (IMS-MS/MS) with electrospray ionization [14].

The characterization of commercial prebiotic galacto-oligosaccharide mixture was done with linear ion-trap mass spectrometer coupled with high-performance anion-exchange chromatography (HPAEC) using electrospray ionization combination with ^{1}H NMR and ^{13}C NMR [15].

The synthesis of a novel sequence of di- and tri-organotin (IV) compounds which contains germanium having the general formula $R_{4-n}SnL_n$ along with characterization was performed by elemental analysis, FT-IR, multinuclear NMR (^{1}H,^{13}C,^{119}Sn) and mass spectrometry by double-focusing mass spectrometer [16].

The characterization of polyisobutylenes was done by various mass spectrometry techniques like tandem mass spectroscopy with MALDI-TOF and ESI-QIT. The primary structure was determined by multistage mass spectrometric analysis, the presence of specific functional groups (e.g., OH or OCH_3) was confirmed, and also differentiation between isomeric functional groups was done [17].

4.3. Peptide and protein sequence/structure analysis

Mass spectroscopy has an important application in analysis of sequence of amino acids in proteins and peptides, that is, analysis of structure of proteins and peptides, and this is employed increasingly. This can be performed by stepwise hydrolysis accompanied with chromatography. Peptides are converted into amino alcohols which are volatile in nature. These amino alcohols derivatized and analyzed in mass spectrometer which aids sequence analysis. However, sequencing of underivatized peptides as in fast atom bombardment mass spectrometry (FABMS) is also employed. New techniques like MALDI and tandem mass spectroscopy are also in trend [18, 19]. Various examples from the literature support the present application.

The peptide sequence was analyzed using combination of gas-phase ion/ion chemistry and tandem mass spectrometry. The quadrupole linear ion trap with electrospray ionization and chemical ionization was also utilized in the analysis to characterize the primary structure of intact proteins [20].

Similarly, RNA polymerase II (Pol II) transcription initiation complex structure was analyzed by cross-linking and mass spectroscopy. They employed linear ion trap quadrupole (LTQ)-orbitrap spectrometer and recorded Fourier transform mass spectrometer (FTMS) spectra at 100,000 resolutions. The cross-linking/MS was used as an integrated structure analysis tool for large multi-protein complexes [21].

A new procedure was revealed which enabled selective sequencing and detection of serine-, threonine- and tyrosine-phosphopeptides at very low level of femto mole in protein digests with electrospray mass spectroscopy (ES-MS) using quadrupole mass spectrometer [22].

Another similar study revealed a method for determination of amino acid sequence of fractions of peptides from apolipoprotein B by tandem mass spectrometry. In this, triple quadrupole mass spectrometer along with LSI-MS was employed [23].

4.4. Clinical studies

Implementation of mass spectroscopy in clinical laboratory resulted in significant advancements. Sometimes greater degree of sensitivity is required when analyte quantity is too low and mass spectroscopy due to its higher sensitivity marks a valuable place in clinical analysis [24]. In any disease condition, the chemistry of body changes which results in the changes in products in body fluids and excretion products can be detected for the diagnosis purpose by chromatographic instrument like gas chromatography equipped with mass spectroscopy [18]. Matrix-assisted laser desorption/ionization mass spectrometry (MALDI-MS) is now in trend that is used to directly analyze and image pharmaceutical compounds in intact tissue [25].

Organic acidurias are an inherited disorder of metabolism in man, and gas chromatography coupled with mass spectroscopy is used to define an organic aciduria involving isovaleric acid. This technique is used in diagnosis and characterization of inborn errors of organic acid metabolism [26]. A simplified method was developed for clinical diagnosis of organic aciduria with gas chromatography–mass spectrometry (GC–MS) using quadrupole mass spectrometer. The urine samples were analyzed from patients, and acids were identified like methylcitric acid, margaric acid and glutaric acid [27].

Matrix-assisted laser desorption/ionization imaging mass spectrometry is emerged as a potent instrument for the investigation of small molecules and proteins in biological systems by *in situ* analysis of tissue sections [28]. A new technique was developed for the identification of proteins on tissue using tryptic digestion followed by matrix-assisted laser desorption/ionization imaging mass spectrometry with tandem mass spectrometry analysis. They used MALDI-TOF for their study [29]. Likewise, the level of antitumor drug SCH 226374 was determined in mouse tumor tissue using MALDI-QqTOF mass spectrometer. In whole brain homogenates, the concentration of drug was determined at nanogram levels with high-performance liquid chromatography/tandem mass spectrometry using triple quadrupole mass spectrometer [25].

Matrix-assisted laser desorption ionization-Fourier transform ion cyclotron resonance (MALDI-FTICR) is also an efficient technique for imaging of drugs and metabolites in tissue. A method based on MALDI-FTICR for imaging of olanzapine in kidney and liver as well as imatinib in glioma was described [30].

There is dramatic amplification in interest and implementation of clinical mass spectrometry for testing of vitamin D. LC–MS/MS method enables to distinguish between vitamin D_2 and vitamin D_3 and also provides information on the vitamin D epimeric form, both of which are not currently possible with existing immunoassays. Mass spectroscopy is the favored method for endocrine disorders analysis, for example, steroid analysis which requires technical competence, skill and experience for the needed improvement [24].

Mass spectroscopy can also be used for the investigation of bile acids in biological fluids. Bile has bile acids as the major constituents which are synthesized in liver and secreted in gall bladder or in intestine. Bile acids have many vital physiological functions like lipid absorption and cholesterol homeostasis. Under normal healthy condition, only small quantities of bile acids are found in peripheral circulation and urine, but in hepatobiliary and intestinal diseases it will get affected. This occurs due to disturbances in synthesis and pharmacokinetics in the body. Hence, the evaluation of bile acid can become useful for investigating the liver or intestinal functions along with the diagnosis of various diseases such as cholestasis, colon and liver cancer. The complex structure and low concentration of bile acids in biological fluids make their analysis technically difficult. For many years, GC–MS has been used but LC–MS has also been used for qualitative analysis of bile acids. A rapid, accurate, sensitive and reproducible method was developed using liquid chromatography–electrospray tandem mass spectrometry (LC–MS/MS) to investigate conjugated and total bile acids in samples of human bile and mixture of bile acid standards. The results coincide with the results obtained by the GC–MS technique. This method has important advantages over others because of the high specificity, sensitivity and selectivity of tandem mass spectrometry [31].

Mass spectroscopy has also application in clinical microbiology. Matrix-assisted laser desorption/ionization time-of-flight mass spectrometry (MALDI-TOF MS) has been freshly adapted for the recognition of whole microorganisms from their colonies on media or directly from cultures in blood and urine. This technique can precisely identify even those bacteria which are difficult by conventional methods and that day is not far away when this technology will complement the conventional microbiologic identification methods. In this technique, a portion of an isolated colony is loaded in the instrument and the comparison of spectrogram is done to a library by a proprietary algorithm to identify the organism. The ease of use, ability to run large numbers of isolates per batch, simplicity of setup, automation, rapid turnaround time and low reagent costs are the major advantages. By optimization, the cost of operation reduces to one-tenth that of conventional method with automated biochemical testing platforms. A Swiss study has done a comparison between MALDI-TOF MS system and conventional methods for the identification of 1371 routine bacteria and yeast isolates. MALDI-TOF MS provided identifications for 98.5% of the isolates, including 93.2% at the species level and 5.3% at the genus level. Of the species-level identifications, 95.1% matched conventional identifications. Important deficiencies in present MALDI-TOF MS platforms include misclassification such as of *Shigella* as *Escherichia coli* and *Streptococcus pneumoniae* as *Streptococcus mitis* and, additionally, poor performance with polymicrobial samples. In some cases, instruments have identified only one organism without indicating the presence of others. Despite the need for improvement, mass spectrometry will become popular in the near future with fast turnaround times, ease of use and potential operational cost savings [32].

4.5. Pharmaceutical analysis

Mass spectroscopy emerged as a powerful tool for various operations in pharmaceutical field mainly in drug development. New methods and instruments in mass spectroscopy are developed at a very high rate. Mass spectroscopy now becomes an irreplaceable tool in all types of drug discoveries due to its high sensitivity, speed, versatility and selectivity [33].

Mass spectroscopy is widely used for detection of impurities in samples. Likewise, the use of LC–MS for multidimensional evaluation of impurities during drug development is described. They used peptide drugs as an example and used ion trap mass spectrometer with electrospray ionization in their method [34]. Similarly, it can also be used for detecting the purity profile of active pharmaceutical ingredients (API), that are, MK-0969, an M3 antagonist; MK-0677, an oral-active growth hormone secretagogue and API-A, a cathepsin K inhibitor. The elucidation of impurity structure was made by utilization of LC–MS using quadrupole ion trap mass spectrometer equipped with an electrospray ionization or an atmospheric pressure chemical ionization (APCI) interface [35]. A protocol for qualitative and quantitative analysis of pharmaceutical compounds by MALDI-TOF mass spectrometry was described. Two drugs lopinavir and ritonavir were analyzed and described that HIV protease inhibitors can successfully be quantified in peripheral blood mononuclear cells using MALDI-TOF mass spectrometry [36].

Matrix-assisted laser desorption/ionization mass spectrometry imaging (MALDI-MSI) emerged as a valuable tool in direct analysis of pharmaceutical formulations. MALDI-MSI can be used for direct analysis of homogeneity of the active drug compound throughout the excipients contained in tablets. A direct analysis in real-time ion source coupled with a time-of-flight mass spectrometer (DART-MS) method for screening of pharmaceutical formulations was developed. A library of compounds were analyzed using mass spectra data collected by DART-MS operated in switching mode at 20, 60 and 90 V settings. This library consisted of 17 commonly encountered drugs in parenteral pharmaceutical formulations, that is, surgical analgesic: fentanyl, hydromorphone and morphine; anesthetic: baclofen, bupivacaine, ketamine, midazolam, ropivacaine and succinylcholine; and a mixture of other drug classes: caffeine, clonidine, dexamethasone, ephedrine, heparin, methadone, oxytocin and phenylephrine [37].

4.6. Forensic applications

In forensic study, sample is in minute quantity; therefore, high sensitivity is required for analysis. Mass spectroscopy coupled with gas chromatography emerged as an indispensable tool in forensic field as well as LC–MS has also wide utility in forensic study. In forensic studies, the use of mass spectroscopy is becoming significant because of increase in the demand to investigate use of illegal drugs through analyzing body fluids and tissues. The sample for forensics in the case of drug abuse is mainly urine, hair and blood. Some of the drugs in routine analysis include opiates, cocaine, marihuana, lysergic acid diethylamide (LSD) and amphetamines. However, cases of murders or death due to poisoning and drug overdose are also the prime targets for these drug candidates' analysis [38].

A liquid chromatographic thermospray tandem mass spectrometric method was developed for quantitative analysis of some drugs having hypnotic, sedative and tranquilizing properties, that is, benzodiazepine, thioxanthene, butyrophenone, methadone and diphenylbutylpiperidine in whole blood. A triple-stage quadrupole mass spectrometer was used in the analysis at a very low detection limit of 0.05–0.5 ng/ml [39]. Similarly, a method was developed for determining common drugs of abuse in body fluids using liquid chromatography-atmospheric pressure chemical ionization mass spectrometry (LC-APCI-MS). Drugs analyzed were opiate agonists (morphine, morphine-3-glucuronide, morphine-6-glucuronide, 6-monoacetylmorphine, codeine, codeine-6-glucuronide, dihydrocodeine, dihydromorphine, buprenorphine, methadone, tramadol, and ibogaine), cocaine and its metabolites (benzoylecgonine and ecgonine methyl ester) and lysergic acid diethylamide in serum, blood, urine and other biological matrices by using single quadrupole instrument [40]. Determination of 11-nor-9-D-tetrahydrocannabinol-9-carboxylic acid (THC-COOH) in urine [41], alfentanil, fentanyl and its derivatives with other opioid drugs like morphine, buprenorphine, codeine, heroin, methadone, naloxone, naltrexone, tramadol, pentazocine, pethidine and others in hair [42] has been done by LC–MS and LC–MS/MS. Analysis of methadone and its metabolites with other illicit drugs like cocaine, phencyclidine, heroin and 6-acetylmorphine in hair by GC–MS is another successful application of mass spectroscopy [43].

4.7. Metabolites analysis

Determination of metabolic pathway and different metabolites of a drug or xenobiotics is very important to assess its different parameters of pharmacokinetics. Drug metabolic reactions can be divided into two parts: 1. Phase I or functionalization reactions and 2. Phase II or conjugation reactions. Both of these transformations involve changes in the molecular weight. These changes can be accurately measured by mass spectrometer.

In structural characterization by mass spectrometry, the exchange of labile hydrogen with deuterium (H/D exchange) in small organic molecules has been widely used and this occurs in solution containing functional groups which have labile hydrogen(s) such as -SH, –OH, –N(R)H, $-NH_2$ and -COOH. During biotransformation, attachment of polar functional groups occurs, which causes changes in the number of exchangeable hydrogens. The number of exchangeable hydrogens in metabolites can give additional information to facilitate structural elucidation. This approach was applied to differentiate sulfoxide and sulfone metabolites from the isomeric mono- and di-hydroxylated metabolites, respectively. For example, the H/D exchange method was used for the drug metabolism studies of denopamine and promethazine in which N- or S-oxide was easily distinguished from the hydroxylated metabolites. A triple-stage quadrupole mass spectrometer equipped with electron impact (EI), FAB, APCI, ESI and thermospray (TSP) systems was utilized in the study [44].

Oxidation of a tertiary amino group to form an N-oxide is an important biotransformation pathway for many drugs and xenobiotics. N-oxide metabolites have the same elemental composition as those metabolites resulting from hydroxylation. Differentiation by mass spectrometry is a challenging task because these analytes exhibit the same m/z for their protonated or deprotonated molecules and their product ion mass spectra are usually very similar. Deoxygenation of N-oxides during atmospheric pressure chemical ionization represents potential way to differentiate N-oxides from hydroxylated metabolites. 6-OH desloratadine and N-oxides can be clearly differentiated as the major fragment ion from N-oxides was due to the loss of an oxygen atom while the prominent fragment ion from 6-OH desloratadine was due to loss of H_2O [45].

Stable isotope-labeled (^2H, ^{13}C, ^{15}N, ^{18}O, ^{34}S and others) xenobiotics can facilitate metabolite detection and identification by mass spectrometry, especially when radiolabeled parent drug is not available [46–48]. Custom-designed isotopic clusters resulting from the mixture of natural and synthetically enriched isotopes can greatly facilitate the detection and identification of metabolites. For example, the detection and identification of ribavirin metabolites in rats was done with the aid of stable isotope labeled drug [49]. Similarly, a fast and sensitive liquid chromatography–tandem mass spectrometry method was developed for simultaneous determination of acetaminophen and its glucuronide and sulfate metabolites (APAP-GLU and APAP-SUL) in small plasma volumes. The tandem triple quadrupole mass spectrometer equipped with an electrospray ionization source was used in the study [50].

5. Conclusion

Mass spectrometry is a very sensitive technique which can analyze even minute quantities of the molecule. This ability is utilized for various purposes like phytochemical, clinical, pharmaceutical and forensic analyses. This technique, not only elucidates the structure of the compounds, but also provides the information of molecular formula and the isotopic abundance of particular molecular formula. The availability of interphases made it possible to hyphenate this sophisticated technique with the different chromatographic techniques. This opened the new horizons of its applicability. The variations and permutation combinations of different ionization techniques with the different analyzers provide the analysis of diversified chemical entities at the femtogram level. The uniqueness of the technique of making fragments of the compound under investigation provides valuable structural information. This is helpful in the study of metabolite, peptide sequencing and macromolecules. This information is directly applicable in pharmaceutical and biomedical analysis. The development of double and triple quad techniques and their application have definitely uplifted the level of research and analysis in biomedical field, and this chapter gives the update on the topic.

Author details

Uttam Singh Baghel[1]*, Atamjit Singh[2], Deeksha Singh[3] and Manish Sinha[2]

*Address all correspondence to: drusb1985@yahoo.com

1 Department of Pharmaceutical Chemistry and Analysis, Kota College of Pharmacy, Kota, India

2 Department of Pharmaceutical Chemistry, Laureate Institute of Pharmacy, Kangra, India

3 Primary Health Centre, Meenabadoda, Sawai Madhopur, India

References

[1] Seger C, Sturm S. Analytical aspects of plant metabolite profiling platforms: Current standings and future aims. Journal of Proteome Research. 2007;**6**:480-497

[2] Safer S, Cicek SS, Pieri V, Schwaiger S, Schneider P, Wissemann V, Stuppner H. Metabolic fingerprinting of *Leontopodium* species (Asteraceae) by means of 1H NMR and HPLC-ESI-MS. Phytochemistry. 2011;**72**:1379-1389

[3] Shellie RA, Marriott PJ. Comprehensive two-dimensional gas chromatography-mass spectrometry analysis of Pelargonium Graveolens essential oil using rapid scanning quadrupole mass spectrometry. The Analyst. 2003;**128**:879-883

[4] Singh R, Kumar A, Giri DD, Bhuvaneshwari K, Pandey KD. Gas chromatography-mass spectrometry analysis and phytochemical screening of methanolic fruit extract of Momordica Charantia. Journal of Recent Advances in Agriculture. 2012;**1**:122-127

[5] Thangavel A, Balakrishnan S, Arumugam A, Duraisamy S, Muthusamy S. Phytochemical screening, gas chromatography-mass spectrometry (GC-MS) analysis of phytochemical constituents and anti-bacterial activity of Aerva Lanata (L.) leaves. African Journal of Pharmacy and Pharmacology. 2014;**8**:126-135

[6] Kannappan P, Bhaskaran KS, Sundaram S, Gopalakrishnan VK. Quantitative phytochemical analysis, in vitro antioxidant potential and gas chromatography-mass spectrometry studies in ethanolic extract of Azolla Microphylla. Asian Journal of Pharmaceutical and Clinical Research. 2016;**9**:318-323

[7] Simons R, Vincken J, Bohin MC, Kuijpers TFM, Verbruggen MA, Gruppen H. Identification of prenylated pterocarpans and other isoflavonoids in *Rhizopus* spp. elicited soya bean seedlings by electrospray ionisation mass spectrometry. Rapid Communications in Mass Spectrometry. 2011;**25**:55-65

[8] Qiu F, Tong Z, Gao J, Wang M, Gong M. Rapid and simultaneous quantification of seven bioactive components in radix Astragali based on pressurized liquid extraction combined with HPLC-ESI-MS/MS analysis. Analytical Methods. 2015;**7**:3054-3062

[9] Fan R, Li N, Jiang X, Yuan F, Gao Y. HPLC–DAD–MS/MS identification and HPLC–ABTS·+ on-line antioxidant activity evaluation of bioactive compounds in liquorice (*Glycyrrhiza uralensis* Fisch.) extract. European Food Research and Technology. 2014;**240**:1035-1048

[10] Fang N, Yu S, Prior RL. LC/MS/MS characterization of phenolic constituents in dried plums. Journal of Agricultural and Food Chemistry. 2002;**50**:3579-3585

[11] Simons R, Vincken JP, Bakx EJ, Verbruggen MA, Gruppen H. A rapid screening method for prenylated flavonoids with ultra-high-performance liquid chromatography/electrospray ionisation mass spectrometry in licorice root extracts. Rapid Communications in Mass Spectrometry. 2009;**23**:3083-3093

[12] Franski R, Bednarek P, Siatkowska D, Wojtaszek P, Stobiecki M. Application of mass spectrometry to structural identification of flavonoid monoglycosides isolated from shoot of lupin (Lupinus Luteus L.). Acta Biochimica Polonica. 1999;**46**:459-473

[13] Zaia J, Costello CE. Tandem mass spectrometry of sulfated heparin-like glycosaminoglycan oligosaccharides. Analytical Chemistry. 2003;**75**:2445-2455

[14] Carevic M, Bezbradica D, Banjanac K, Milivojevic A, Fanuel M, Rogniaux H, Ropartz D, Velickovic D. Structural elucidation of enzymatically synthesized galacto-oligosaccharides using ion-mobility spectrometry–tandem mass spectrometry. Journal of Agricultural and Food Chemistry. 2016;**64**:3609–3615

[15] Coulier L, Timmermans J, Bas R, Dool RVD, Haaksman I, Klarenbeek B, Slaghek T, Dongen WV. In-depth characterization of prebiotic galacto-oligosaccharides by a combination of analytical techniques. Journal of Agricultural and Food Chemistry. 2009;**57**:8488-8495

[16] Khosa MK, Mazhar M, Ali S, Dastgir S. Synthesis, characterisation, and structural elucidation by spectral investigation (FT-IR, multinuclear NMR, mass spectrometry) of biologically active organotin(IV) compounds containing germanium. Turkish Journal of Chemistry. 2010;**34**:387-398

[17] Wollyung KM, Wesdemiotis C, Nagy A, Kennedy JP. Synthesis and mass spectrometry characterization of centrally and terminally amine-functionalized polyisobutylenes. Polymer Chemistry. 2005;**43**:946-958

[18] Beckett AH, Stenlake JB. (2007), Practical pharmaceutical chemistry. 4th ed. India:CBS publishers and distributors; 2007. 493 p.

[19] Yergey AL, Coorssen JR, Backlund PS, Blank PS, Humphrey GA, Zimmerberg J, Campbell JM, Vestal ML. De novo sequencing of peptides using MALDI/TOF-TOF. Journal of the American Society for Mass Spectrometry. 2002;**13**:784-791

[20] Syka JEP, Coon JJ, Schroeder MJ, Shabanowitz J, Hunt DF. Peptide and protein sequence analysis by electron transfer dissociation mass spectrometry. Proceedings of the National Academy of Sciences of the United States of America. 2004;**101**:9528-9533

[21] Chen ZA, Jawhari A, Fischer L, Buchen C, Tahir S, Kamenski T, Rasmussen M, Lariviere L, Bukowski-Wills JC, Nilges M, Cramer P, Rappsilber J. Architecture of the RNA polymerase II-TFIIF complex revealed by cross-linking and mass spectrometry. The EMBO Journal. 2010;**29**:717-726

[22] Carr SA, Huddleston MJ, Annan RS. Selective detection and sequencing of phosphopeptides at the femtomole level by mass spectrometry. Analytical Biochemistry. 1996;**239**:180-192

[23] Hunt DF, Yates III JR, Shabanowitz J, Winston S, Hauer CR. Protein sequencing by tandem mass spectrometry. Proceedings of the National Academy of Sciences of the United States of America. 1986;83:6233-6237.

[24] Strathann FG, Hoofnagle AN. Current and future applications of mass spectrometry to the clinical laboratory. American Journal of Clinical Pathology. 2011;**136**:609-616

[25] Reyzer ML, Hsieh Y, Ng K, Korfmacher WA, Caprioli RM. Direct analysis of drug candidates in tissue by matrix-assisted laser desorption/ionization mass spectrometry. Journal of Mass Spectrometry. 2003;**38**:1081-1092

[26] Chalmers RA, Lawson AM. Organic acids in man: The analytical chemistry, biochemistry and diagnosis of the organic acidurias. Analytical Biochemistry. 1982;**127**:458-460

[27] Nakagawa K, Kawana S, Hasegawa Y, Yamaguchi S. Simplified method for the chemical diagnosis of organic aciduria using GC/MS. Journal of Chromatography B. 2010;**878**:942-948

[28] Zaima N, Hayasaka T, Goto-inoue N, Setou M. Matrix-assisted laser desorption/ionization imaging mass spectrometry. International Journal of Molecular Sciences. 2010;**11**:5040-5055

[29] Groseclose MR, Andersson M, Hardesty WM, Caprioli RM. Identification of proteins directly from tissue: *in situ* tryptic digestions coupled with imaging mass spectrometry. Journal of Mass Spectrometry. 2007;**42**:254-262

[30] Cornett DS, Frappier SL, Caprioli RM. MALDI-FTICR imaging mass spectrometry of drugs and metabolites in tissue. Analytical Chemistry. 2008;**80**:5648-5653

[31] Perwaiz S, Tuchweber B, Mignault D, Gilat T, Yousef IM. Determination of bile acids in biological fluids by liquid chromatography-electrospray tandem mass spectrometry. Journal of Lipid Research. 2001;**42**:114-119

[32] Mitsuma SF, Mansour MK, Dekker JP, Kim J, Rahman MZ, Tweed-Kent A, Schuetz P. Promising new assays and technologies for the diagnosis and management of infectious diseases. Clinical Infectious Diseases. 2013;**56**:996-1002

[33] Feng WY. Mass spectrometry in drug discovery: A current review. Current Drug Discovery Technologies. 2004;**1**:295-312

[34] Ermer J. The use of hyphenated LC–MS technique for characterisation of impurity profiles during drug development. Journal of Pharmaceutical and Biomedical Analysis. 1998;**18**:707-714

[35] Zhou L, Mao B, Reamer R, Novak T, Ge Z. Impurity profile tracking for active pharmaceutical ingredients: Case reports. Journal of Pharmaceutical and Biomedical Analysis. 2007;**44**:421-429

[36] Van Kampen JJ, Burgers PC, De Groot R, Luider TM. Qualitative and quantitative analysis of pharmaceutical compounds by MALDI-TOF mass spectrometry. Analytical Chemistry. 2006;**78**:5403-5411

[37] Poklis JL, Mohs AJ, Wolf CE, Poklis A, Peace MR. Identification of drugs in parenteral pharmaceutical preparations from a quality assurance and a diversion program by direct analysis in real-time AccuTOFTM-mass spectrometry (DART-MS). Journal of Analytical Toxicology. 2016;**40**:608-616

[38] Foltz RL. Recent applications of mass spectrometry in forensic toxicology. International Journal of Mass Spectrometry and Ion Processes. 1992;**118/119**:237-263

[39] Verweij AMA, Hordijk ML, Lipman PJL. Liquid chromatographic-thermospray tandem mass spectrometric quantitative analysis of some drugs with hypnotic, sedative and tranquillising property in whole blood. Journal of Chromatography B: Biomedical Sciences and Applications. 1996;**686**:27-34

[40] Bogusz MJ, Maier RD, Kruger KD, Kohls U. Determination of common drugs of abuse in body fluids using one isolation procedure and liquid chromatography--atmospheric-pressure chemical-ionization mass spectrometry. Journal of Analytical Toxicology. 1998;**22**:549-558

[41] Breindahl T, Andreasen K. Determination of 11-nor-delta-9-tetrahydrocannabinol-9-carboxylic acid in urine using high-performance liquid chromatography and electrospray

ionization mass spectrometry. Journal of Chromatography B: Biomedical Sciences and Applications. 1999;**732**:155-164

[42] Gergov M, Nokua P, Vuori E, Ojanpera I. Simultaneous screening and quantification of 25 opioid drugs in post-mortem blood and urine by liquid chromatography–tandem mass spectrometry. Forensic Science International. 2009;**186**:36-43

[43] Goldberger BA, Darraj AG, Caplan YH, Cone EJ. Detection of methadone, methadone metabolites, and other illicit drugs of abuse in hair of methadone-treatment subjects. Journal of Analytical Toxicology. 1998;**22**:526-530

[44] Ohashi N, Furuuchi S, Yoshikawa M. Usefulness of the hydrogen--deuterium exchange method in the study of drug metabolism using liquid chromatography-tandem mass spectrometry. Journal of Pharmaceutical and Biomedical Analysis. 1998;**18**:325-334

[45] Ramanathan R, AD S, Alvarez N, Blumenkrantz N, Chowdhury SK, Alton K, Patrick J. Liquid chromatography/mass spectrometry methods for distinguishing N-oxides from hydroxylated compounds. Analytical Chemistry. 2000;**72**:1352-1359

[46] Branfma AR, McComish MF, Bruni RJ, Callahan MM, Robertson R, Yesair DW. Characterization of diaminouracil metabolites of caffeine in human urine. Drug Metabolism & Disposition. 1983;**11**:206-210

[47] Baillie TA, Halpin RA, Matuszewski BK, Geer LA, Chavez-Eng CM, Dean D, Braun M, Doss G, Jones A, Marks T, Melillo D, Vyas KP. Mechanistic studies on the reversible metabolism of rofecoxib to 5-hydroxyrofecoxib in the rat: Evidence for transient ring opening of a substituted 2-furanone derivative using stable isotope-labeling techniques. Drug Metabolism Disposition. 2001;**29**:1614-1628

[48] Weidolf L, Covey TR. Studies on the metabolism of omeprazole in the rat using liquid chromatography/ion spray mass spectrometry and the isotope cluster technique with [34S]omeprazole. Rapid Communications in Mass Spectrometry. 1992;**6**:192-196

[49] Chowdhury SK, editor. Identification and Quantification of Drugs, Metabolites and Metabolizing Enzymes by LC-MS. 1st ed. Amsterdam: Elsevier; 2005. 277 p

[50] Gicquel T, Aubert J, Lepage S, Fromentry B, Morel I. Quantitative analysis of acetaminophen and its primary metabolites in small plasma volumes by liquid chromatography-tandem mass spectrometry. Journal of Analytical Toxicology. 2013;**37**:110-116

Chapter 7

Metal Complexes of Pharmaceutical Substances

Tünde Jurca, Eleonora Marian, Laura Graţiela Vicaş,
Mariana Eugenia Mureşan and Luminiţa Fritea

Additional information is available at the end of the chapter

http://dx.doi.org/10.5772/65390

Abstract

Significant progresses have been made in the inorganic and organic chemistry up to the present concerning the synthesis, characterization, and application of the metal complexes of pharmaceutical substances. From the wide range of fields in which these coordination compounds find their application, many efforts were focused on the study of their importance in the biological processes. The coordination complexes of many pharmaceutical substances having different pharmacological effects e.g., pyrazinamide (PZA), nicotinamide (NAM), nicotinic acid (NIC), theophylline (TEO), captopril (CPL), tolbutamide (TBA), clonidine (CLN), guanfacine (GUAF), etc. with transition metals were synthesized and used in order to improve their pharmacological and pharmacotechnical properties and also for the drug analysis and control. Several techniques such as Fourier transform infrared spectroscopy (FTIR), Raman spectroscopy, surface-enhanced Raman spectroscopy (SERS), X-ray spectroscopy, mass spectrometry, ultraviolet-visible (UV-Vis) spectrophotometry, electron paramagnetic resonance (EPR) spectroscopy, X-ray diffraction, elemental analysis, electrochemical methods, thermal methods, and scanning electron microscopy were used for the physicochemical characterization of the complex composition. A significant interest in the development of metal complex-based drugs with unique research and therapeutic and diagnostic opportunities is currently observed in the medicinal inorganic chemistry area.

Keywords: coordination complexes, transition metals, pharmaceutical substances, characterization methods, medicinal chemistry

1. Introduction

The coordination complexes have been studied since 1798 starting with the Tassaert studies, and till nowadays significant progresses have been made in the inorganic and organic chemistry concerning the synthesis, characterization, and application of this large group of metal complexes. Concerning their structure, complexes were considered those compounds which do not fit within the classical theory of valence, meaning that the combination ratio of the elements exceeded their valences. This coordination theory elaborated by Alfred Werner indicated that the secondary valences of the elements are involved in the formation of the second-order combinations leading to the actual representation of the complexes formed by the first coordination sphere marked between brackets [central atom (ligand)] and the second coordination sphere (ionization sphere) coming outside of the brackets. The central atom can be any chemical element; meanwhile, the ligands can be ions, atoms, or neutral molecules, which can act as donors [1]. Neutral molecules or mono-/polyatomic anions which have one or more unshared electron pairs can act as mono-/polydentate ligands, the latter ones form complexes with cyclic structure known as chelates. A large number of pharmaceutical substances behave in vivo or in vitro conditions as ligands and chelating agents [2].

The number and the large structural variety of these complexes could not allow a rigorous systematization, even though some attempts by using certain classification criteria have been made such as the number of the central atoms, the charge of the complex ion, the type of ligands, and the coordination number. The coordination compounds were classified into Werner complexes, complexes with metal-metal bonds, metal carbonyls, clusters, complexes with macrocyclic ligands, molecular complexes (adducts, clathrates), chelates, and metal-organic complexes [1].

Natural metal complexes consisting of a central metal atom or ion (especially of the 3D transition metals) are involved in a plenty of biological mechanisms among which photosynthesis, transport of oxygen in blood, coordination of some metabolic processes, pathological states, enzymatic reactions, etc., even though the metallic ions represent only 3% of the body composition. Many biomolecules (amino acids, peptides, carboxylic acids, etc.) can form metal complexes with different stabilities having biomedical importance. Some drugs have a certain therapeutic effect (e.g., antimicrobial, diuretic, antidepressant) due to the complexation of the metallic ion (Cu^{2+}, Zn^{2+}, Fe^{2+}, Mg^{2+}, etc.) essential for a certain biochemical process. Metal complexes and products containing oligoelements are widely used in therapy due to their pharmacodynamic properties, bioavailability enhancement, and toxicity decrease of some metal ions [1].

The main aspects concerning the formation of complexes between pharmaceutical substances and various ligands are supported by several observations. According to the biological, physiological, and pathophysiological role of metal ions and ligands with pharmacological effect, metal ions present a great importance in carrying out the vital functions of living organisms acting as complexes or chelates and also in the analysis and control methods of drug substances by forming complexes that can be detected by spectral techniques. The use of ligands, chelating agents, or complexes in medicine and biology concerns several purposes such as antidotes in poisoning with metal ions or hydrocyanic acid or cyanides; introduction in the living organisms of some essential metal ions found to be deficient; depriving bacteria, viruses, or microbial enzyme systems of micronutrients essential for their work; or providing toxic metals for the pathogenic agents [2].

Many coordination complexes have been used in medicine containing metals such as platinum (cisplatin as anticancer chemotherapy drug), gold (as auranofin used for rheumatoid arthritis), technetium and rhenium (as radiopharmaceuticals used in imaging and radiotherapy), ruthenium (as anticancer drug), gadolinium, cobalt, lithium, bismuth, iron, calcium, lanthanum, gallium, tin, arsenic, rhodium, copper, zinc, aluminum, lutetium, vanadium, manganese, etc. [3, 4]. Only a reduced number of Co(III) complexes can be mentioned as having biochemical properties: vitamin B_{12}, a natural organometallic complex of Co(III) with glyoxime. Other important examples are the series of Co(III) complexes containing N- and O-donor ligands based on a chelating Schiff base (imidazole, methylimidazole) with efficiency in the treatment of epithelial herpetic keratitis (the molecular target is supposed to be a virus protease containing histidine), adenovirus keratoconjunctivitis, and human immunodeficiency virus type 1. $[Co(NH_3)_6]Cl_3$ presents potent antiviral activity (against Sindbis virus). Some studies reported also the antibacterial activity of Co(II) and Co(III) complexes against *Bacillus subtilis, Enterobacter aeruginosa, Escherichia coli, Staphylococcus aureus,* etc. [3].

It was demonstrated that the antibacterial activity was increased upon chelation making the ligand a more powerful agent [5, 6]. The complexation of derivatives of sterically hindered *o*-diphenols and *o*-aminophenols with Cu(II), Co(II), Ni(II), and Zn(II) ions exhibited antioxidant, antiviral, and antimicrobial activity with low toxicity. Their synthesis, their separation as crystalline powders, the composition, and physicochemical characteristics of the complexes were also studied in Ref. [7].

Metal complexes have become an emerging tool in drug discovery being widely used as therapeutic compounds to treat several human diseases such as carcinomas, lymphomas, infection control, diabetes, anti-inflammatory, and neurological disorders [8, 9]. Due to various implications and applications of complexes (especially the chelates) in the biomedical field, many aspects are required to be studied such as their nature, their stability, the factors determining their formation and stability, and possibilities for preventing some reactions and for releasing a metal ion from a complex; all these are necessary in order to understand how they act in biological processes [2].

2. Physicochemical characterization of metal complexes of some pharmaceuticals

The coordination complexes of a wide range of pharmaceutical substances [pyrazinamide (PZA), nicotinamide (NAM), nicotinic acid (NIC), tolbutamide (TBA), theophylline (TEO), captopril (CFL), clonidine (CLN), and guanfacine (GUAF)] with transition metals [Cu(II), Cd(II), Ni(II), Mn(II), Zn(II), and Co(II)] were synthesized and then characterized by using various techniques such as elemental analysis, spectral, electrochemical, thermal, and microscopic methods.

2.1. Metal complexes of pyrazinamide

Pyrazinamide (PZA) (pyrazine carboxamide) is a nicotinamide analogue used as a first-line drug to treat tuberculosis. The complexes of PZA with Cu(II) were assessed by different techniques such as elemental analysis, spectral methods [Fourier transform infrared spectroscopy (FTIR), FT-Raman spectrometry, mass spectrometry], and scanning electron microscopy (SEM) coupled with X-ray spectroscopy [energy-dispersive spectroscopy (EDS)] [1, 10–13].

The elemental analysis indicated that the combination ratio of metal:ligand (Me:L) is 1:2 for $[Cu(PZA)_2]Cl_2$ and $[Cu(PZA)_2](C_6H_5COO)_2$ complexes. The mass spectra of the complex of PZA with Co(II) benzoate revealed the identity and the purity of PZA and of the complex fragments confirming its structure (**Figure 1**) [1, 10].

The FTIR spectra of the complexes highlighted that –C=O groups and nitrogen from the pyrazine ring are implied in the coordination process (**Table 1**) [11]. Comparing the Raman spectra of PZA and of $[Cu(PZA)_2]Cl_2$, another analytical evidence for the complex formation is obtained. The appearance of new band characteristic for the Me:L bonds can be observed analyzing in detail the spectral region of low values of wave number (**Figure 2**) [1, 11, 12].

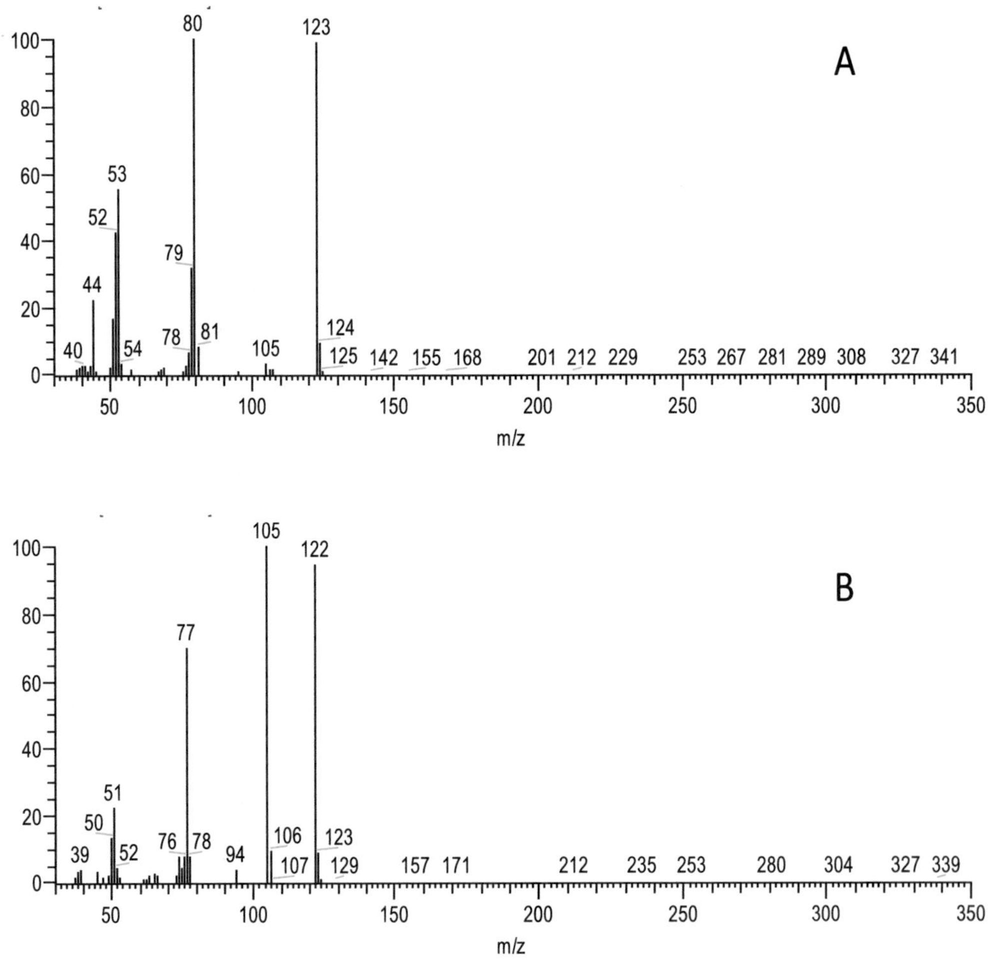

Figure 1. The mass spectra of PZA (A) and $[Cu(PZA)_2](C_6H_5COO)_2$ (B) [1, 10]. Reprinted with permission of Revista de Chimie and of Editura Universității din Oradea.

http://dx.doi.org/10.5772/65390

PZA	$[Cu(PZA)_2](C_6H_5COO)_2$	$[Cu(PZA)_2]Cl_2$	Assignment
3410s	3610w	3430s	ν_{as} NH_2
3140m	3170m	3110m	ν_s NH_2
3080	3065m	3070m	νCH
1705s	1915w	1700s	νC=O(1)
1600m	1590m	1590m	δNH_2(2)
1570	1585m	1585m	ν ring
1530	1545s	1510w	ν ring
1375s	1380s	1385s	νCN(III)
1150w	1180w	1170m	δCH
1090m	1085w	1080w	ϱ NH_2
870w	850w	870m	δ ring
665w	680m	670w	ϱ NH_2

Note: v, very; s, strong; m, medium; w, weak; ν, stretching; δ, in plane bending; ϱ, rocking.
Source: Reprinted with permission of Farmacia and of Editura Universității din Oradea.

Table 1. Assignment of the characteristic IR bands of the metal complexes of PZA [1, 14].

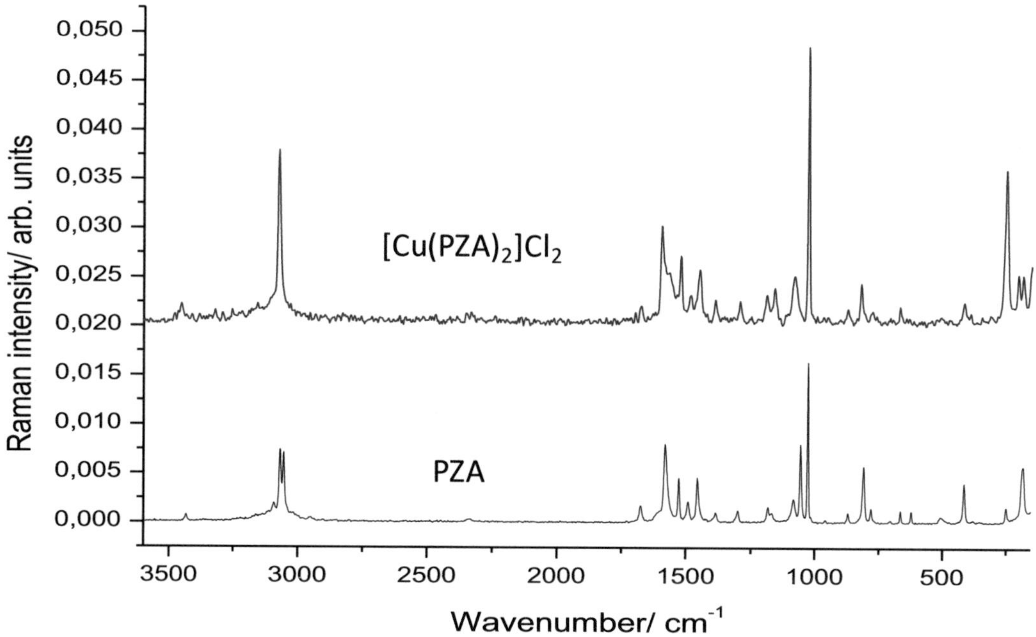

Figure 2. Raman spectra of PZA and of $[Cu(PZA)_2]Cl_2$ [1, 12]. Reprinted with permission of Studia Universitatis Babes-Bolyai and of Editura Universității din Oradea.

The morphology and the crystal structure of the two complexes were revealed by the SEM images and EDS spectra (**Figures 3** and **4**). The first complex, $[Cu(PZA)_2](C_6H_5COO)_2$, presented irregular conglomeration with different shapes and dimensions (**Figure 3A**); meanwhile, the second one, $Cu(PZA)_2]Cl_2$, presented acicular and elongated particles with an average size of about 1.5 microns (**Figure 4A**) [1, 14].

2.2. Metal complexes of nicotinamide

Nicotinamide (NAM) (3-pyridine carboxylic acid amide) is the amide of nicotinic acid playing an important role in the biosynthesis of pyridine nucleotides, and it is a reactive moiety of the coenzyme nicotinamide adenine dinucleotide, a soluble electron carrier in biochemical reactions. The NAM complexes with transition metals [Cu(II), Cd(II), Hg(II)] were synthesized and characterized by using elemental analysis, UV-Vis, and FTIR spectroscopy [1, 15, 16]. The spectral data confirmed tetradentate coordination of NAM with Hg(II), Cd(II), and hexadentate coordination with Cu(II). In the FTIR spectra of these complexes, it can be observed that the characteristic bands of NAM are slightly shifted after coordination (**Table 2**) [16]. The slight shifting of the bands from NAM complexes with Hg may be explained by the stereochemistry of $HgCl_2$, which is less bulky than $Cu(C_6H_5COO)_2$ and $Cd(SCN)_2$.

2.3. Metal complexes of nicotinic acid

Nicotinic acid (NIC) (pyridine-3-carboxylic acid) known as vitamin B_3, niacin, has two important pharmacological properties: peripheral vasodilator and hypocholesterolemic drug. Its complexes with Co(II) and Cu(II) were synthesized and characterized by elemental analysis and spectral methods [FTIR spectroscopy, Raman spectroscopy, and surface-enhanced

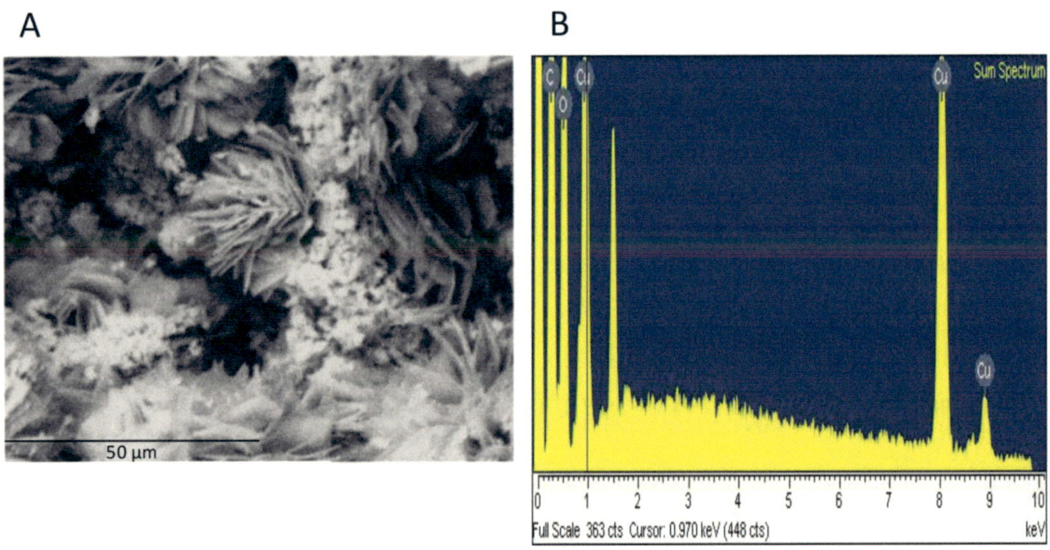

Figure 3. SEM image (A) and EDS spectrum (B) of $[Cu(PZA)_2](C_6H_5COO)_2$ [1, 14]. Reprinted with permission of Farmacia and of Editura Universității din Oradea.

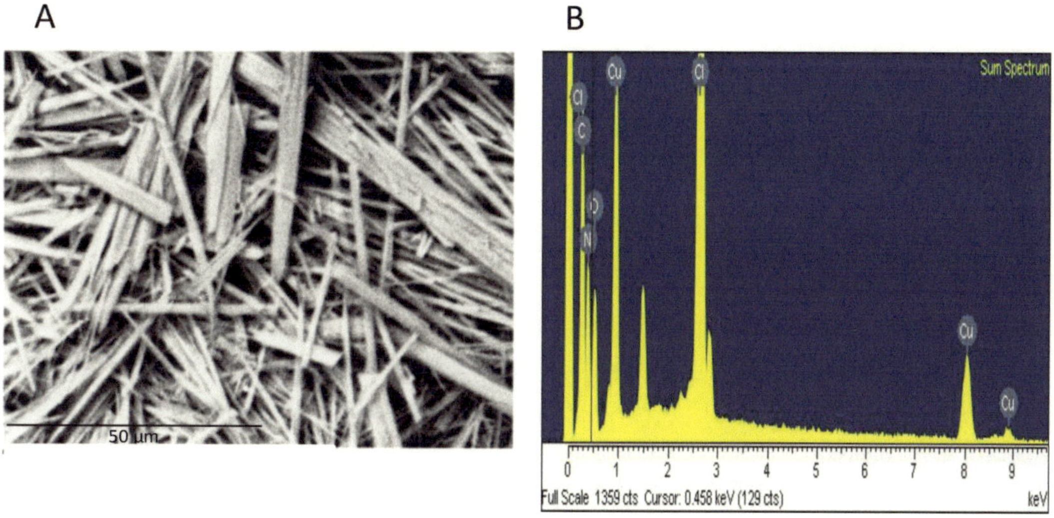

Figure 4. SEM image (A) and EDS spectrum (B) of $[Cu(PZA)_2]Cl_2$ [1, 14]. Reprinted with permission of Farmacia and of Editura Universității din Oradea.

Raman spectroscopy (SERS)] (**Figure 5**). The significant differences observed from the spectral data of the metal complexes can be attributed to the coordination process with the metal ions: the stretching vibrations ν(C—C) from the pyridine ring (1500–1600 cm^{-1}) and ν(ring) of NIC (1037 cm^{-1}) are shifted; meanwhile, the vibration band γ(CH) of the ring at 811 cm^{-1} is shifted and also splitted indicating the ring deformation during the coordination process. There appear new bands corresponding to the Me:L bonds (at about 500 cm^{-1}) (**Table 3**). The spectral results confirmed the monodentate coordination of NIC with Cu(II) and Co(II) [17].

2.4. Metal complexes of guanfacine

Guanfacine (GUAF) (*N*-(diaminomethylidene)-2-(2,6-dichlorophenyl)acetamide), used as antihypertensive drug, is able to form colored complexes (combination ratio Me:L 1:2) with Mn(II) and Cd(II) having different spectral characteristics. These complexes were analyzed by using spectral techniques such as FTIR and Raman spectroscopy. The imine group vibration from the FTIR data of GUAF ($\nu_{C=N}$ at 1700 cm^{-1}) was shifted (Δ = 10–60 cm^{-1}) in the spectra of GUAF complexes with Mn(II) ($\nu_{C=N}$ at 1710 cm^{-1}) and Cd(II) ($\nu_{C=N}$ at 1760 cm^{-1}) showing the imine group involvement in the complex formation. The formation of new bonds Me:L was observed at around 500 cm^{-1} in the case of the two mentioned complexes. Significant differences appeared in the Raman spectra of the complexes in the region 1100–1250 cm^{-1} due to the electronic delocalization from $NH{=}C{-}NH_2$ (**Figure 6**). After coordination, in the case of both complexes, two distinct bands were revealed at 1212 cm^{-1} for $NH{-}C{-}NH_2$ and at 1174 cm^{-1} for $NH{=}C{-}NH_2$. The spectral data indicated that GUAF is coordinated by nitrogen atoms, and the results confirmed a tetradentate coordination of Cd(II) complexes [18].

NAM	$[Hg(NAM)_2]Cl_2$	$[Cu(NAM)_2]$ $(C_6H_5COO)_2 2H_2O$	$[Cd(NAM)_2](SCN)_2$	Assignment
			3531w	(OH)
3364vs	3363s	3369s	3479s	$\nu_{as}(NH_2)$
3167s	3171s	3207vs	3176m	$\nu_s(NH_2)$
3065sh	3071sh	3071sh	3072w	ν(CH)
1654s	1654vs	1668vs	1667vs	ν(CO)
1622ws	1623ws	1632ws	1618m	$\delta(NH_2)$
1577ws	1577vs	1596s	1577s	ν(CN) + ν(CC)
1449m	1449m	1490s	1485m	β(CH)
1395ws	1400m	1377vs	1400s	ν(CN) amide
1297m	1296m	1301w	1331m	ν(CC)
1178m	1179m	1193w	1204s	ν(ring) NAM
1141m	1142s	1153sh	1153m	ν (CN)
1120m	1119s	1116m	1112m	$\rho(NH_2)$
1022m	1024w	1025m	1040m	υ(CNS)
–	919m	928w	937w	γ(CH) ring
–	844s	853m	840m	$\nu_{as}(C–CH_3)$
771m	786s	775w	770s	γ(CH) ring
698m	700ws	719w	719w	$\omega(NH_2)$
684m	686ws	687s	687ws	δ(ring)NAM
633s	641s	655w	657ws	γ(NH)

Note: v, very; s, strong; m, medium; w, weak; sh, shoulder; sp, splitting; ν, stretching; β, in-plane bending; γ, out-of-plane bending; δ, in-plane bending; ω, out-of-plane wag.

Source: Reprinted with permission of Revista de Chimie.

Table 2. Assignment of the characteristic IR bands of the metal complexes of NAM [16].

2.5. Metal complexes of theophylline

Theophylline (TEO) (3,7-dihydro-1,3-dimethyl,1H-purine-2,6-dione) also known as 1,3-dimethyl-xanthine belongs to the class of peripheral and cerebral vasodilator drugs. Metal complexes of TEO were synthesized having the general formula: $[Me_n(TEO)_x]A_m \cdot yH_2O$, where Me = Cu(II), Co(II), Cd(II), Zn(II), and Ni(II) and A = CH_3COO^-, $C_6H_5COO^-$; n = 1, x = 1 or 2, m = 2, and y = 2 or 4. The combination ratio was determined by using elemental analysis and conductometric titration; meanwhile, the number of water molecules was determined by using thermal analysis [2, 19–22].

http://dx.doi.org/10.5772/65390

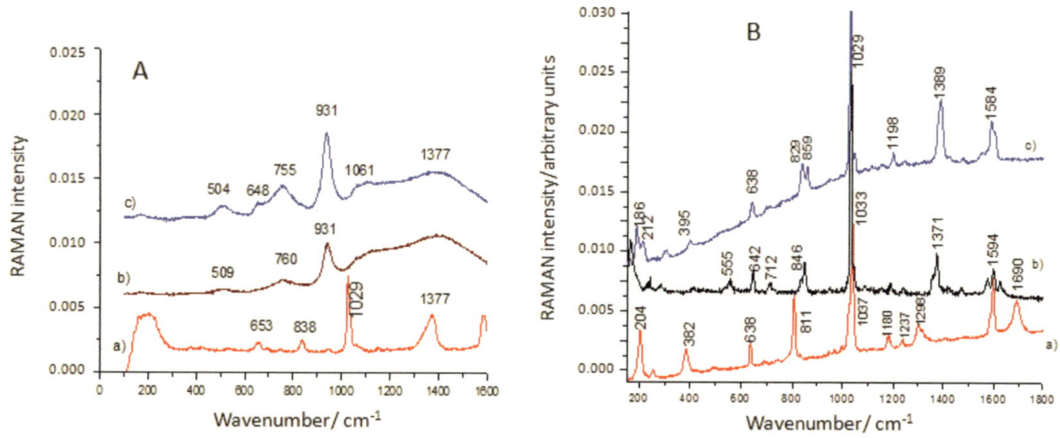

Figure 5. Raman (A) and SERS (B) spectra of NIC (a) and its complex with Cu(II) (b) and Co(II) (c) [17]. Reprinted with permission of Revista de Chimie.

Assignment (cm^{-1})	NIC Raman (cm^{-1})	NIC SERS (cm^{-1})	$Cu(NIC)_2(CH_3COO)_2$ Raman (cm^{-1})	$Cu(NIC)_2(CH_3COO)_2$ SERS (cm^{-1})	$Co(NIC)_2(CH_3COO)_2$ Raman (cm^{-1})	$Co(NIC)_2(CH_3COO)_2$ SERS (cm^{-1})
ν(OH) acid	–	–	–	–	–	–
ν(CH)	–	–	–	–	–	–
νC=O	1690m	–	–	–	–	–
ν(ring) NIC	1594m	1594m	1594m	–	1584m	–
ν(CN)	–	–	–	–	–	–
	–	1377m	1371m	1377wv	1389s	1377wv
ν(CC)	1298m	–	–	–	–	–
δ(CN)	1180m	–	–	–	1198m	–
ν(ring)NIC	1037vs	1029s	1033vs	1060wv	1029vs	1061wv
δ(OH)acid	–	–	–	931m	–	931m
γ(CH) ring	811m	838wv	846m	760wv	829m	755wv
δ(CH) ring	638m	653wv	642m	–	638m	648wv
νMe-O	–	–	555wv	509wv	–	504wv

Note: v, very; s, strong; m, medium; w, weak; ν, stretching; γ, out-of-plane bending; δ, in-plane bending.
Source: Reprinted with permission of Revista de Chimie.

Table 3. Assignment of the characteristic Raman and SERS bands of the metal complexes of NIC [17].

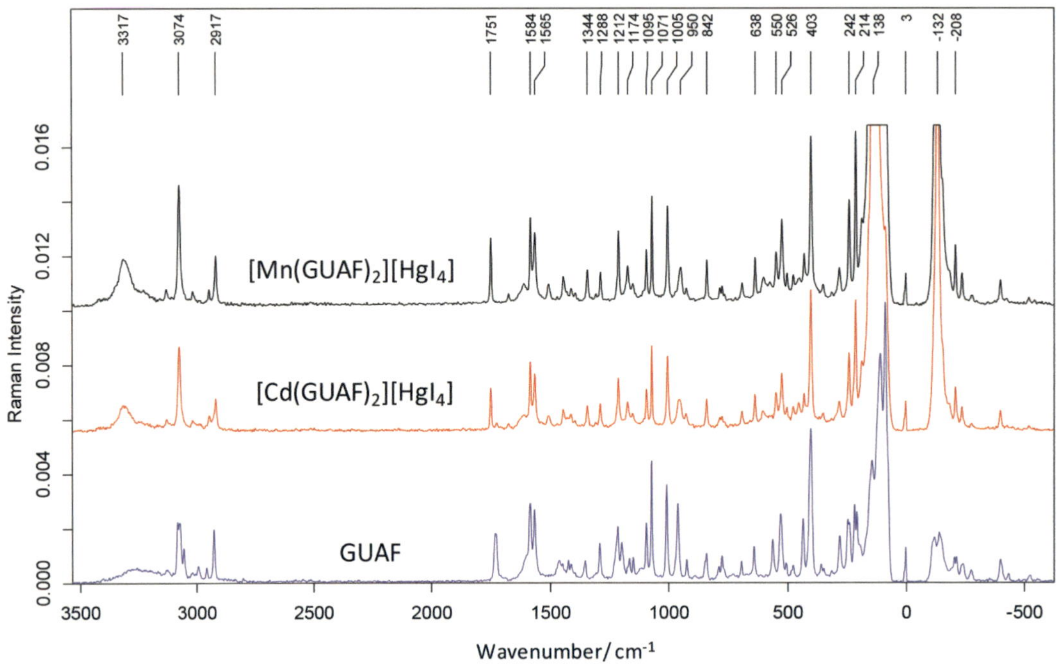

Figure 6. Raman spectra of GUAF and of its metal complexes [19].

The combination ratio Me:TEO is 1:2 for the complexes having the acetate anion. The complex $[Cu(TEO)_2](CH_3COO)_2$ has a high thermal instability even at 40°C, its thermal decomposition being already started. On the thermal curves, eight stages of decomposition, all scarcely separable, can be observed. The first five were weakly endothermic, and three were strongly exothermic. The X-ray diffractogram revealed that this complex crystallizes in the monoclinic system. The microscopic analysis showed a mixture of particles with different shapes: acicular, flake, irregular, and lamellar (**Figure 7**) [2, 19–22].

In the case of $[Cd(TEO)_2](CH_3COO)_2$, the last stage of thermal decomposition was not achieved in the investigated temperature range; therefore, heating was required up to a higher temperature (850°C) when constant weight was reached corresponding to the cadmium oxide. The complex crystallizes in the monoclinic system, and it presents microcrystals with parallelepiped shape (**Figure 8**) [2, 19, 21, 22].

The thermal decomposition of $[Co(TEO)_2](CH_3COO)_2$ takes place in four stages: one endothermic and three exothermic. It presents monoclinic crystal system, and the microcrystals have a tabular form (**Figure 9**). The complex $[Zn(TEO)_2](CH_3COO)_2$ presented similar properties as $[Cd(TEO)_2](CH_3COO)_2$ complex [2, 19, 21, 22].

The endothermic peak at 272°C, which is characteristic for TEO decomposition, is not found in the differential scanning calorimetry (DSC) curves of the complexes, being a credible argument for the complex synthesis and not as a simple mechanical mixture (**Figure 10**) [2, 19].

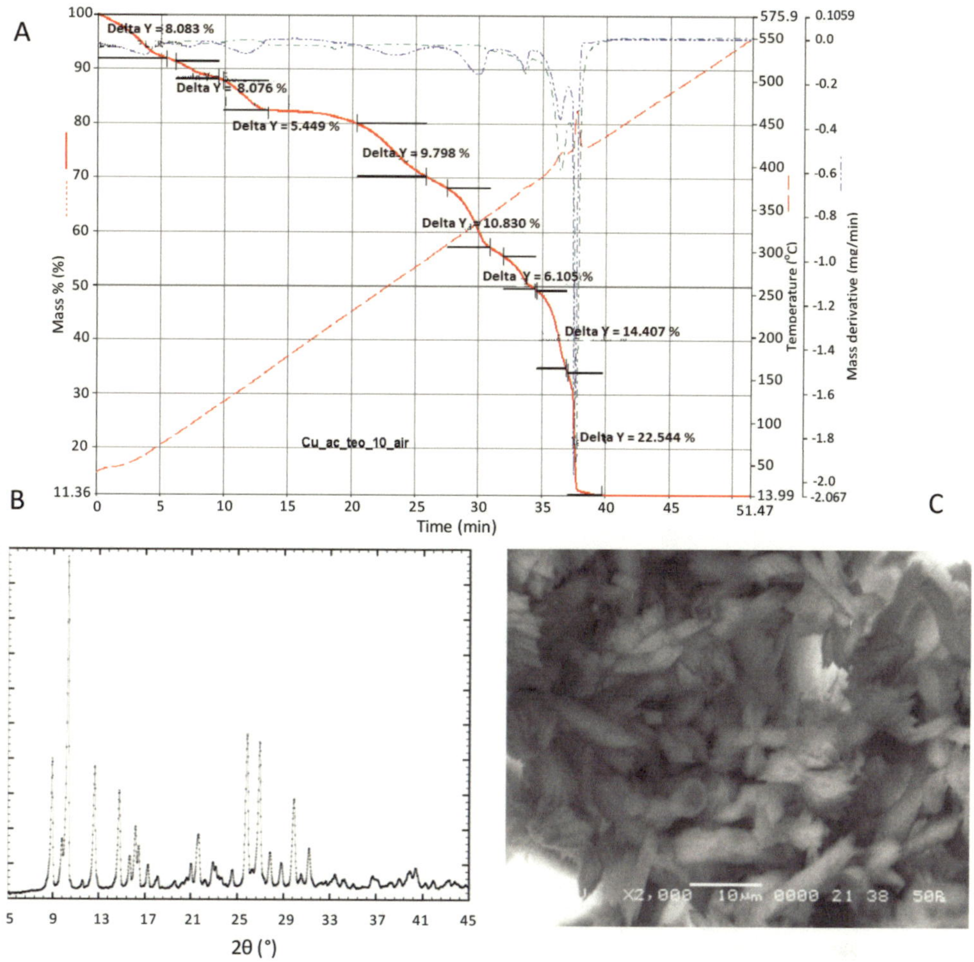

Figure 7. Thermogramms TG, DTG, and DTA (A); X-ray diffractogram; (B) and SEM images (C) of $[Cu(TEO)_2](CH_3COO)_2$ [2, 19, 21, 22]. Reprinted with permission of Revista de Chimie, Farmacia and of Editura Politehnica Timişoara.

The FTIR data also indicated the complex formation: the disappearance of the symmetric vibration band of —C=O from TEO at 1717 cm^{-1} in the complexes spectra indicating that this bond is involved in the formation of Me:TEO coordinative bond; the deformation vibration of Me:N bond found at 570–685 cm^{-1}, the appearance of symmetric and asymmetric stretching vibrations of the —COOH group (1260–1250 and 1535–1530 cm^{-1}), and the possibility of coordinating also the water of crystallization (appearance of large bands at 3050–3500 cm^{-1}) (**Figure 11**) [2, 19, 21, 22].

The combination ratio Me:TEO is 1:1 for the complexes having the benzoate anion: $[Co(TEO)](C_6H_5COO)_2{\cdot}2H_2O$, $[Ni(TEO)](C_6H_5COO)_2{\cdot}2H_2O$, and $[Cu(TEO)](C_6H_5COO)_2{\cdot}2H_2O$. Their thermal decomposition takes place in four stages, the first one being the stage of loss of water of crystallization (**Figure 12A**). The FTIR data are similar with those of the complexes mentioned above (having the acetate group as anion); in addition, the specific vibration band

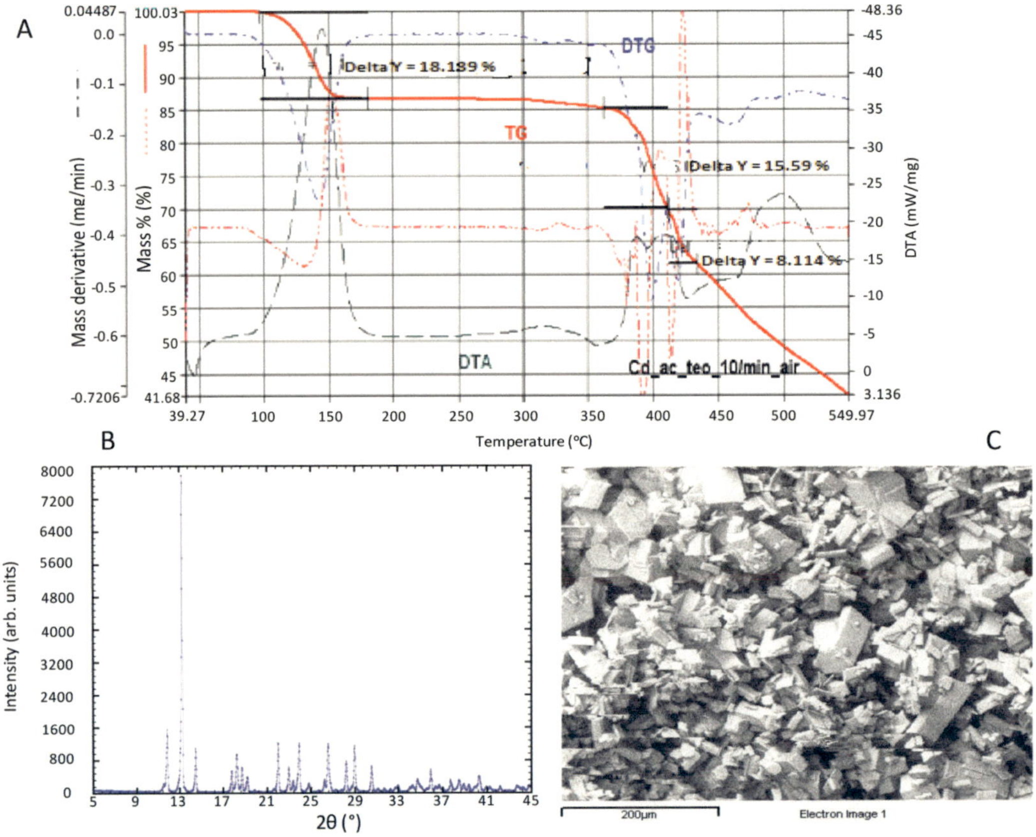

Figure 8. Thermogramms TG, DTG, and DTA (A); X-ray diffractogram; (B) and SEM images (C) of $[Cd(TEO)_2](CH_3COO)_2$ [2, 19, 21, 22]. Reprinted with permission of Revista de Chimie, Farmacia and of Editura Politehnica Timişoara.

of the aromatic ring (1438, 1442, 1440 cm^{-1}) appears. The microscopic image of [Co(TEO)] $(C_6H_5COO)_2$ showed the acicular shape of the particles (**Figure 12B**) [2, 20].

2.6. Metal complexes of captopril

The chemical structure of captopril (CPL), a dipeptide derivative of *L*-alanine-*L*-proline with antihypertensive effect, contains bonds such as —C=O and $—N(—CH_2)_2$ with donor atoms capable of forming Me:L bonds. The interaction between the metal ions such as Mn(II), Co(II), Zn(II), Ni(II), and Cd(II) with N and O atoms from the peptide (which act as donors) leads to the formation of stable chelate cycles. These complexes were characterized by elemental analysis obtaining the results presented in **Table 4** [23, 24].

CPL forms complexes with transition metals mentioned above in the presence of tetraiodomercurate anion, $[HgI_4]^{2-}$. The formation and the structure of these complexes are observed in the data of the elemental analysis and in the UV and IR spectra of the complexes with changes of the wavelength values and of absorbance due to the presence of Me:CPL bonds. In the IR spectra of the complexes, a diminution of the band at 1748 cm^{-1} of —C=O from the carboxyl group, in comparison with the IR spectrum of CPL, was observed. A wider band appeared

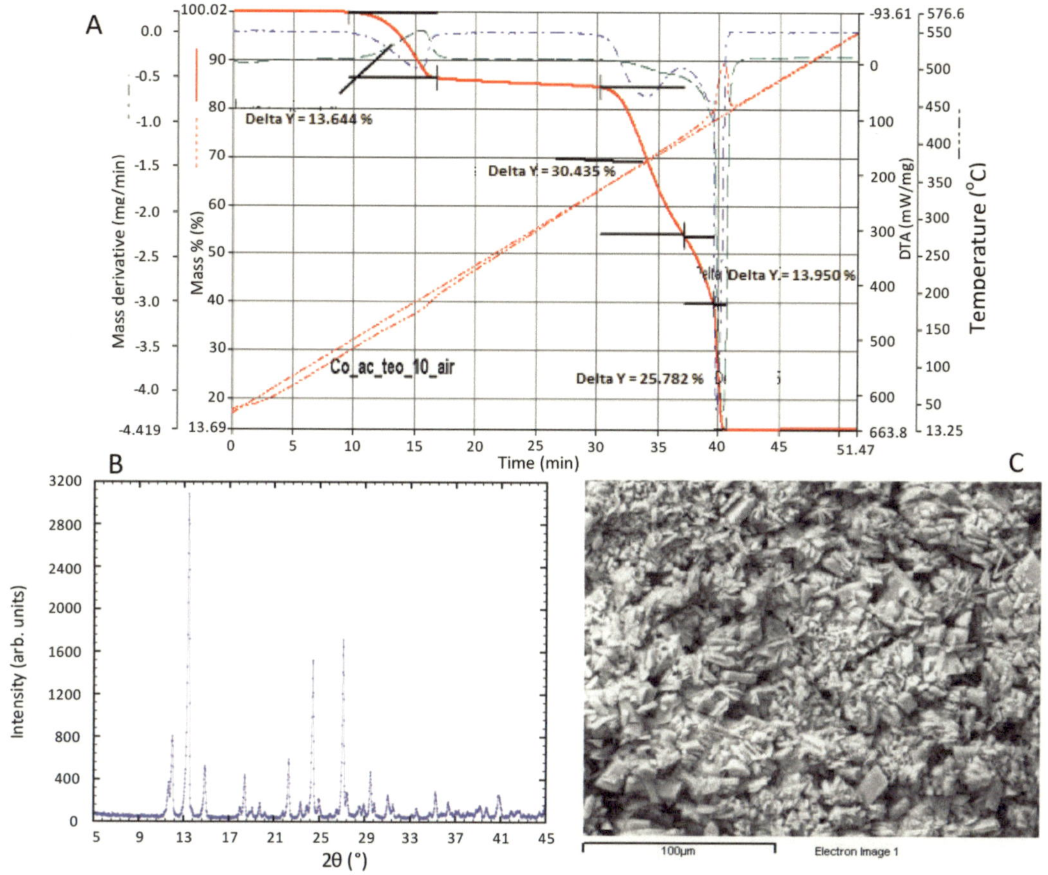

Figure 9. Thermogramms TG, DTG, and DTA (A); X-ray diffractogram; (B) and SEM images (C) of $[Co(TEO)_2](CH_3COO)_2$ [2, 19, 21, 22]. Reprinted with permission of Revista de Chimie, Farmacia and of Editura Politehnica Timişoara.

at 1600 cm^{-1} due to the overlapping of the bands corresponding to —C=O from the amide group. In addition, a new band is observed at 1450 cm^{-1} due to —C=O from the carboxyl group (—COO$^-$). In the IR spectra of the Zn:CPL complex, the band corresponding to —C=O from carboxyl group decreased. It is possible that the reaction with some metals was not completely performed or some degradation products of CPL may be involved in the complexation reaction. In the case of the complex $Cu_2^{II}CPL_2(H_2O)_2$, the IR spectra have indicated the participation of —COOH, —C=O, and —SH groups in coordination along with H_2O included in the inner coordination sphere [23, 24].

The UV spectra of the complexes were compared to the UV spectrum of CPL in dimethylformamide establishing the parameters presented in **Table 5** ($A^{\%}_{1\,cm}$ = 190 for 2.5 μg% CPL) [23, 24].

2.7. Metal complexes of tolbutamide

Tolbutamide (TBA) (*N*-*p*-tolylsulfonyl-*N′*-*n*-butylcarbamide) is the first generation of sulfonylurea oral hypoglycemic drug. Three complexes of TBA with Cu(II) were synthesized,

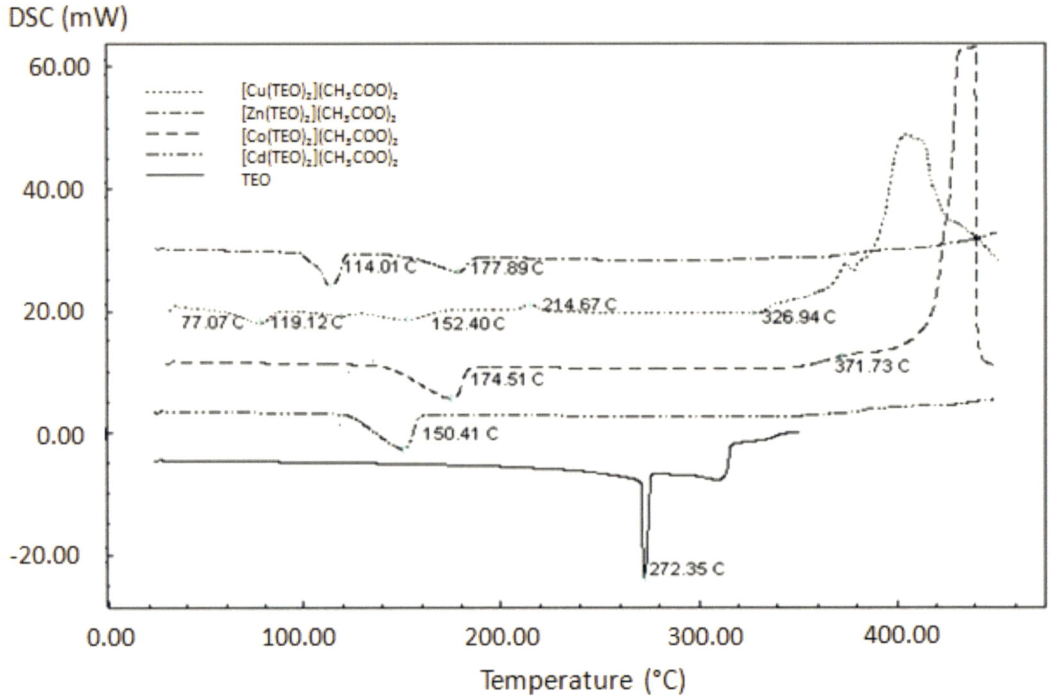

Figure 10. DSC thermogramms of metal complexes of TEO [2, 19]. Reprinted with permission of Revista de Chimie and of Editura Politehnica Timişoara.

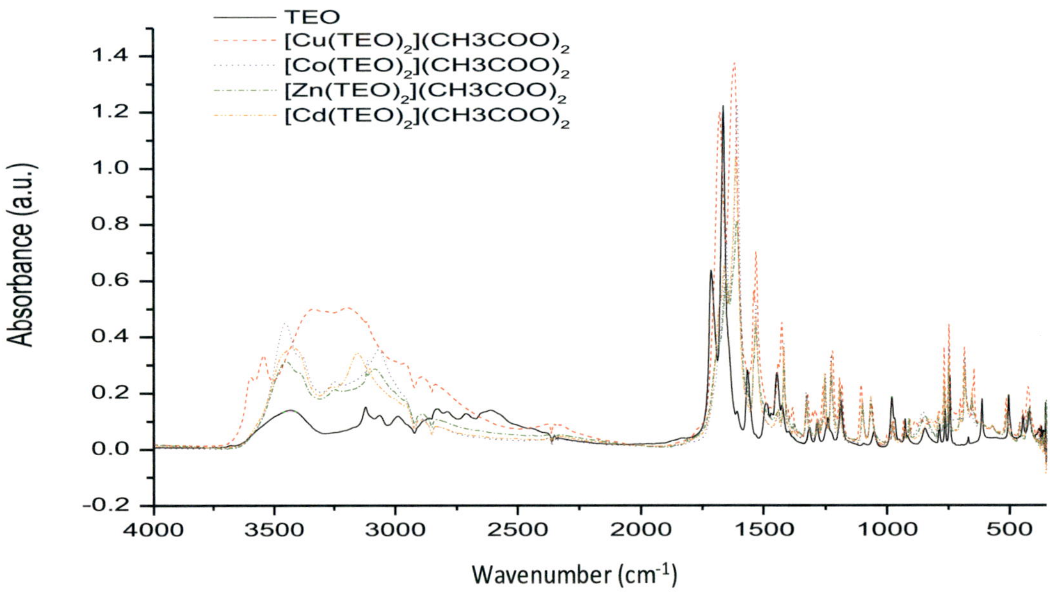

Figure 11. FTIR spectra of TEO and of its metal complexes (acetate anion).

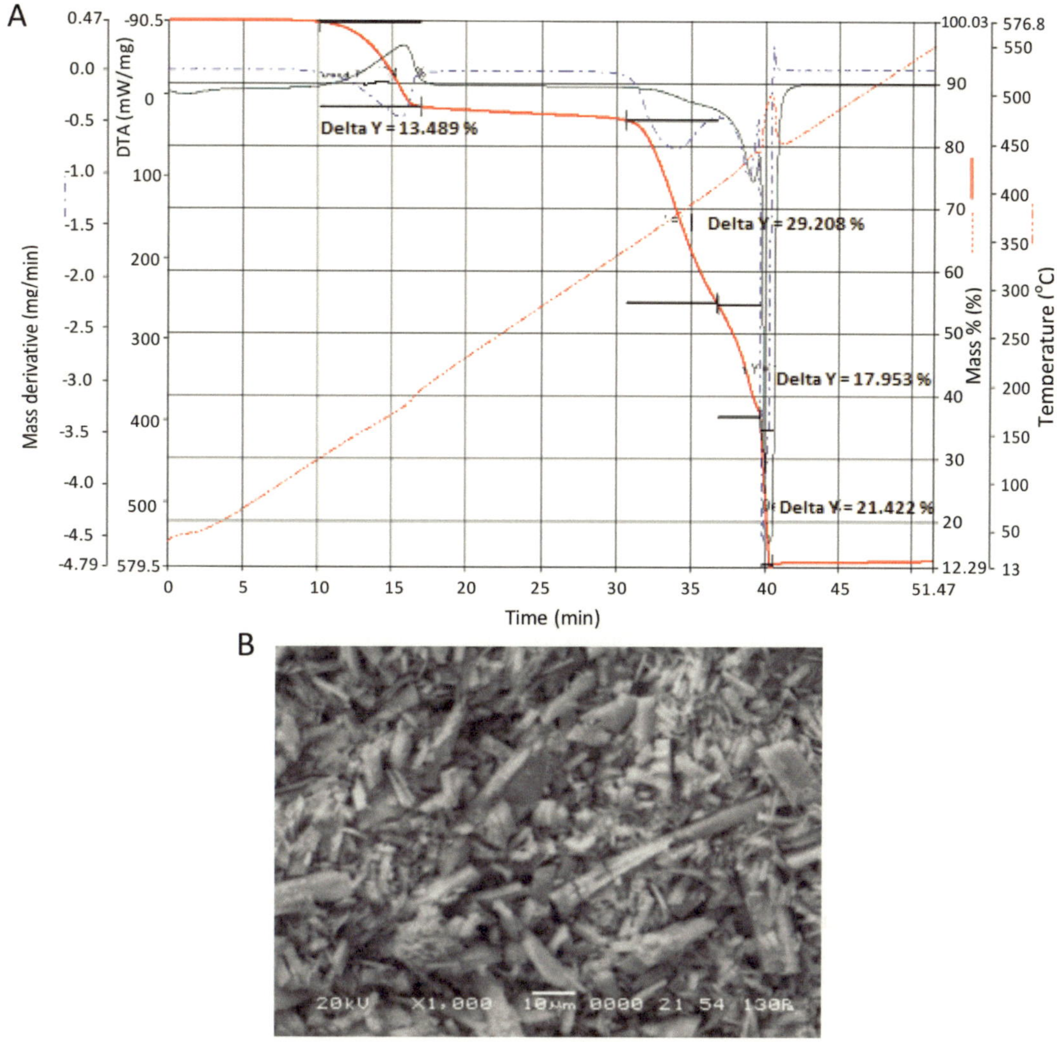

Figure 12. Thermogramms TG, DTG, and DTA (A) and SEM images (B) of $[Co(TEO)](C_6H_5COO)_2$ [2, 20]. Reprinted with permission of Revista de Chimie and of Editura Politehnica Timişoara.

$[Cu(TBA)_2](SCN)_2$, $[Cu(TBA)_2]Cl_2 \cdot 2H_2O$, and $[Cu(TBA)_2][Hg(SCN)_4] \cdot H_2O$ and then were characterized by elemental analysis, FTIR spectroscopy, electron paramagnetic resonance (EPR) spectroscopy, and thermal methods establishing the combination ratio Cu:TBA 1:2, the presence of water of crystallization, and the coordination system. The FTIR spectral studies indicated that the three mentioned complexes were coordinated through the carbonyl group. The EPR spectra showed that the Cu^{2+} ions were disposed in an octahedral vicinity of axial symmetry with a different hyperfine structure of the three complexes [25–27].

The molecular formula and weight	Color	Melting point (°C)	C% Found/ calculated	H% Found/ calculated	N% Found/ calculated	S% Found/ calculated
$[Cd(CPL)_2][HgI_4]$ M = 1255.18	White	210	17.09/17.21	2.56/2.39	1.881/2.23	5.579/5.099
$[Zn(CPL)_2][HgI_4]$ M = 1244.18	White	165	16.94/17.8	3.019/2.48	1.866/2.31	5.547/5.29
$[Ni(CPL)_2][HgI_4]$ M = 1199.48	Greenish yellow	170	18.07/18.01	3.022/2.5	1.955/2.33	6.083/5.33
$[Co(CPL)_2][HgI_4]$ M = 1199.68	Light pink	180	17.84/18.01	2.866/2.51	1.938/2.31	6.148/5.43
$[Mn(CPL)_2][HgI_4]$ M = 1195.68	White crystals	182	18.04/18.06	2.648/2.50	1.946/2.34	5.85/5.35

Source: Reprinted with permission of Farmacia and of Editura Universităţii din Oradea.

Table 4. Physicochemical characterization of the metal complexes of CPL [23, 24].

Complex	λ (nm)	$A^{\%}_{1\,cm}$	Concentration (μg %)
CPL–Cd	300	270	5
CPL–Zn	300	400	5
CPL–Ni	300	475	4
CPL–Co	300	250	8
CPL–Mn	321	520	4.5

Source: Reprinted with permission of Editura Universităţii din Oradea.

Table 5. The parameters of the metal complexes of CPL from UV spectra [23].

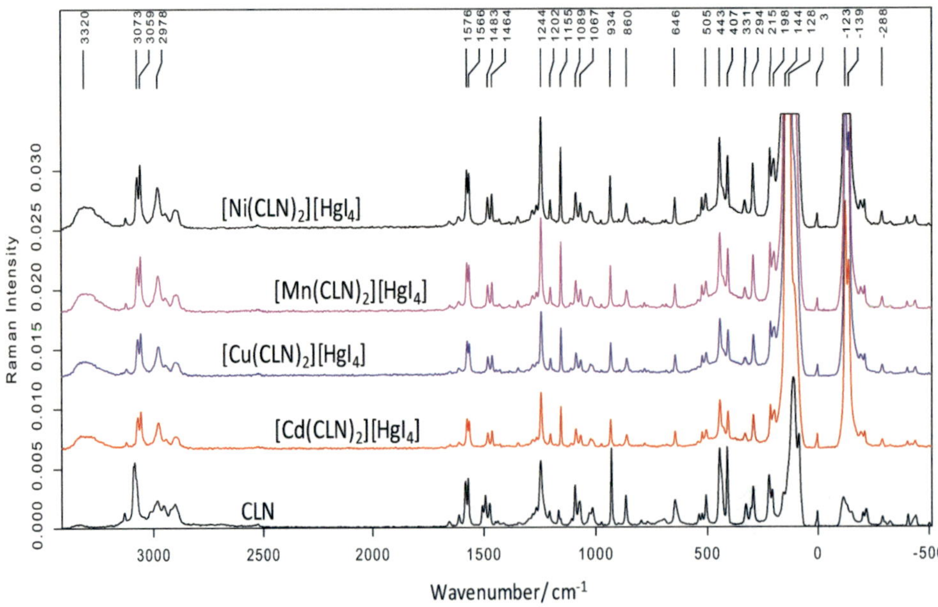

Figure 13. Raman spectra of CLN and of its metal complexes.

2.8. Metal complexes of clonidine

Clonidine (CLN) (2[(2,6-dichlorophenyl)imino]-imidazolidine) is a centrally acting α_2 adrenergic agonist used as antihypertensive drug. Metal complexes of CLN such as $[Me(CLN)_2][HgI_4]$ where Me = Cd(II), Mn(II), Ni(II), and Cu(II) were synthesized and studied by elemental analysis, FTIR spectroscopy, Raman spectroscopy (**Figure 13**), and EPR spectroscopy confirming the structure and the changes in the complex conformation [28].

3. Biomedical significance of metal complexes with pharmaceuticals

The study of the complexes structure and of their biological importance represented the major research interest toward the use of organic drugs as ligands in coordination chemistry for their application in the biomedical field.

The molecules of the pharmaceutical substances have one or more unshared electron pairs that can function as ligands. In fact, many of the basic components of living organisms (amino acids, peptides, proteins, hormones, lipids, carbohydrates, etc.) may function as ligands because they contain donor atoms in their molecules such as nitrogen, oxygen, sulfur, and phosphorus. It is well known that many molecules of drug substances act as ligands both in vitro and in vivo conditions. It is noteworthy to mention that in vivo these ligands will compete for a particular metal ion with a variety of other ligands determining that the extrapolation of this in vitro behavior should be done with moderation. It should always be taken into consideration that the therapeutic effect will be mainly influenced by the conformation of the drug ligands molecules and by their ability to combine with receptors.

Thus, the use of these metal complexes in the biomedical field can be realized by various purposes such as the introduction in the body of deficient metal ions, the use of the ligands as antidotes in various intoxications with metals, and the acquirement of pharmacotherapy effects by blocking metal ions essential for some enzymatic systems. Metal ions are of great importance not only in the vital functions of living organisms, but also they can be intensively used in analysis and control methods for pharmaceutical substances by forming complexes that can be detected by using different physicochemical methods such as spectroscopy, chromatography, microscopy, etc.

4. Conclusions

Transition metal complexes find their application in catalysis, material synthesis, photochemistry, therapy, and diagnostics. Various chemical, optical, and magnetic properties of the metal complexes of some pharmaceutical substances (pyrazinamide, nicotinamide, nicotinic acid, tolbutamide, theophylline, captopril, clonidine, and guanfacine) have been studied by using a wide range of techniques. The spectral methods such as Fourier transform infrared spectroscopy, Raman spectroscopy, surface-enhanced Raman spectroscopy, X-ray spectroscopy, mass spectrometry, ultraviolet-visible spectrophotometry, electron paramagnetic resonance spectroscopy, and X-ray diffraction provided information about the complexes and ligand structure. Other techniques such as elemental analysis, electrochemical, and thermal methods were also employed for the assessment of the complexation ratio. The scanning electron microscopy images revealed the morphology of the metal complexes underlying their crystalline or amorphous character. Many studies were conducted concerning the synthesis and the investigation of metal complexes in which the pharmaceutical substances play the role of ligand highlighting their increasing clinical and commercial importance.

Author details

Tünde Jurca, Eleonora Marian, Laura Graţiela Vicaş, Mariana Eugenia Mureşan and Luminiţa Fritea*

*Address all correspondence to: fritea_luminita@yahoo.com

Faculty of Medicine and Pharmacy, University of Oradea, Oradea, Romania

References

[1] Jurca T Combinatii complexe ale metalelor tranzitionale cu substante medicamentoase. Oradea: Editura Universitatii din Oradea; 2008. 205 p. ISBN: 978-973-759-585-0.

[2] Marian E, Complecsi ai unor metale tranzitionale cu substante medicamentoase. Timişoara: Editura Politehnica; 2009. 189 p. ISBN: 978-973-625-876-3.

[3] Chang EL, Simmers C, Knight DA. Cobalt complexes as antiviral and antibacterial agents. Pharmaceuticals. 2010;**3**:1711–1728. DOI: 10.3390/ph3061711

[4] Nagy L, Csintalan G, Kalman E, Sipos P, Szventnik A. Applications of metal ions and their complexes in medicine I. Acta Pharmaceutica Hungarica. 2003;**73**(4):221–236.

[5] Warra AA, Transition metal complexes and their application in drugs and cosmetics – A review. Journal of Chemical and Pharmaceutical Research. 2011;**3**(4):951–958.

[6] Storr T, Thompson KH, Orvig C. Design of targeting ligands in medicinal inorganic chemistry. Chemical Society Review. 2006;**35**(6):534–544.

[7] Loginova NV, Kovalchuk TV, Polozov GI, Osipovich, NP, Chernyavskaya AA, Sorokin VL, Shadyro OI. Redox-active antimicrobial metal complexes with sterically hindered o-diphenol and o-aminophenol derivatives. In: Blanc G, Moreau D, editors. Biometals: Molecular Structures, Binding Properties. Hauppauge ed. New York: Nova Science Publisher; 2010. pp. 59–90.

[8] Pattan SR, Pawar SB, Vetal SS, Gharate UD, Bhawar SB. The scope of metal complexes in drug design - a review. Indian Drugs. 2012;**49**(11):5–12.

[9] Chiang L, Jones MR, Ferreira CL, Storr T. Multifunctional ligands in medicinal inorganic chemistry–current trends and future directions. Current Topics in Medicinal Chemistry. 2012;**12**(3):122–144.

[10] Jurca T, Bănică F, Bungău S. Studiul spectrelor de masă ale pirazinamidei şi combinaţiilor sale complexe cu benzoatul de cupru (II). Revista de Chimie. 2006;**57**(8):859–861.

[11] Jurca T, Banica F, Cavalu S, Simon V, Cinta Panzaru S. FT-IR and Raman studies of pyrazinamide and its complex combination with Cu(II). Studia Universitatis Babes-Bolyai Physica. 2004;**69**(3):63–67.

[12] Jurca T, Cavalu S, Canta Pinzaru S, Simon V. Vibrational and EPR analysis of new complexes of pyrazinamide with Cu(II). Studia Universitatis Babes-Bolyai. 2003;**68**(2):457–460.

[13] Simon V, Jurca T, Simon S. Transition metal effect on the structure of pyrazinamide complexes. International Journal of Modern Physics B. 2004;**18**:1–8.

[14] Jurca T, Marian E. Scanning electronic microscopy study of new pyrazinamide compounds with metallic ions. Farmacia. 2009;**57**(2):247–254.

[15] Cavalu S, Jurca T, Damian G. Synthesis and spectroscopic studies of metallic nicotinamide complexes. Farmacia. 2003;**51**(3):92–99.

[16] Jurca T, Cavalu S, Vicaş L, Damian G, Bănică F. ATR FT-IR studies of transitional nicotinamide metallic complexes. Revista de Chimie. 2004;**55**(7):492–494.

[17] Jurca T, Marian E, Cavalu S. Raman and SERS investigation of new synthesized nicotinic acid metallic complexes. Revista de Chimie. 2009;**60**(3):320–322.

[18] Jurca T, Marian E, Banica F. Interaction of metal ions with guanfacine. Revista de Chimie. 2010;**61**(5):487–490.

[19] Marian E, Jurca T, Kacso I, Borodi G, Bratu I, Structure investigations of some complexes of theophylline with transitional metals. Revista de Chimie. 2009;**60**(6):599–604.

[20] Marian E, Jurca T, Banica F, Tita B, Tita D, Scanning electronic microscopy study of new theophyline compounds with metallic ions. Revista de Chimie. 2008;**59**(5):495–498.

[21] Marian E, Jurca T, Banica F, Morgovan C, Bratu I. Thermal analysis and Raman spectrometry of some complexes of theophylline with transitional metals. Revista de Chimie. 2010;**61**(6):569–574.

[22] Marian E, Cavalu S, Jurca T, Banica F, Bratu I. Synthesis, spectroscopic behavior and scanning electron microscopy of new complexes of theophylline with some transitional metals ions. Farmacia. 2010;**58**(6):749–757.

[23] Vicaş LG, Inhibitorii enzimei de conversie: Date analitice şi farmacologice. Oradea: Editura Universităţii din Oradea; 2006. 126 p. ISBN: 973-613-808-9.

[24] Jurca T, Vicaş L, Complexes of the ACE – Inhibitor captopril. Farmacia. 2010;**58**(2):198–202.

[25] Marian E, Jurca T, Banica F, Vicas L. Sinteza si caracterizarea fizico-chimica a combinatiilor complexe ale tolbutamidei cu ionii de cupru (II). Revista de Medicină şi Farmacie Târgu-Mureş. 2004;**50**(2):171–175.

[26] Marian E, Jurca T, Simon V, Banica F, Cavalu S. Spectroscopic behaviour of copper complexes of tolbutamine. Studia Universitatis Babes-Bolyai Physica. 2004;**69**(2):71–75.

[27] Marian E, Jurca T, Simon V, Vicas L, Bănică F, IR and EPR investigation of copper complexes of tolbutamide. Timişoara Medical Journal. 2005;**55**(5):130–132.

[28] Jurca T, Cavalu S, Cinta-Panzaru S, Simon V. Synthesis and spectroscopic characterisation of guanfacine with transition metals. In: The 10th European Conference on the Spectroscopy of Biological Molecules; 30.08–4.09; Szeged, Hungary; 2003. Book of Abstracts, p. 116. ISBN: 963-482-614-8.

Chapter 8

Quantum Dots for Pharmaceutical and Biomedical Analysis

Hayriye Eda Şatana Kara and Nusret Ertaş

Additional information is available at the end of the chapter

http://dx.doi.org/10.5772/intechopen.70034

Abstract

Quantum dots (QDs) are luminescent semiconductor nanocrystals that have extraordinary luminescence emission properties. Their semiconductor properties are different from bulk material because of the quantum confinement effects. These properties allow the use of QDs as a luminescent probe for pharmaceutical and biomedical analysis. Herein, we want to mention the synthesis, surface modification, characterization, and application of QDs. The aim of this chapter is to compile and discuss the advantages and disadvantages of QDs and their usage areas.

Keywords: quantum dots, luminescence, fluorescence, chemiluminescence, phosphorescence, pharmaceutics

1. Introduction

The semiconductor nanoparticles known as quantum dots (QDs) are one of the most relevant developments in the nanotechnology area. Therefore, they are finding new important fields of application in pharmaceutical, biomedical, and food analysis and biomonitoring. QDs are zero-dimensional materials composed of II–VI groups (e.g., CdSe, CdTe, CdS, and ZnS) or III–V elements (e.g., InAs) [1–4]. Colloidal semiconducting QDs have spherical shape, and their radii are in between 2 and 10 nm in diameter, which is less than or equal to the excitation Bohr radius [5–8]. A decrease in the crystal size causes emission at longer wavelength due to increase of the Stokes shift. At such small sizes, these nanoparticles behave differently from the bulk form because of quantum confinement effect, which is responsible for the optoelectronic properties of QDs such as narrow spectral band and high quantum yield (QY).

In the last decades, QDs have gained great interests as luminescent probes for the determination of pharmaceuticals [9–12] in different sample matrices, *in vitro* bioimaging [13–18] and *in vivo* applications [19–21], as well as computing, light-emitting devices, and photodetector devices. Because of their unique optical properties, including good optical properties, stability against photobleaching and chemical reaction, broad excitation bands, sharp and symmetric emission bands, size control luminescence, as well as high photoluminescence QY, QDs are used as an alternative to organic and inorganic fluorophores [22–24].

QDs can be modified by different molecules such as polymers and biomolecules in order to make them water soluble and biocompatible. Modification of QDs with biomolecules (e.g., DNA, enzyme, antibody, antigen) [25–27] and metal ions [28–31] has formed an important field of sensor applications [32–34] for the analysis of ions [35–38], biomacromolecules, pharmaceuticals, and small molecules [12, 39–41]. In addition, the surface modification of QDs can increase their luminescent QYs, prevent them from chemical instability and aggregation, and give a special feature to interact with target molecules.

2. Structural and optical properties

2.1. Structural properties

Generally, QDs are composed of core, shell, and surface-coating parts, which gain high photoluminescence QY, surface activation, and stability to chemicals and photons [42, 43]. The core is composed of few monolayers of a semiconductor material, i.e., CdSe, CdTe, fluorescence emission, as well as excitation wavelengths, depends on the composition of the core. Shell part surrounds and stabilizes the core. Shell is also effective on the fluorescence QY, decay kinetics, and photostability of QDs. The organic capping determines its stability, biological functionality, and solubility [44]. Coating part at initially prepared QDs is hydrophobic, whereas nowadays hydrophilic polymers or molecules are used. These amphiphilic polymers increase the water solubility of QDs and allow incorporating ionizable functional groups. Both shell and capping are covered to the particle surface and optimize these characters. Typical QDs are core or core-shell structures. The passivation shell is chemical coating, and coated nanoparticles are called core-shell systems. Core (for example, CdTe) or core-shell (for example, CdSe/ZnS and CdTe/CdS) QDs are functionalized with different coatings. In core-shell system, the band gap of the shell is higher than the band gap of core [45–50]. Additionally, a slight red shift in absorption and emission is observed because of tunneling of charge carriers into the shell.

2.2. Characterization

Definition of size, structure, and shape of synthesized QDs is important. The characterization of QDs is evaluated by high-resolution transmission electron microscopy (HR-TEM), scanning electron microscopy (SEM), atomic force microscopy (AFM), X-ray fluorescence (XRF), X-ray diffraction (XRD), and Fourier transform infrared spectroscopy (FT-IR) methods. The size of QDs is generally detected by TEM and SEM [3, 51–53]. In addition to these methods, Brus equation is also used to calculate the diameter of QDs [39]. The optical characterization is made by UV-visible (UV-vis), fluorescence, Raman, and nuclear magnetic resonance spectroscopy (NMR) [54, 55].

2.3. Optical properties of QDs

Although QDs are composed of semiconductor materials, their small size results in spectroscopic properties that are radically different from bulk forms. The electron of the valence band moves to conductance band when QDs absorb the photon. Absorption occurs as long as the energy of photon is higher than the bandgap energy of QDs; thus, excitons can be created with a wide range of energies within the core. The higher energy excitons relax to the lowest bandgap energy before emitting a photon. Therefore, excitation spectrum is broad, whereas the emission spectrum is narrow (**Figure 1**).

QDs are artificial atoms with typical dimensions ranging from 2 to 10 nm. QDs can be designed to have different emission wavelength by adjusting their size (**Figure 2**). The emission is adjusted by the particle diameter in the visible area and by particle composition in the longer wavelength. As diameter of QDs gets bigger, red shift is observed in the emitted light [56, 57]. The properties of QDs are changed by constructing the properties of the electron and hole. The electrons and holes of QDs also present discrete energy levels. As nanoparticles size get smaller, the band gap will be larger, and energy difference between the highest valence band and the lowest conduction band will be increased. As a result, the high energy is required to excite the dots, and therefore, more energy is emitted when nanoparticles return to ground state.

QDs have broad absorption band and a narrow symmetrical emission band; therefore, overlap with other emission colors is minimal. The wavelengths of absorption and emission are tunable by particle size as mentioned before. The broad absorption bands allow selection of the excitation wavelength, and consequently, excitation and emission wavelengths can be separated [48, 58–61]. Physicochemical and optical properties of QDs are summarized in **Table 1**.

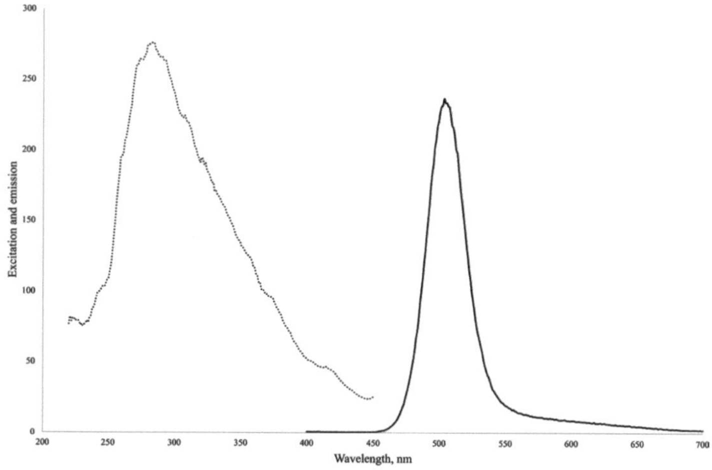

Figure 1. Excitation (dot line) and emission (line) spectra of MPA-capped CdTe QDs.

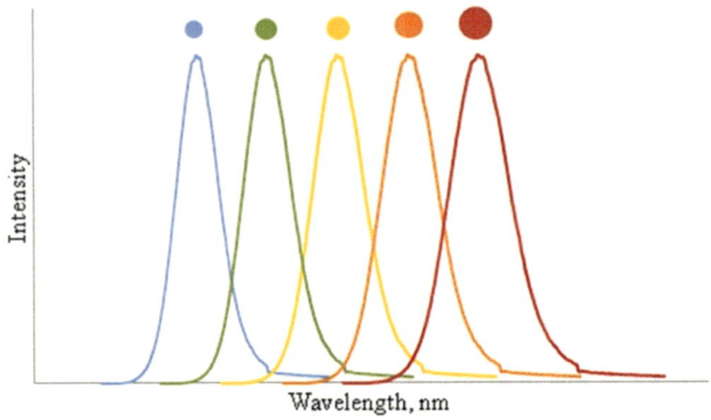

Figure 2. Photoluminescence spectra of the QDs by changing the size of the particle.

Property	Quantum dots
Size	2–10 nm
Thermal stability	High, depends on shell
Photostability	High, stable fluorophores due to their inorganic composition
Chemical stability	More resistant to degradation
Brightness	10–20 times more than organic dyes
Absorption spectra	Broader absorption spectra enables selection of excitation wavelength
Molar absorption coefficient	10^5–10^6 M^{-1} cm^{-1}
Emission spectra	A narrow (30–90 nm), symmetric, sharply defined emission peak
Quantum yield	0.1–0.8
Stokes shift	Large stokes shift
Lifetime	Longer lifetime helps to eliminate background signal
Excitation by single or multiple sources	Ideal for the same source and multicolor experiments
Solubility	Depends on surface modification
Sensitivity	High S/N ratio
Applicability to single-molecule analysis	Good
Bioimaging	Better contrast with electron microscope

Table 1. Properties of QDs.

2.4. Synthesis and surface chemistry

Synthesis step of semiconductor nanocrystals plays a critical role. The solvent selection is crucial because this step is not only controlling nanocrystal size but also changing their polarity, solubility in aqueous or organic medium, functionality, and applicability. QDs can be synthesized in the organic or aqueous medium. When compared to organic way, water-based QDs synthesis is less toxic, useful for biological applications, and cheaper.

In order to prepare QDs, different types of materials and methods have been reported. In these methods, colloidal synthesis is commonly used technique for the preparation of QDs. During early 1990s, Murray et al. [62] first reported the synthesis for highly monodisperse, size-tunable QDs. In this method, trioctylphosphine (TOP) and trioctylphosphine oxide (TOPO) are used for the synthetic approach for II–VI QDs such as CdSe, CdS, and CdTe. Highly luminescent Cd-E (E = Se/Te/S) nanocrystals were synthesized by using dimethylcadmium ($Cd(CH_3)_2$) as QDs precursors in the coordinating solvent (TOPO) at high temperature (300°C). However, nowadays, cadmium oxide (CdO) is used instead of dimethylcadmium due to its toxicity. The commonly used method for synthesis of colloidal nanocrystal is based on core growth process, starting from organometallic precursor in a mixed solvent including organic surfactants and coordinating solvents. Briefly, reaction medium is heated to high temperature, and precursors are injected while solution stirring. At this step, precursors transform into monomers. As the process continues, nanocrystal growth starts with a nucleation process, and the color of solutions changes from yellow to red. Coordinating solvent caps the nanocrystal surface and stabilizes its surface, moreover, changes its solubility in organic and aqua media and prevents aggregation. Organic surfactants are used to avoid aggregation and give water soluble character [4, 63–66]. At nanocrystal-growth process, not only solvent but also temperature, pH, and growth time are important. Generally, when the heated and unheated forms of QDs are compared, heated QDs show emission, whereas unheated QDs do not have an emission band. Minor variations in pH values affect the diameter of nanocrystals.

Coordinating solvents can be hydrophobic or hydrophilic. Hydrophobic coatings are not suitable for aqueous assay especially biological analysis; therefore, hydrophilic capping agent must be introduced for this purpose. Various approaches have been developed to make water soluble. The hydrophobic layer can be changed with acidic and hydrophilic groups. Hydrophobic part reacts with the hydrophobic surface of QDs, whereas the hydrophilic part on the outer end will give water soluble character. The stability of water dispersed QDs is generally achieved by using charged organic molecules or polymers such as mercaptopropionic acid (MPA), mercaptoacetic acid (MAA), mercaptosuccinic acid (MSA), and cysteine [67–69].

Doping of QDs with transition-metal ions such as Mn^{2+} and Cu^{2+} has been widely studied to enrich the features of nanocrystals. These advantages include stability, large Stokes shift, and longer emission lifetime which allow phosphorescent measurement [70–72]. In these doped QDs, Mn-doped ZnS which have orange-yellow emission is widely used as a phosphorescent probe for analysis of different kinds of analytes [73–75]. The purification step is needed to remove unreacted precursors and other chemicals. A widely used method for this purpose is precipitation of QDs in methanol or ethanol and centrifugation.

Additionally, lithography [76, 77], epitaxy [76, 78], electrochemical assembly [79–81], plasma synthesis [82–86], biological synthesis [87, 88], gamma-irradiation [89–91], and microwave-assisted synthesis [92, 93] ways are also used for the synthesis of QDs.

3. Applications of QDs

Surface properties of QDs affect the luminescence character. The chemical or physical interaction of analytes and QDs can influence the optical properties of the QDs. Depending on this change, QDs have been widely applied to detect different kinds of analytes including ions, pharmaceuticals, small molecules, and biological macromolecules. In these analytes, direct analysis of pharmaceutical and biological samples is difficult due to interference molecules. However, chemical surface modifications of QDs with functional groups or biomolecules enhance the selectivity and favorable luminescence features. Most of the detection methods are based on the fluorescence properties of these QDs. Besides, in recent years, increasing number of work on making use of the inherent phosphorescent properties of QDs is in the literatures. In most QDs applications, the detection is based on quenching of luminescence signal, while new methods are developed on signal enhancing.

3.1. Fluorescence-based measurements

The luminescence properties of QDs are used for qualitative or quantitative analysis of different kinds of analytes. Initial studies are generally based on measuring the enhancement or quenching of luminescence signal of QDs resulting from the interaction of the QDs and the analyte. This surface interaction generally is not specific and allows interacting with simple and small molecules. Nonspecifically binding is a major problem especially in biomedical applications due to the interaction of a variety of biomolecules and structures including nucleic acids, proteins, or matrix compounds. In order to increase selectivity, conjugation of QDs with polymers, antibodies, amino acids, and proteins has been proposed and applied [94, 95].

In a pioneering work, Cd-based QDs have been reported for optical sensing of small molecules and ions. Many studies in this field, focusing on interactions between QDs and analyte, showed that the luminescence response was affected by the surface capping ligands. For example, the addition of Cd ions to a basic solution, including CdS nanoparticle, has caused increasing the luminescence QY of the nanoparticle. This effect was attributed to the formation of a $Cd(OH)_2$ shell on the CdS core which eliminates the nonradiative pathways. Furthermore, addition of Zn^{2+} and Cd^{2+} ions to basic CdS or ZnS colloidal solutions caused similar photoluminescence-activation effect [96–98]. In addition, organic capping agents such as mercapto acids and mercaptoamines have been used for the modification of QDs surface [99]. Modification strategies are based on not only intensity enhancing but also quenching and emission wavelength shifting. Quenching mechanism depends on the interaction of quencher and nanoparticle and includes different deactivation pathways such as electrotransfer process, nonradiative pathways, inner-filter effect, and complex formation. Quenching occurs by two different mechanisms called dynamic (collisional) and static quenching. In dynamic quenching, the quencher and fluorophore come into contact during the lifetime of the excited state, and the fluorophore returns to the ground state without emission. In static quenching, fluorophore and quencher form a nonfluorescent complex at the ground state. These quenching systems can be differentiated by their different dependence on temperature and viscosity. At higher temperatures, dynamic quenching increases due to faster diffusion. On the contrary, in the static quenching, dissociation of weakly bound complexes causes decreasing of quenching. In some cases, the fluorophore can be quenched by collisions and complex formation with the same quencher at the same time [100].

Literature survey shows that a great number of fluorescence methods have been developed for analysis of pharmaceuticals and biomolecules (**Table 2**). These methods are generally based on quenching of fluorescence intensity of QDs (**Figure 3**). Thioglycolic acid (TGA)–modified water-soluble CdSe QDs were synthesized and used for determination of paraoxon, which is acetylcholinesterase inhibitor. In this study, multilayers of chitosan, TGA-capped CdSe QDs, and organophosphorus hydrolase polyelectrolytes were incorporated into layer-by-layer architecture. The size of the nanoparticles was determined by HR-TEM and found 3.4 nm. The presence of paraoxon in the sample solution caused decreasing fluorescence emission of QDs, attributed to an interaction of the analyte with QDs and to change surface conformation [101].

QD material	QD coating	Analyte	Matrix	Detection limit	Measuring signal	Ref.
CdSe	TGA	Paraoxon	–	–	Fluorescence quenching	[101]
CdTe-Mn doped	TGA	Glutathione	Tablet	0.06 μM	Fluorescence quenching/ enhancement	[102]
CdTe	TGA	Doxycycline	Honey, human serum	1.1×10^{-7} M	Fluorescence quenching	[103]
CdS	L-Cysteine	Ceftriaxone	Urine	1.3×10^{-9} M	Fluorescence quenching	[104]
CdSe	–	Spironolactone	Tablet	4.8×10^{-7} M	Fluorescence quenching	[105]
CdSe/CdS	TGA	Sparfloxacin	Tablet	0.1391 μg mL^{-1}	Fluorescence quenching	[106]
CdS	MPA	Penicillamine	Tablet	0.1123 μg mL^{-1}	Fluorescence enhancement	[107]
CdTe	MPA	Rifampicin, Rifaximin	Urine, tablet	1.5 mg mL^{-1}, 1.0 mg mL^{-1}	Fluorescence quenching	[108]
C	PEG 2000	Ceftazidime, Cefepime	Tablet	4.7×10^{-3} μg mL^{-1}, 5.1×10^{-3} μg mL^{-1}	Fluorescence quenching	[10]

Table 2. QD-based fluorescent probes for determination of pharmaceuticals.

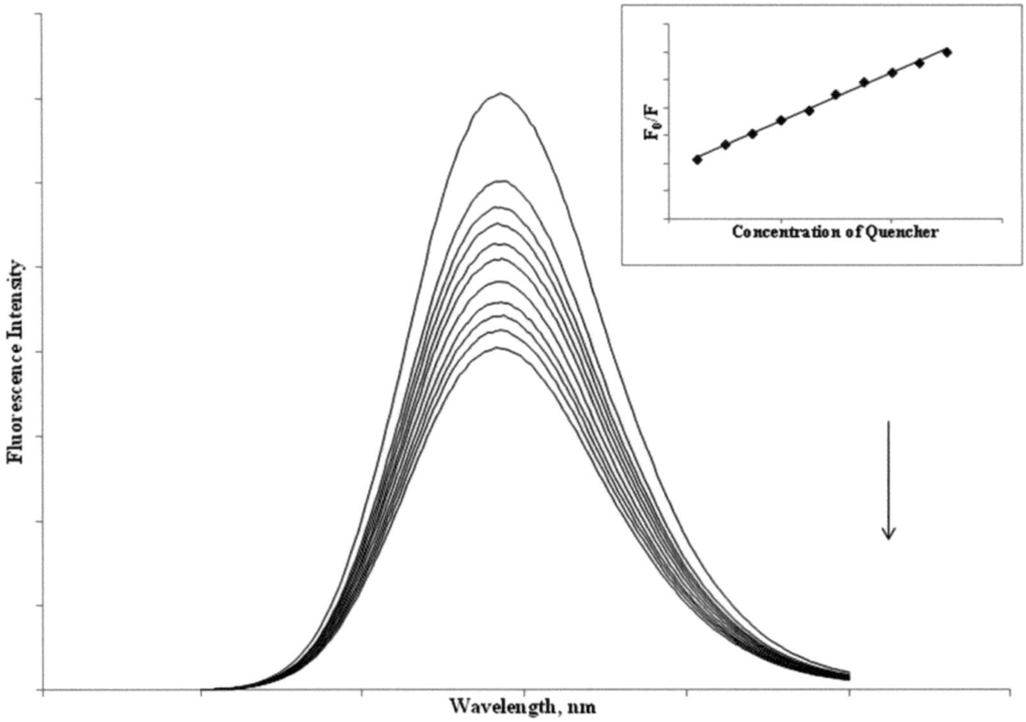

Figure 3. The general aspect of fluorescence spectra of QDs at different concentration of quencher. The inset is the calibration curve of F_0/F versus the concentration of quencher.

Yu et al. have developed a fluorescence switch sensor consisting of Mn-doped CdTe QDs-methyl viologen (MV^{2+}) nanohybrids to analyze bioactive peptide glutathione (GSH). Characterization of QDs was performed by TEM, XRD, and FT-IR methods. The results obtained from these studies showed that prepared QDs were monodisperse and have spherical shape with sizes 20 nm and hexagonal crystalline structure. In addition, TGA molecules and QDs conjugated successfully. In the sensor, MV^{2+} has two quaternary ammonium groups which link TGA on the surface of QDs through electrostatic interaction. Because of the electron transfer between QDs and quencher, the fluorescence signal is quenched. As the addition of GSH, the peptide can effectively replace TGA ligands on the surface of QDs, and fluorescence intensity is again recovered. Fluorescence recovery level of QDs is depended to the amount of GSH [102].

Synthesized by sonochemical technique, TGA-capped CdTe is used as a fluorescent probe for determination of doxycycline (DC), a member of tetracycline antibiotics, in honey and human serum. Prepared QDs were identified by FT-IR. In here, the peak related to SH group of TGA was not observed in the spectrum of TGA-QDs hybrids due to interaction between thiol and CdTe QDs. Furthermore, the TEM images showed that size distribution and shape were 4–6 nm and spherical, respectively. Similar to previous studies, the fluorescence intensity of QDs was quenched by adding of DC. To identify the mechanism of quenching process,

Stern-Volmer equation has been plotted at a different temperature. Obtained results showed that the quenching mechanism was dynamic [103].

L-Cysteine is also widely used surface modification agent which gives hydrophilic character to QDs. L-Cysteine–coated CdS QDs have prepared, characterized, and used to analyze ceftriaxone (CFX) in biological samples [104]. Optical characterization was identified by fluorimetry and UV-vis spectrometry. QDs have a broad absorption band in pH 7 buffer solution. Moreover, symmetric and narrow emission peak was at around 556 nm with excitation wavelength 359 nm. Recorded FT-IR spectra indicated that characteristic S-H band was absent in the L-cysteine–capped QDs. Also, the band at 3175 cm^{-1} belongs to CFX was not seen in the spectrum of L-cysteine/QDs/CFX complex. Prepared and characterized QDs were used in determination of CFX in urine. The effects of reaction time, temperature, and pH were identified, and 10 min, 25°C, and pH 7 were selected as optimum conditions.

Liang et al. have developed a method based on the quenching of the fluorescence of CdSe QDs by spironolactone in organic media. CdSe QDs were prepared from CdO as a precursor and stearic acid. After heating step under Ar flow, trioctylphosphine oxide and hexadecylamine were added, heated again at high temperature, and Se solution was injected. Obtained dried nanoparticles were redispersed in hexane. The emission spectra of QDs were recorded after titration with spironolactone [105].

Core-shell nanoparticles were also used for sparfloxacin in tablets. Water-soluble CdSe/CdS QDs modified with TGA have been synthesized and acted as a fluorescent probe. Hyperchromicity and forming of new absorption band were observed when the spectra of CdS and CdSe/CdS QDs were compared. This result indicated that CdS coated the surface of CdSe. After optimization of working conditions, TGA-capped CdSe/CdS QDs was used for the determination of sparfloxacin. Quenching mechanism was found static according to results of Stern-Volmer equation [106].

Besides these applications, fluorescence enhancement method has been developed for the determination of pharmaceuticals. Pawar et al. has used MPA-modified CdS nanoparticles as a turn-on probe to determine penicillamine. The obtained results showed that QDs aggregated after addition of the drug, the average size of QDs increased, and the fluorescence intensity of QDs was enhanced due to the interaction between QDs and penicillamine [107].

Nowadays, innovations and applications related to QDs are continuously increased. One of these is the using of QDs in automated systems. Rifampicin and rifaximin, which are complex macrocyclic antibiotics, have been analyzed by automated QDs-based analytical method using flow system. The automated flow system allows repeatable handling solutions, automation, miniaturization, lower reagent consumption, and waste. In this method, in the initial status, carrier ($Na_2CO_3/NaHCO_3$ buffer solution) flowed through the flow cell, and then valves of sample solutions and QDs were switched on. The analytical signal of QDs-sample mixture was recorded. A blank signal which is only QDs signal was also recorded before any sample analysis [108].

Most inorganic QDs such as CdX nanoparticles consist of highly toxic heavy metal ions, and this could be a major concern for *in-vivo* applications. Therefore, novel fluorescence carbon-based nanoparticles have found wide using area in this field. They have some advantages over traditional QDs, for example, free of heavy metal ions, low toxicity, and excellent biocompatibility [10, 109]. Due to their small molecular mass, carbon nanoparticles easily enter the living cell and allow *in-vivo* monitoring [110]. Huang et al. have developed a sensitive method for the determination of ceftazidime and cefepime in their pharmaceutical forms based on fluorescence quenching of poly(ethylene glycol) (PEG) 2000-capped carbon QDs. The chemical oxidation method was used for preparation of nanocrystals. In this method, sawdust as a carbon source and nitric acid were mixed and heated. Then, pH of the solution was adjusted to neutral by adding NaOH and centrifuged. In the last step, for surface modification, PEG 2000 was added and heated in a microwave oven. Obtained carbon nanoparticles were characterized by TEM, and size of them was found in between 5 and 8 nm. Developed QDs were used for determination of drugs and method was sensitive, selective, and with high recovery value [10].

Functionalized semiconductor QDs have been used as fluorescence labels for biological detection and imaging. For example, avidin, highly positively charged tetramer, and functionalized CdSe/ZnS QDs were used in the detection of biotin [111]. Similarly, different sizes CdSe/ZnS QDs conjugated with different antibodies have been applied for simultaneous detection of four toxins (Shiga-like toxin 1, staphylococcal enterotoxin B, cholera toxin, and ricin) [112]. In another example, determination of 17β-estradiol has been done by using biotin-labeled antirabbit IgG and streptavidin conjugated by QDs [113]. As mentioned above, carbon-based QDs have low toxicity and biocompatibility. Li et al. have developed glucose sensor based on combining electrostatic attraction between anionic fluorescent carbon QDs that bear polar carboxy and hydroxy groups and cationic boronic acid–substituted bipyridinium salt. The interaction between them results in the formation of a ground state complex leading to a decrease in the fluorescence intensity. When glucose is added to the medium, the tetrahedral anionic glucoboranate esters are formed which effectively neutralize the net charge of the cationic bipyridinium and recover the fluorescence intensity of QDs [114].

3.2. Fluorescence resonance energy transfer (FRET)

Fluorescence resonance energy transfer (FRET) is a powerful technique describing energy transfer between two light-sensitive molecules. In here, an exited donor chromophore group transfers energy to an acceptor chromophore through the nonradiative coupling. The efficiency of this energy transfer depends on the distance between both fluorophores, inversely proportional to the sixth power of the distance between acceptor and donor; therefore, FRET is more sensitive to small changes in distance [115, 116]. Photoemission properties and high QY of QDs allow efficient energy transfer when compared with the organic dyes. In addition, distinguishing the emission of the donor from acceptor is much easier due to narrow and symmetric emission spectrum of QDs. This technique is a useful tool to analyze biomolecules such as proteins and DNA. For this purpose, a FRET-based bioassay was developed by using QDs (donor) functionalized with a label (i.e., protein, antibody) which recognizes and binds to target. When FRET process occurs, the bioconjugate captures the dye-labeled analog (acceptor), and fluorescence quenching takes place. In addition of target molecule to the medium, analyte competes with its analog and binds to the label. The acceptor is displaced from QD surface; thus, fluorescence enhancement is observed. Besides, different approaches have been developed toward the use of QDs in FRET assays, such as analyte displaces fluorescent ligands, analyte cleaves donor-acceptor linkage, analyte changes the conformation of acceptor-donor linkage, and analyte mediates donor-acceptor binding [117].

Several studies can be found that develop FRET-based assays for the detection of pharmaceuticals and biomolecules (**Table 3**). Generally, QDs and organic dye are used as donor and acceptor, respectively. However, the opposite situation, in which dye acted as a donor and QDs as acceptors, has been reported. Briefly, QD as a donor in an exited state transfers its excitation energy to an acceptor non-fluorescent dye in a nonradiative fashion. When the analyte introduces the medium, analyte replaces the dye, and fluorescence emission is recovered (**Figure 4**).

Similarly in this approach, amantadine has been determined in pharmaceutical form by using FRET mechanism. The optical sensor was designed by using water-soluble β-cyclodextrin (CD)-functionalized CdTe QDs and Rhodamine B (RB). The interior of the β-CD is not hydrophobic but considerably less hydrophilic than the aqueous environment and thus able to host other hydrophobic molecules. Therefore, RB could enter the cavity of β-CD by hydrophobic interaction, and the FRET process occurred between QDs (donor) and RB (acceptor). When amantadine introduced the system, it replaced RB in the cavity; the process of FRET was switched off. The authors also have used the developed sensor for *in-vivo* imaging. For this purpose, functionalized QDs with amantadine in the cavity were incubated with HepG2 cells and observed in the cytoplasm by fluorescence microscope [118].

Analyte	Donor	Acceptor	Detection limit	Ref.
Amantadine	β-CD-functionalized CdTe	Rhodamine B	8.82×10^{-6} mol/L	[118]
Atrovastatin/linezolid	β-CD-conjugated CdSe/silica atrovastatin and linezolid	Atrovastatin/linezolid	–	[119]
Vitamin B12	MPA functionalized CdS	Vitamin B12	6.91 μg mL^{-1}	[120]
DNA	Streptavidin-coated QDs	Cyanine dye		[121]
Biomarkers	Streptavidin labeled lanthanide complexes (europium and terbium)	Biotin coated CdSe/ZnS	0.10–0.63 pM	[122]
Estradiol	Lanthanide (III), europium and terbium, chelate	Protein-coupled CdTe QDs	–	[123]

Table 3. FRET-based fluorescent probes for determination of pharmaceuticals.

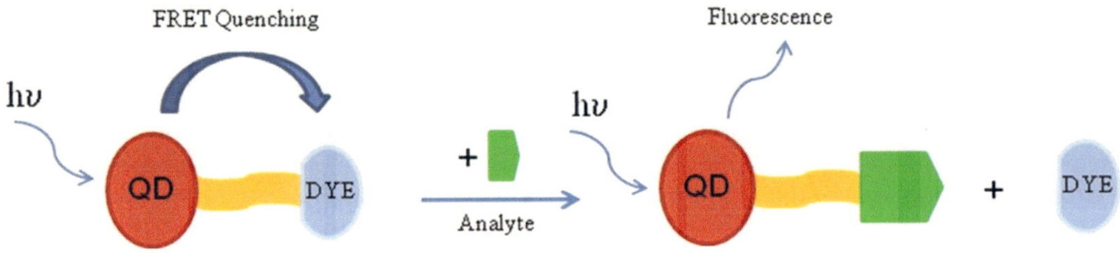

Figure 4. Schematic diagram of the quantum-dot based FRET sensor. QDs (donor) and non-fluorescent organic dye (acceptor).

Antony et al. [119] also used β-CD-conjugated CdSe/silica nanoparticles for determination of atrovastatin and linezolid. The FRET system occurred between CdSe/SiO_2 nanoparticles and the drugs encapsulated in the CD cavity. Coating and conjugation of prepared QDs were identified by FT-IR. FTIR spectra of CdSe and CdSe/SiO_2-β-CD complex were recorded and vibration bands at 1031.92 and 1117.29 cm^{-1} appeared due to Si—O—Si bond. The Fröster distance between the encapsulated drugs and the CdSe nanoparticles was calculated and found below 3 nm.

More simple FRET method has been developed for analysis of Vitamin B12 in human serum, urine, and pharmaceutical forms. Herein, the MPA functionalized CdS QDs synthesized from cadmium chloride and sodium sulfide in aqueous media by a chemical method. FRET-based quenching mechanism was due to photoinduced charge transfer from QDs to drug. For investigation, the quenching mechanism, UV-vis absorption, and fluorescence spectra of QDs in the absence and presence of vitamin B12 were examined. Obtained results show that energy transfer from CdS QDs to vitamin B12 could occur with high probability resulting in the fluorescence quenching of QDs. Under the optimized conditions, the relationship between the fluorescence intensity of QDs and concentration of vitamin B12 was linear, and limit of detection was found as 6.91 μg mL^{-1} [120].

Zhang et al. [121] has developed a method for detection of DNA using commercially available streptavidin-coated QDs. In this method, nanoparticle and cyanine dye were donor and acceptor, respectively. Cyanine-labeled DNA was assembled onto the QDs surface by specific streptavidin-biotin binding. The binding of molecules to QDs resulted in the formation of QD/DNA/Dye complexes. FRET was occurred between QDs and dye by excitation with the appropriate wavelength.

The use of QDs as energy acceptors in FRET-based techniques is not so common. QDs were inadequate acceptors when compared with molecular dyes because of a longer lifetime of QDs [95]. In addition to this, the donor used has to emit luminescence at a wavelength shorter than that of the QDs to allow the FRET process. A method based on this technique has been developed by Geissler et al. Herein, FRET process was realized between visible-emitting lanthanide complexes of europium and terbium, streptavidin labeled (as a donor) to CdSe/ZnS biotin-coated QDs (as acceptor). Developed method has been used for determination of five biomarkers [122]. Similarly, determination of estradiol was examined by using luminescent energy transfer between protein-coupled CdTe QDs and lanthanide (III), europium and terbium, chelate [123].

3.3. Chemiluminescence

Chemiluminescence (CL) is typically defined as the emission of light, as the result of a chemical reaction. Generally, chemiluminescent reactions show weak luminescence due to low QY. Therefore, it is necessary to enhance the CL intensity for analytical applications. For this reason, QDs have attracted great attention due to their properties as mentioned before such as brightness and reactivity. In addition, the use of QDs as chemiluminescent probe can give an advantage such as the emission at wide range wavelengths without light source [47, 124]. Nowadays, high-quality semiconductor QDs (core or core-shell) can be easily synthesized and have been used in CL systems such as CdTe, CdSe, and CdSe/CdS. Besides, doped QDs are also used in CL assays because of their catalytic features. The advances of using QDs in CL not only expand conventional usage of them but also give an opportunity to develop new nanomaterials.

There are three possible mechanisms that could be explained for the enhancement of CL by QDs as explained by Frigerio et al. [47]: (i) as emitter species after direct oxidation; direct oxidation happens when QDs is an only luminescent compound in the system; (ii) as a catalyst of a reaction involving others luminophores; when more than one luminophores exist in the system, the final emitter is the luminophores due to the catalytic effect of QDs; and (iii) as emitter species after chemiluminescence resonance energy transfer (CRET); the difference from catalytic effect, the final emitter is QDs.

Chen et al. have used MPA-capped CdTe QDs in H_2O_2-HCO_3 CL system and applied for the determination of ascorbic acid in serum [125]. The chemical process followed; peroxymonocarbonate (HCO_4^-) was formed by reaction of hydrogen peroxide and sodium hydrogen carbonate. This unstable compound decomposed and caused to form superoxide ion radical ($\cdot O_2^-$) and finally singlet oxygen molecules with emission after several chemical steps. Radical scavenger, ascorbic acid, was used to study the emitting species. The proposed reaction mechanism based on the presence of four emitters: $^1O_2$, $(O_2)_2^*$, $(CO_2)_2^*$, and CdTe* in the system. Authors also explained that the CL emission intensity depended on sizes of QDs, bigger nanoparticles decreased CL intensity.

In another work, Khateee et al. [126] used a flow-injection analysis system to investigate $KMnO_4$-morin sensitized with CdS QDs and applied it to environmental water samples and pharmaceutical forms for determination of nalidixic acid. In addition, luminescence intensity was enhanced not only by adding L-cysteine–capped CdS QDs but also nalidixic acid. Possible CL mechanism was based on oxidation of morin and CdS by $KMnO_4$ in acidic media. Moreover, obtained UV-vis and luminescence spectrum showed that transmission of the energy of excited morin to CdS QDs can occur. According to the spectral knowledge, the addition of nalidixic acid to $KMnO_4$-morin-CdS QDs system cannot generate new luminophore species. The final emitter species in the mentioned CL system is exited CdS QDs.

The same group also used a similar QDs system for the baclofen analysis in water samples and pharmaceutical forms. Various oxidants in basic and acidic aqueous medium were examined, and the best results were obtained with $KMnO_4$ in acidic media. In addition, $Na_2S_2O_3$ significantly enhanced the CL intensity of $KMnO_4$ L-cysteine–caped CdS QDs, while adding of baclofen caused inhibition of intensity. There are two emissions bands observed, attributed to CdS QDs (at around 520 nm) and exited manganese (at around 725 nm). The process of CL is that (i) $KMnO_4$ oxidizes the L-cysteine to produce excited L-cysteine, (ii) excited L-cysteine transforms its energy to CdS QDs, and (iii) excited QDs produce the emission. Furthermore, the inhibition effect of baclofen was explained by incorporation of baclofen and $KMnO_4$. The consumption of $KMnO_4$ by baclofen leads to decrease in the amount of excited CdS QDs and then CL emission [127].

QDs can be used in CL system as a catalyst, due to the redox properties of both conduction and valence bands. Imani-Nabiyyi and Sorouraddin showed that the CL emission was enhanced by combination of cysteine-capped CdTe QDs and luminol in the presence of KIO_4. The amplified CL was effectively quenched in the presence of naphazoline. According to spectroscopic and chemical investigations, weak CL emission was observed with the reaction between periodate and luminol in alkaline conditions; however, adding of QDs caused increasing of the CL emission. Based on these data, this phenomenon was explained by author that QDs could interact with the reactants catalytically and caused to form reactive oxygen species which reacts with luminol in order to give emission [128].

3.4. Chemiluminescence resonance energy transfer (CRET)

CRET is a nonradiative transfer of energy between chemiluminescent donors to a fluorophore acceptor. An essential condition is that there should be an overlap between CL emission spectrum and the absorption spectrum of the fluorescent acceptor. QDs are well-suited fluorescent acceptors due to their broad excitation spectra. In CRET, QDs are the final emitters, which can be confirmed by the emitted spectra. However, sometimes, it is also possible that direct oxidation and CRET take place simultaneously; thus with CRET, it is difficult to define that the excited forms of QDs are formed by a resonance energy transfer process or a redox process.

The water-soluble MPA-capped CdTe QDs as sensitizers are used for the chemiluminometric determination of the anti-diabetic drugs gliclazide and glipizide in their pharmaceutical formulations. Both glipizide and gliclazide quenched the CL emission of the Ce(IV)–SO_3^{2-}-CdTe QD system, probably due to radical scavenging activity [129].

Golub et al. demonstrated CRET system for highly sensitive detection of DNA by the labeling of the probe-analyte complex with a hemin/G-quadruplex nanostructure [130]. The emission of CdS QDs was observed by stimulation with hemin/G-quadruplex-catalyzed luminol-H_2O_2 system. The detection limit for DNA is 2 nmol L^{-1}.

Similar nanostructure was modified with glucose oxidase and conjugated to CdSe/ZnS QDs for the CL detection of glucose. The glucose oxidase catalyzed the oxidation of glucose to compose gluconic acid and H_2O_2. Then, in the presence of luminol catalyzed by hemin/G-quadruplex generated strong CL, which initiated a CRET process to the CdSe/ZnS QDs. Quantitative determination of glucose can be realized from the luminescence intensity of the QDs. The detection limit of glucose was calculated to 5 mmol L^{-1} [131].

3.5. Phosphorescence

Phosphorescence is the radiative transition from the lowest excited triplet state, T_1, to the (singlet) ground state, S_0. On the contrary to fluorescence (singlet-singlet transition), phosphorescence is a spin-forbidden process [100]. In order to obtain phosphorescence, the phosphorophore is excited by electromagnetic radiation of the appropriate wavelength. If S energy levels and T energy levels are close, some of the excited molecules can drop into the T state through an intersystem crossing. The intersystem crossing quantum efficiency can be enhanced by different approaches such as cryogenic conditions, micelle, and heavy atom effect. Phosphorescence techniques have advantages over the fluorescence methods such as selectivity, sensitivity, longer emission lifetime, and a wider gap between excitation and emission spectra. The longer lifetime of the triplet excited state allows using an appropriate delay time so that possible spectral interferences coming from system and light scattering can be avoided [132].

The optical, electrical, and magnetic character of QDs can be modified by using different types of dopants. Compared with traditional QDs such as CdSe, ZnS is a more attractive host nanoparticle for doping to form new type of QDs. Doping Mn^{2+} into ZnS QDs provides unique phosphorescence properties. Mn-doped ZnS QDs exhibit phosphorescence emission, which is produced by the energy transfer from the band gap of ZnS to Mn^{2+} dopant and the subsequent transition from the triplet state to the ground state of the Mn^{2+} involved in the ZnS host lattice [133]. Similar to fluorescent methods, mechanism of the system is quenching of phosphorescence emission of QDs. Adding quencher to QDs solution causes decrease of phosphorescence intensity due to adsorption onto the surface of QDs. When added, analyte interacts with the quencher, the new complex molecule is formed, and phosphorescence intensity is recovered due to removal of quencher from the surface. This type of QDs has been used in the phosphorescent sensing of drugs and biomolecules without any sample pretreatment [12].

General synthesis process of Mn-doped ZnS was explained by He et al. [134]. Briefly, capping agent such as L-cysteine and MPA, $ZnSO_4$, and $MnCl_2$ were added to a flask. pH of the mixture was adjusted to 11 with NaOH. Then, after air was removed with argon purging at room temperature, Na_2S was quickly added to the solution. The mixture was stirred, and then, the solution was aged at 50°C under open air to form capped Mn-doped ZnS QDs. The heating step is vital for synthesis. For example, the phosphorescence spectrum of L-cysteine–capped Mn-doped ZnS QDs exhibited a maximum phosphorescence emission peak at 590 nm when excited at 290 nm. This peak was not observed without the aging step; however, after aging step, the peak appeared [39].

He et al. report Mn-doped ZnS QDs for the RTP detection of enoxacin in biological fluids. The fluorescence spectra of the Mn-doped ZnS QDs show two emission bands, at 435 and 590 nm, while the phosphorescence spectra exhibit only a single-emission peak at 590 nm. The emission at 590 nm presents typical characteristics of an RTP and shows a long decay time of 2 ms because of intersystem crossing. Reported Mn-doped ZnS QDs-based RTP method was not need the using of oxygen scavenger and other inducers and allowed the detection of enoxacin in biological fluids without interference from autofluorescence and the scattering light of the matrix [134].

In phosphorescence study, not only uncapped but also capped QDs are used. For this purpose, widely used capping agents are MPA and L-cysteine. For example, MPA-capped Mn-doped ZnS QDs/CTAB nanohybrids were prepared through electrostatic self-assembly and applied to detection of rutin [135]. Cetyltrimethyl ammonium bromide (CTAB) is a cationic surfactant and has high stability to chemicals, heat, light, pressure, and pH; therefore, CTAB-based nanohybrid also shows highly stable features. Besides, adding of CTAB to QDs causes enhancement of phosphorescence intensity. Quantitative determination of rutin was done by using of linearity of RTP quenching value of QDs and rutin concentration.

N-acetyl-L-cysteine (NAC) and L-cysteine–capped Mn-doped ZnS QDs (NAC-Mn/ZnS QDs and L-cysteine-Mn/ZnS QDs) were prepared by hydrothermal methods and used for determination of L-ascorbic in the human serum sample. The characterization of QDs was made by TEM. Both NAC and L-cysteine–capped QDs were of spherical shape with size 8–10 nm. FT-IR spectra of NAC-capped QDs showed that the band of sulfhydryl group disappeared, and the band of carboxyl group was shifted. When it comes to L-cysteine spectra, their S—H vibration band disappeared. These results indicated that NAC and L-cysteine capped the QDs successfully. The proposed method was selective and sensitive. The Stern-Volmer plot and phosphorescence decay of nanohybrid QDs indicated the dynamic quenching mechanism [136].

Similar QDs system was applied to the investigation of the interaction of anticancer drug and DNA. Herein, L-cysteine capped Mn-doped ZnS QDs/idarubicin (IDA) nanohybrid was used as a phosphorescent probe. IDA was adsorbed on the surface of Mn-doped ZnS QDs and quenched of phosphorescence signal. With the addition of ds-DNA, IDA interacts with DNA, desorbed from the surface of the QDs, and the phosphorescence signal is increased. The quenching mechanism of phosphorescence of QDs by IDA was a combined dynamic and static quenching [12].

Same mechanism was used to investigate anticancer drug sanguinarine and DNA interaction [53]. Sanguinarine can adsorb on the surface of Mn-doped QDs and quench the phosphorescence emission. When the G-quadruplex-sanguinarine complex formed, the phosphorescence intensities of the QDs sensors would be restored.

The macromolecules such as DNA and ATP are also used for capping agents. An ultrasonic-assisted approach was developed for the synthesis of adenosine triphosphate (ATP)–capped Mn-doped ZnS QDs. The prepared QDs were combined Mg^{2+}-ATP-arginine ternary system and used a phosphorescent probe to detect arginine and methylated arginine [137]. The supramolecular interactions of Mg^{2+} and arginine with ATP have been investigated. Arginine and Mg^{2+} acted as a cofactor, interacted specifically, and catalyzed the hydrolysis of ATP. The binding of ATP-capped Mn-doped ZnS QDs to arginine in the presence of Mg^{2+} caused to quenching of the phosphorescence intensity of the QDs, which allowed detection of arginine with a detection limit of 0.23 mM.

http://dx.doi.org/10.5772/intechopen.70034

Phosphorescent QDs have been used as a probe in numerous bioanalysis such as for nucleic acid or protein detection. Gong et al. developed riboflavin (RF)-modulated MPA-capped Mn-doped ZnS QDs and utilized as RTP sensor for DNA detection. As an electron acceptor, RF could quench the RTP emission of QDs via photo-induced electron transfer (PIET) and form Mn-doped ZnS QDs/RF nanohybrids by electrostatic attraction. RF also effectively interacted with DNA in groove-binding mode. In Mn-doped ZnS QDs/RF nanohybrids system, adding of DNA to medium caused the removal of RF from the surface of QDs due to interaction with DNA double helix. Therefore, releasing the RTP of Mn-doped ZnS QDs was observed. The degree of recovery of Mn-doped ZnS QDs depended on DNA concentration. The developed QD-based RTP sensor acted in a turn-on mode and offered high sensitivity to DNA [138].

Another study for detecting DNA is based on self-assembly of nanohybrids from octa(3-aminopropyl) octasilsequioxane octahydrochloride (OA-POSS) and MPA-capped Mn-doped ZnS QDs (MPA-1) [139]. OA-POSS has eight quaternary ammonium groups on each corner and acts as cubic linkers between MPA-1 through electrostatic interaction. MPA-1 and OA-POSS form spherical nanohybrids (1/OA-POSS) in aqueous solution with these linkers. DNA possesses negative charge in phosphate groups and competes with MPA-1 for forming complexes with OA-POSS. This competition led to the decrease of the emission intensity of 1/OAPOSS nanohybrids and allows developing a method for quantitative determination of DNA.

4. Conclusions

A pervasive trend in the pharmaceutical and biomedical analysis is the development of ultrasensitive and high-throughput technologies for the rapid detection and quantification of drugs, proteins, and nucleic acid. QDs have an important role in this field. QDs have unique structural and surface properties such as stability, tunable size, wide spectrum band, and large surface-to-volume ratio that have enabled a new avenue of research to be opened. QD-based nanotechnology will be constantly expanding its applications due to their continued development of specialized nanoparticles. Chemical-surface modifications of the QDs allow enhancing the selectivity of the systems and to profit from their favorable emission features. Moreover, different approaches such as the combination of the nanoparticles with energy-transfer processes and phosphorescence detection are helping to open new research areas. These intelligent, multifunctional, low-toxic or nontoxic nanoparticles are achievements for the future.

Author details

Hayriye Eda Şatana Kara* and Nusret Ertaş

*Address all correspondence to: eda@gazi.edu.tr

Department of Analytical Chemistry, Faculty of Pharmacy, Gazi University, Ankara, Turkey

References

[1] Costa-Fernandez JM, Pereiro R, Sanz-Medel A. The use of luminescent quantum dots for optical sensing. Trends in Analytical Chemistry. 2006;**25**:207-218

[2] Niemeyer CM. Nanoparticles, proteins and nucleic acids: Biotechnology meets materials science. Angewandte Chemie International Edition. 2001;**40**:4128-4158

[3] Murphy CJ, Coffer JL. Quantum dots: A primer. Applied Spectroscopy. 2002;**56**:16A-27A

[4] Kuang H, Zhao Y, Ma W, Xu L, Wang L, Xu C. Recent developments in analytical applications of quantum dots. Trends in Analytical Chemistry. 2011;**30**:1620-1635

[5] Ozkan M. Quantum dots and other nanoparticles: What can they offer to drug discovery?. Drug Discovery Today. 2004;**9**:1065-1071

[6] Yoffe AD. Semiconductor quantum dots and related systems: Electronic optical, luminescence and related properties of low dimensional systems. Advances in Physics. 2001;**50**:1-208

[7] Efros AL, Rosen M. The Electronic structure of semiconductor nanocrystals. Annual Review of Material Sciences. 2000;**30**:475-521

[8] Gill R, Zayats M, Willner I. Semiconductor quantum dots for bioanalysis. Angewandte Chemie International Edition. 2008;**47**:7602-7625

[9] Ma J, Wang F, Fan M, Wang T, Bian W, Shuang S. Phosphorescence detection of hydrochlorothiazide using Mn-doped ZnS quantum dots. Analytical Methods. 2013;**5**:6094-6099

[10] Huang Y, Zhang Y, Yan Z, Liao S. Assay of ceftazidime and cefepime based on fluorescence quenching of carbon quantum dots. Luminescence. 2015;**30**:1133-1138

[11] Lopez JJ, Molina-Garcia L, Rodrigues SSM, Santos JLM, Ortega-Barrales P. Automated determination of rifamycins making use of MPA-CdTe. Journal of Luminescence. 2016;**175**:158-164

[12] Ertas N, Satana Kara HE. L-Cysteine capped Mn-doped ZnS quantum dots as a room temperature phosphorescence sensor for in-vitro binding assay of idarubicin and DNA. Biosensors and Bioelectronics. 2015;**70**:345-350

[13] Chan WC, Nie S. Quantum dot bioconjugates for ultrasensitive nonisotopic detection. Science. 1998;**281**:2016-2018

[14] Hoshino A, Hanaki KI, Suzuki K, Yamamoto K. Applications of T-lymphoma labeled with fluorescent quantum dots to cell tracing markers in mouse body. Biochemical and Biophysical Research Communications. 2004;**314**:46-53

[15] Parak WJ, Gerion D, Pellegrino T, Zanchet D, Micheel C, Williams SC, et al. Biological applications of colloidal nanocrystals. Nanotechnology. 2003;**14**:R15-R27

[16] Jaiswal JK, Mattoussi H, Mauro JM, Simon SM. Long-term multiple color imaging of live cells using quantum dot bioconjugates. Nature Biotechnology. 2003;**21**:47-51

[17] Dahan M, Levi S, Luccardini C, Rostaing P, Riveau B, Triller A. Diffusion dynamics of glycine receptors revealed by single-quantum dot tracking. Science. 2003;**302**:442-445

[18] Wu X, Liu H, Liu J, Haley NK, Treadway JA, Larson JP, et al. Immunofluorescent labeling of cancer marker Her2 and other cellular targets with semiconductor quantum dots. Nature Biotechnology. 2003;**21**:41-46

[19] Larson DR, Zipfel WR, Williams RM, Clarck SW, Bruchez MP, Wise FW, et al. Water-soluble quantum dots for multiphoton fluorescence imaging *in vivo*. Science. 2003;**300**:1434-1436

[20] Dubertret B, Skourides P, Norris DJ, Noireaux V, Brivanlou AH, Libchaber A. *In vivo* imaging of quantum dots encapsulated in phospholipid micelles. Science. 2002;**298**:1759-1762

[21] Akerman ME, Chan WC, Laakkonen P, Bhatia SN, Ruoslahti E. Nanocrystal targeting *in vivo*. Proceedings of the National Academy of Science of the United States of America. 2002;**99**:12617-12621

[22] Resch-Genger U, Grabolle M, Cavaliere-Jaricot S, Nitschke R, Nann T. Quantum dots versus organic dyes as fluorescent labels. Nature Methods. 2008;**5**:763-775

[23] Wang F, Tan WB, Zhang Y, Fan X, Wang M. Luminescent nanomaterials for biological labeling. Nanotechnology. 2006;**17**:R1-R13

[24] Pham W, Choi YD, Weissleder R, Tung CH. Developing a peptide based near-infrared molecular probe for protease sensing. Bioconjugate Chemistry. 2004;**15**:1403-1407

[25] Dybiec M, Chornokur G, Ostapenko S, Wolcott A, Zhang JZ, Zajac A, et al. Photoluminescence spectroscopy of bioconjugated CdSe/ZnS quantum dots. Applied Physics Letters. 2007;**90**:263112

[26] Gerion D, Chen F, Kannan B, Fu AH, Parak WJ, Chen DJ, et al. Room-temperature single-nucleotide polymorphism and multiallele DNA detection using fluorescent nanocrystals and microarrays. Analytical Chemistry. 2005;**75**:4766-4772

[27] Baron R, Willner B, Willner I. Biomolecule-nanoparticle hybrids as functional units for nanotechnology. Chemical Communications. 2007;**4**:323-332

[28] Ma H, Liu X, Wang X, Li X, Yang C, Iqbal A, et al. Sensitive fluorescent light-up probe for enzymatic determination of glucose using carbon dots modified with MnO_2 nanosheets. Microchimica ACTA. 2017;**184**:177-185

[29] Wei F, Yu H, Hu C, Cai Z, Yang J, Liu L, et al. A simple method for copper ion determination in beverages using Mn-modified CdTe/CdS quantum dots as a fluorescence probe. Journal of Nanoscience and Nanotechnology. 2016;**16**:12282-12287

[30] Uppa Y, Kulchat S, Ngamdee K, Pradublai K, Tuntulani T, Ngeontae W. Silver ion modulated CdS quantum dots for highly selective detection of trace Hg^{2+}. Journal of Luminescence. 2016;**178**:437-445

[31] Zhang H, Kang S, Wang G, Zhang Y, Zhaou H. Fluorescence determination of nitrite in water using prawn-shell derived nitrogen-doped carbon nanodots as fluorophores. ACS Sensors. 2016;**1**:875-881

[32] Shojaei TR, Salleh MAM, Sijam K, Rahim RA, Mohsenifar A, Safarnejad R, et al. Detection of Citrus Tristeza virus by using fluorescence resonance energy transfer-based biosensor. Spectrochimica ACTA Part A-Molecular and Biomolecular Spectroscopy. 2016;**169**:216-222

[33] Ahmed F, Arshi N, Dwivedi S, Koo BH, Azam A, Alsharaeh E. Low-temperature growth of ZnO nanotubes for fluorescence quenching detection of DNA. Journal of Materials Science-Materials in Medicine. 2016;**27**:189

[34] Pratiwi FW, Hsia CH, Kuo CW, Yang SM, Hwu YK, Chen P. Construction of single fluorophore ratiometric pH sensors using dual-emission Mn^{2+}-doped quantum dots. Biosensors & Bioelectronics. 2016;**84**:133-140

[35] Chen JL, Zhu CQ. Functionalized cadmium sulfide quantum dots as fluorescence probe for silver ion determination. Analytica Chimica Acta. 2005;**546**:147-153

[36] Li J, Bao D, Hong X, Li D, Li J, Bai Y, et al. Luminescent CdTe quantum dots and nanorods as metal in probes. Colloids and Surfaces A: Physicochemical and Engineering Aspects. 2005;**257-258**:267-271

[37] Xie HY, Liang JG, Zhang ZL, Liu Y, He ZK, Pan DW. Luminescent CdSe-ZnS quantum dots as selective Cu^{2+} probe. Spectrochimica Acta Part A: Molecular and Biomolecular Spectroscopy. 2004;**60**:2527-2530

[38] Chen Y, Rosenzweig Z. Luminescent CdS Quantum dots as selective ion probes. Analytical Chemistry. 2002;**74**:5132-5138

[39] Kara Satana HE, Demirhan B, Demirhan Er B. Mn-doped ZnS quantum dots as a room-temperature phosphorescent probe for analysis of glutamic in food stuff. Turkish Journal of Chemistry. 2016;**40**:762-771

[40] Peng CF, Li ZK, Zhu YY, Chen W, Yuan Y, Liu LQ, et al. Simultaneous and sensitive determination of multiplex chemical residues based on multicolor quantum dot probes. Biosensors and Bioelectronics. 2009;**24**:3657-3662

[41] Peng H, Zhang L, Kjallman THM, Soeller C, Travas-Sejdic J. DNA hybridization detection with blue luminescent quantum dots and dye-labelled single-stranded DNA. Journal of American Chemical Society. 2007;**129**:3048-3049

[42] Bawja N, Mehra NK, Jain K, Jian NK. Pharmaceutical and biomedical applications of quantum dots, Artificial Cells. Nanomedicine and Biotechnology. 2016;**44**:758-768

[43] Moriyama S, Fuse T, Suzuki M, Aoyagi Y, Ishibashi K. Four-electron shell structures and an interacting two-electron system in carbon nanotube quantum dots. Physical Review Letters. 2005;**94**:186806

[44] Santos BS, Farias PMA, Fontes A. Handbook of self assembled semiconductor nanostructures for novel devices in photonics and electronics. In: Henini M, editor. Semiconductor Quantum Dots for Biological Applications. Netherlands: Elsevier; 2008. pp. 773-799

[45] Peng X, Schlamp MC, Kadavanich AV, Alivisatos AP. Epitaxial growth of highly luminescent CdSe/CdS core/shell nanocrystals with photostability and electronic accessibility. Journal of American Chemical Society. 1997;**119**:7019-7029

[46] Talapin DV, Rogach AL, Kornowski A, Haase M, Weller H. Highly luminescent monodisperse CdSe and CdSe/ZnS nanocrystals synthesized in a hexadecylamine–trioctylphosphine oxide–trioctylphospine mixture. Nano Letters. 2001;**1**:207-211

[47] Frigerio C, Ribeiro DSM, Rodrigues SSM, Abreu VLRG, Barbosa JAC, Prior JAV, et al. Application of quantum dots as analytical tools in automated chemical analysis: A review. Analytica Chimica Acta. 2012;**735**:9-22

[48] Gerion D, Pinaud F, Williams SC, Parak WJ, Zanchet, D, Weiss, S,et al. Synthesis and properties of biocompatible water-soluble silica coated CdSe/ZnS semiconductor quantum dots. The Journal of Physical Chemistry. 2001;**105**:8861-8871

[49] Bruchez M, Moronne M, Gin P, Weiss S, Alivisatos AP. Semiconductor nanocrystals as fluorescent biological labels. Science. 1998;**281**:2013-2016

[50] Kim S, Bawendi MG. Oligomeric ligands for luminescent and stable nanocrystal quantum dots. Journal of American Chemical Society. 2003;**125**:14652-14653

[51] Yamashita K, Nakate T, Okimoto K, Ohike A, Tokunaga Y, Ibuki R, et al. Establishment of new preparation method for solid dispersion. International Journal of Pharmaceutics. 2003;**267**:79-91

[52] Miao Y, Zhang Z, Gong Y, Yan G. Phosphorescent quantum dots/doxorubicin nanohybrids based on photo-induced electron transfer for detection of DNA. Biosensors and Bioelectronics. 2014;**59**:300-306

[53] Wen LN, Xie MX. Competitive binding assay for G-quadruplex DNA and sanguinarine based on room temperature phosphorescence of Mn-doped ZnS quantum dots. Journal of Photochemistry and Photobiology A: Chemistry. 2014;**279**:24-31

[54] Banin U, Lee JC, Guzelian AA, Kadavanich AV, Alivisatos AP. Exchange interaction in InAs nanocrystal quantum dots. Superlattice and Microstructures. 1997;**22**:559-568

[55] Rajh T, Micic OI, Nozik OJ. Synthesis and characterization of surface-modified colloidal cadmium telluride quantum dots. Journal of Physical Chemistry. 1993;**97**:11999-12003

[56] Chan WCW, Maxwell DJ, Gao X, Bailey RE, Han M, Nie S. Luminescent quantum dots for multiplexed biological detection and imaging. Current Opinion in Biotechnology. 2002;**13**:40-46

[57] Norris DJ, Bawendi MG. Measurement and assignment of the size dependent optical spectrum in CdSe nanocrystals. Physical Review B Condensed Matter and Materials Physics. 1996;**53**:16338-16346

[58] Pinaud F, Michalet X, Bentolila LA, Tsay JM, Doose S, Li JJ, et al. Advances in fluorescence imaging with quantum dot bio-probes. Biomaterials. 2006;**27**:1679-1687

[59] Yu WW, Qu L, Guo W, Peng X. Experimental determination of the extinction coefficient of CdTe, CdSe, and CdS nanocrystals. Chemistry of Materials. 2003;**15**:2854-2860

[60] Kuçur E, Boldt FM, Cavaliere-Jaricot S, Ziegler J, Nann T. Quantitative analysis of the CdSe nanocrystal concentration by comparative techniques. Analytical Chemistry. 2007;**79**:8987-8993

[61] Rueda D, Walter NG. Fluorescent energy transfer readout of an aptazyme-based biosensor. Methods in Molecular Biology. 2006;**335**:289-310

[62] Murray CB, Norris DJ, Bawendi MG. Synthesis and characterization of nearly monodisperse CdE (E = sulphur, selenium, tellurium) semiconductor nanocrystallites. Journal of American Chemical Society. 1993;**115**:8706-8715

[63] Modani S, Kharwade M, Nijhawan M. Quantum dots: A novelty of medical field with multiple applications. International Journal of Current Pharmaceutical Research. 2013;**5**:55-59

[64] Pandurangan DK, Mounika KS. Quantum dot aptamers-an emerging technology with wide scope in pharmacy. International Journal of Pharmacy and Pharmaceutical Sciences. 2012;**4**:24-31

[65] Bandyopadhyay S, Menon L, Kouklin N, Zeng H, Sellmyer DJ. Electrochemically self-assembled quantum dot arrays. Journal of Electronic Materials. 1999;**28**:515-519

[66] Murray CB, Kagan CR, Bawendi MG. Synthesis and characterization of monodisperse nanocrystals and close-packed nanocrystal assemblies. Annual Review of Materials Research. 2000;**30**(1):545-610

[67] Tada H, Higuchi H, Wanatabe TM, Ohuchi N. *In vivo* real-time tracking of single quantum dots conjugated with monoclonal anti-HER2 antibody in tumors of mice. Cancer Research. 2007;**67**(3):1138-1144

[68] Obonyo O, Fisher E, Edwards M, Douroumis D. Quantum dots synthesis and biological applications as imaging and drug delivery systems. Critical Reviews in Biotechnology. 2010;**30**(4):283-301

[69] Lakshmeesha R, Afrasim M, Gowda DV, Santhosh TR, Abhilasha TP, Gritta S, Osmani RAM. Quantum dots: A new-fangled loom in drug delivery and therapeutics. World Journal of Pharmaceutical Sciences. 2016;**4**(4):125-134

[70] Norris DJ, Yao N, Charnock FT, Kennedy TA. High-quality manganese-doped ZnSe nanocrystals. Nano Letters. 2001;**1**:3-7

[71] Yang H, Santra S, Holloway PH. Syntheses and applications of Mn-doped II-VI semiconductor nanocrystals. Journal of Nanoscience and Nanotechnology. 2005;**5**:1364-1375

[72] Sotelo-Gonzalez E, Fernandez-Arguelles MT, Costa-Fernandez JM, Sanz-Medel A. Mn-doped ZnS quantum dots for the determination of acetone by phosphorescence attenuation. Analytica Chimica Acta. 2012;**712**:120-126

[73] Wei B, Jing M, Wenrong G, Dongtao Lu, Meng F, Yanli W, et al. Phosphorescence detection of L-ascorbic acid with surface-attached N-acetyl-L-cysteine and L-cysteine Mn-doped ZnS quantum dots. Talanta. 2013;**116**:794-800

[74] Demirhan Er B, Demirhan B, Satana Kara HE. Room-temperature phosphorescence determination of melamine in dairy product using L-cysteine capped Mn-doped zinc sulfide (ZnS) quantum dots. Journal of Dairy Science. 2015;**98**:2992-3000

[75] Wu P, He Y, Wang HF, Yan XP. Conjugation of glucose oxidase onto Mn-doped ZnS quantum dots for phosphorescent sensing of glucose in biological fluids. Analytical Chemistry. 2010;**82**:1427-1433

[76] Mishra S, Tripathy P, Sinha SP. Advancements in the field of quantum dots. International Journal of Advancements in Research & Technology. 2012;**1**:1-5

[77] Henini M, Sanguinetti S, Brusaferri L, Grilli E, Guzzi M, Upward MD, et al. Structural and optical characterization of self-assembled InAs-GaAs quantum dots grown on high index surfaces. Microelectronics Journal. 1997;**28**:933-938

[78] Franchi S, Trevisi G, Seravalli L, Frigeri P. Quantum dot nanostructures and molecular beam epitaxy. Progress in Crystal Growth and Characterization of Materials. 2003;**47**:166-195

[79] Kim YT, Han JH, Hong BH, Kwon YU. Electrochemical synthesis of CdSe quantum-dot arrays on a graphene basal plane using mesoporous silica thin-film templates. Advanced Materials. 2010;**22**:515-518

[80] Huang H, Zhu JJ. The electrochemical applications of quantum dots. Analyst. 2013;**138**:5855-5865

[81] Penner RM. Hybrid electrochemical/chemical synthesis of quantum dots. Accounts of Chemical Research. 2000;**33**:78-86

[82] Mangolini L, Thimsen E, Kortshagen U. High-yield plasma synthesis of luminescent silicon nanocrystals. Nano Letters. 2005;**5**(4):655-659

[83] Sankaran RM, Holunga D, Flagan RC, Giapis KP. Synthesis of blue luminescent Si nanoparticles using atmospheric-pressure micro discharges. Nano Letters. 2005;**5**:537-541

[84] Kortshagen U. Nonthermal plasma synthesis of semiconductor nanocrystals. Journal of Physics D: Applied Physics. 2009;**42**:113001

[85] Mangolini L, Kortshagen U. Plasma-assisted synthesis of silicon nanocrystal inks. Advanced Materials. 2007;**19**:2513-2519

[86] Pereira RN, Almeida AJ. Doped semiconductor nanoparticles synthesized in gas-phase plasmas. Journal of Physics D: Applied Physics. 2015;**48**:314005

[87] Mao C, Flynn CE, Hayhurst A, Sweeney R, Qi J, Georgiou G, et al. Viral assembly of oriented quantum dot nanowires. Proceedings of the National Academy of Sciences. 2003;**100**(12):6946-6951

[88] Borovaya MN, Burlaka O, Yemets A, Blume Y. Biosynthesis of quantum dots and their potential applications in biology and biomedicine. In: Fesenko O, Yatsenko L, editors. Nanoplasmonics, Nano-Optics, Nanocomposites, and Surface Studies. Switzerland: Springer International Publishing; 2015

[89] Bekasova OD, Revina AA, Rusanov AL, Kornienko ES, Kurganov B. Effect of gamma-ray irradiation on the size and properties of CdS quantum dots in reverse micelles. Radiation Physics and Chemistry. 2013;**92**:87-92

[90] Jovanović S, Marković Z, Budimir M, Spitalsky Z, Vidoeski B, Marković BT. Effects of low gamma irradiation dose on the photoluminescence properties of graphene quantum dots. Optical and Quantum Electronics. 2016;**48**:259

[91] Jovanovic SP, Syrgiannis Z, Markovic ZM, Bonasera A, Kepic DP, Budimir MD, et al.. Modification of structural and luminescence properties of graphene quantum dots by gamma irradiation and their application in a photodynamic therapy. ACS Applied Material Interfaces. 2015;**7**:25865-25874

[92] Xuan T, Wang X, Zhu G, Li H, Pan L, Sun Z. One-step microwave-assisted synthesis of water soluble CdSe quantum dots for white light-emitting diodes with excellent color rending. Journal of Alloys and Compounds. 2013;**558**:105-108

[93] Moghaddam MM, Baghbanzadeh M, Keilbachb A, Kappe CO. Microwave-assisted synthesis of CdSe quantum dots: Can the electromagnetic field influence the formation and quality of the resulting nanocrystals. Nanoscale. 2012;**4**:7435-7442

[94] Kairdolf BA, Qian X, Nie S. Bioconjugated Nanoparticles for biosensing, *in vivo* imaging, and medical diagnostics. Analytical Chemistry. 2017;**89**:1015-1031

[95] Esteve-Turrillas FA, Abad-Fuentes A. Applications of quantum dots as probes in immunosensing of small-sized analytes. Biosensors and Bioelectronics. 2013;**41**:12-29

[96] Spanhel L, Haase M, Weller H, Henglein A. Photochemistry of colloidal semiconductors. 20. Surface modification and stability of strong luminescing CdS particles. Journal of American Chemical Society. 1987;**109**:5649-5655

[97] Sooklal K, Cullum BS, Angel SM, Murphy CJ. Photophysical properties of ZnS nanoclusters with spatially localized Mn^{2+}. Journal of Physical Chemistry. 1996;**100**:4551-4555

[98] Moore DE, Patel K. Q-CdS photoluminescence activation on Zn^{2+} and Cd^{2+} salt introduction. Langmuir. 2001;**17**(8):2541-2544

[99] Gao M, Kirstein S, Mohwald H. Strongly photoluminescent CdTe nanocrystals by proper surface modification. Journal of Physical Chemistry B. 1988;**102**:8360-8363

[100] JR Lakowicz. Principles of Fluorescence Spectroscopy. 2nd ed. New York: Plenum Press; 1999

[101] Constantine CA, Gattás-Asfura KM, Mello SV, Crespo G, Rastogi V, Cheng TC, et al. Layer-by-layer films of chitosan, organophosphorus hydrolase and thioglycolic acid-capped CdSe quantum dots for the detection of paraoxon. Journal of Physical Chemistry. 2003;**107**:13762-13764

[102] Yu L, Li L, Ding Y, Lu Y. A Fluorescent switch sensor for glutathione detection based on Mn-doped CdTe quantum dots-methyl viologen nanohybrids. Journal of Fluorescence. 2016;**26**:651-660

[103] Tashkhourian J, Absalan G, Jafari M, Zare S. A rapid and sensitive assay for determination of doxycycline using thioglycolic acid-capped cadmium telluride quantum dots. Spectrochimica Acta Part A: Molecular and Biomolecular Spectroscopy. 2016;**152**:119-125

[104] Samadi N, Narimani S. An ultrasensitive and selective method for the determination of Ceftriaxone using cysteine-capped cadmium sulfide fluorescence. Spectrochimica Acta Part A: Molecular and Biomolecular Spectroscopy. 2016;**163**:8-12

[105] Liang J, Huang S, Zeng D, He Z, Ji X, Ai X, et al. CdSe quantum dots as luminescent probes for spironolactone determination. Talanta. 2006;**69**:126-130

[106] Hou M, Yan X, Xiong L. Determination of sparfloxacin with CdSe/CdS quantum dots as fluorescent probes. Journal of Luminescence. 2015;**157**:58-62

[107] Pawar SP, Gore AH, Walekar LS, Anbhule PV, Patil SR, Kolekar GB. Turn-on fluorescence probe for selective and sensitive detection of d-penicillamine by CdS quantum dots in aqueous media: Application to pharmaceutical formulation. Sensors and Actuators B. 2015;**209**:911-918

[108] Jimenez-Lopez J, Molina-Garcia L, Rodrigues SSM, Santos JLM, Ortega-Barrales P, Ruiz-Medina A. Automated determination of rifamycins making use of MPA–CdTe quantum dots. Journal of Luminescence. 2016;**175**:158-164

[109] Lin L, Rong M, Luo F, Chen D, Wang Y, Chen X. Luminescent graphene quantum dots as new fluorescent materials for environmental and biological applications. Trends in Analytical Chemistry. 2014;**54**:83-102

[110] Cao L, Wang X, Meziani MJ, Lu F, Wang H, Luo PG, et al. Carbon dots for multiphoton bioimaging. Journal of American Chemical Society. 2007;**129**:11318-11319

[111] Goldman ER, Balighian ED, Mattoussi H, Kuno MK, Mauro JM, Tran PT, et al. Avidin: A natural bridge for quantum dot-antibody conjugates. Journal of American Chemical Society. 2002;**124**(22):6378-6382

[112] Goldman ER, Clapp AR, Anderson GP, Uyeda HT, Mauro JM, Medintz IL, et al. Multiplexed toxin analysis using four colors of quantum dot fluororeagents. Analytical Chemistry. 2004;**76**:684-688

[113] Sun M, Du L, Gao S, Bao Y, Wang S. Determination of 17β-oestradiol by fluorescence immunoassay with streptavidin-conjugated quantum dots as label. Steroids. 2010;**75**:400-403

[114] Li YH, Huang J, Liang RP, Qiu JD. Fluorescent graphene quantum dots with a boronic acid appended bipyridinium salt to sense monosaccharides in aqueous solution. Chemical Communications. 2013;**49**:5180-5182

[115] Harris D. Quantitative Chemical Analysis. 8th ed. New York: W. H. Freeman and Co; 2010

[116] Clapp AR, Medintz IL, Mattouss H. Förster resonance energy transfer investigations using quantum-dot fluorophores. Chemical Physics and Physical Chemistry. 2006;**7**:47-57

[117] Coto-Garcia AM, Sotelo-Gonzalez E, Fernandez-Argüelles MT, Pereiro R, Costa-Fernandez JM, Sanz-Medel A. Nanoparticles as fluorescent labels for optical imaging and sensing in genomics and proteomics. Analytical and Bioanalytical Chemistry. 2011;**399**:29-42

[118] Ai X, Niu L, Li Y, Yang F, Su X. A novel β-cyclodextrin-QDs optical biosensor for the determination of amantadine and its application in cell imaging. Talanta. 2012;**99**:409-414

[119] Antony EJ, Shibu A, Ramasamy S, Paulraj MS, Enoch IVMV. Loading of atorvastatin and linezolid in β-cyclodextrin–conjugated cadmium selenide/silica nanoparticles: A spectroscopic study. Materials Science and Engineering C. 2016;**65**:194-198

[120] Gore AH, Kale MB, Anbhule PV, Patil SR, Kolekar GB. A novel FRET probe for selective and sensitive determination of vitamin B12 by functionalized CdS QDs in aqueous media: Applications to pharmaceutical and biomedical analysis. RSC Advances. 2014;**4**:683-692

[121] Zhang CY, Johnson LW. Quantum dot-based fluorescence resonance energy transfer with improved FRET efficiency in capillary flows. Analytical Chemistry. 2006;**78**:5532-5537

[122] Geissler D, Charbonniere LJ, Ziessel RF, Butlin NG, Löhmannsröben HG, Hildebrandt N. Quantum dot biosensors for ultrasensitive multiplexed diagnostics. Angewandte Chemie International Edition. 2010;**49**:1396-1401

[123] Harma H, Soukka T, Shavel A, Gaponik N, Weller H. Luminescent energy transfer between cadmium telluride nanoparticle and lanthanide (III) chelate in competitive bioaffinity assays of biotin and estradiol. Analytica Chimica Acta. 2007;**604**:177-183

[124] Chen H, Lin L, Li H, Lin JM. Quantum dots-enhanced chemiluminescence: Mechanism and application. Coordination Chemistry Reviews. 2014;**263-264**:86-100

[125] Chen H, Lin L, Lin Z, Guo G, Lin JM. Chemiluminescence arising from the decomposition of peroxymonocarbonate and enhanced by CdTe quantum dots. The Journal of Physical Chemistry A. 2010;**114**:10049-10058

[126] Khateee A, Lotfi R, Hasanzadeh A, Iranifamb M, Joo SW. A flow injection chemiluminescence method for determination of nalidixic acid based on $KMnO_4$–morin sensitized with CdS quantum dots. Spectrochimica Acta Part A: Molecular and Biomolecular Spectroscopy. 2016;**154**:243-251

[127] Khataeea A, Hasanzadeha A, Iranifamb M, Jooc SW. A novel flow-injection chemiluminescence method for determination of baclofen using L-cysteine capped CdS quantum dots. Sensors and Actuators B. 2015;**215**:272-282

[128] Imani-Nabiyyi A, Sorouraddin MH. Determination of naphazoline hydrochloride in biological and pharmaceutical samples by a quantum dot-assisted chemiluminescence system using response-surface methodology. Luminescence. 2014;**29**:994-1002

[129] Fortesa PR, Frigerio C, Silvestre CIC, Santos JLM, Limab JLFC, Zagattoa EAG. Cadmium telluride nanocrystals as luminescent sensitizers in flow analysis. Talanta. 2011;**84**:1314-1317

[130] Golub E, Niazov A, Freeman R, Zatsepin M, Willner I. Photoelectrochemical biosensors without external irradiation: Probing enzyme activities and DNA sensing using hemin/G-quadruplex-stimulated chemiluminescence resonance energy transfer (CRET) generation of photocurrents. The Journal of Physical Chemistry C. 2012; **116**:13827-13834

[131] Niazov A, Freeman R, Girsh J, Willner I. Following glucose oxidase activity by chemiluminescence and chemiluminescence resonance energy transfer (CRET) processes involving enzyme-DNAzyme conjugates. Sensors. 2011;**11**:10388-10397

[132] Kuijt J, Ariese F, Brinkman UAT, Gooijer C. Room temperature phosphorescence in the liquid state as a tool in analytical chemistry. Analytica Chimica Acta. 2003;**488**:135-171

[133] Bi L, Dong XT, Yu YH. Room-temperature phosphorescence sensor based on manganese doped zinc sulfide quantum dots for detection of urea. Journal of Luminescence. 2014;**153**:356-360

[134] He Y, Wang HF, Yan XP. Exploring Mn-doped ZnS quantum dots for the room-temperature phosphorescence detection of enoxacin in biological fluids. Analytical Chemistry. 2008;**80**:3832-3837

[135] Miao Y, Zhang Z, Gong Y, Zhang Q, Yan G. Self-assembly of manganese doped zinc sulfide quantum dots/CTAB nanohybrids for detection of rutin. Biosensors and Bioelectronics. 2014;**52**:271-276

[136] Bian W, Mab J, Guo W, Lu D, Fan M, Wei Y, et al. Phosphorescence detection of L-ascorbic acid with surface-attached N-acetyl-L-cysteine and L-cysteine Mn-doped ZnS quantum dots. Talanta. 2013;**116**:794-800

[137] Ren HB, Yan XP. Ultrasonic assisted synthesis of adenosine triphosphate capped manganese-doped ZnS quantum dots for selective room temperature phosphorescence detection of arginine and methylated arginine in urine based on supramolecular Mg^{2+}-adenosine triphosphate-arginine ternary system. Talanta. 2012;**97**:16-22

[138] Gong Y, Fan Z. Room-temperature phosphorescence turn-on detection of DNA based on riboflavin-modulated manganese doped zinc sulfide quantum dots. Journal of Fluorescence. 2016;**26**:385-393

[139] He Y, Wang HF, Yan XP. Self-assembly of Mn-doped ZnS quantum dots/octa(3-aminopropyl)octasilsequioxane octahydrochloride nanohybrids for optosensing DNA. Chemistry-A European Journal. 2009;**15**:5436-5440

Section 2

Applications of Spectrophotometric Methods in Pharmaceutical and Biomedical Analyses

Ion-Pair Spectrophotometry in Pharmaceutical and Biomedical Analysis: Challenges and Perspectives

Marinela Florea and Mihaela Ilie

Additional information is available at the end of the chapter

http://dx.doi.org/10.5772/intechopen.69778

Abstract

Experimental and theoretical studies of the mechanisms that underlay ion-pair formation, their properties and applications in various fields have been and still are focused by researchers since the introduction of the concept in 1926, by Bjerrum. Ion pairs are distinct chemical entities, electrically neutral, formed between ions of opposite charge and held together by Coulomb forces, without formation of a covalent bond. Investigation methods used are various, from classical conductometric measurements to up-to-date methods, such as spectrophotometry, chromatography and capillary electrophoresis. In the pharmaceutical field, ion pairs were used to develop methods of separation, identification and assay for the active substances in complex matrices, to obtain pharmaceutical formulations with controlled release and to explain the mechanisms of transport and action for certain drugs. The chapter is an attempt to describe new trends in the spectrophotometry of ion pairs and their applications in the pharmaceutical field. The development of the concept and types of ion pairs are first presented; further, examples of applications using molecular absorption, fluorimetry and resonance light scattering spectrophotometry are presented. Based on the literature data and the authors' experience in the field, challenges and perspectives in the ion-pair spectrophotometry are also considered.

Keywords: ion-pair spectrophotometry, pharmaceuticals, UV-Vis absorption, fluorescence, resonance light scattering, resonance Rayleigh scattering

1. Introduction

Ion-pair spectrophotometry refers to analysis methods based on the optical properties of the ion pairs. Infrared, nuclear magnetic resonance and Raman spectrometry are the methods generally used to investigate the structure of the ion pairs and molecular and atomic absorption, fluorimetry and resonance light scattering are used as assay methods.

Known also as *ionic associations* or *ionic association complexes*, ion pairs are pairs of oppositely charged ions held together by Coulomb attraction without formation of a covalent bond [1]. The lifetime of an ion pair was determined to be of at least 10^{-5} seconds, equivalent to about 10^8 molecular vibrations, demonstrating that ion pairs can be considered as independent species [2].

The inclusion of a substance in an ionic association causes changes in its physical-chemical properties without changing its structure, because an ion pair is electrically neutral and has increased lipophilicity compared with the free ions in its composition [3]. The optimum experimental conditions for a quantitative ion-pair equilibrium (solvent, pH, ionic strength) are easily settled. The selectivity of the methods can be increased by selecting the optimum reagent (counterion, ion-pair forming reagent) and the subsequent extraction of the ion pair in an organic phase.

The significant number of the scientific papers published indicates the appropriateness of ion pairing in solving important issues in the pharmaceutical field, especially in analytical chemistry, biochemistry and pharmaceutical technology. Ion pairs are used for the development of new pharmaceutical forms with controlled release, especially for peptides [3–5]. In this case, one of the main advantages is the unmodified pharmaco-toxicological profile of the active substance after ion pairing, because it does not suffer structural changes. The stability [6] and bioavailability [7, 8] of the drugs can be improved. A series of kinetic studies proposed the ion-pair formation as an absorption mechanism for the pharmaceutical substances [9, 10]. Investigations on DNA stability in various matrices [11], protein determination [12] and synthesis of ion-pair receptors based on biological models [13] are important applications of ion-pair equilibrium in biochemistry. Ion-pair–based assay methods are proficient in both isolation, identification and quantification of certain substances of biomedical interest [14]. Over time, titrimetric [15, 16], gravimetric [17, 18], electrometric [19, 20], spectrophotometric [21, 22] and chromatographic [23, 24] methods based on the ion-pair formation were developed.

Classical and modern as well, ion-pair–based spectrophotometric methods had a dynamic evolution over the time. On the one hand, this is due to the elucidation of the mechanisms underlying the formation of ion pairs, and thus, the setting of the experimental conditions, which allow the obtaining of ion pairs for all types of substances, is easier; in this regard, computational chemistry is a very useful tool. On the other hand, the synthesis of new pharmacologically active molecules at very low concentrations requires sensitive analysis methods. Among them, less used spectrophotometric techniques, such as resonance light scattering, have found interesting applications when ion pairing was taken into account.

2. Fundamentals of ion pair

2.1. History of the ion-pair concept

The ion-pair equilibrium has been first considered for inorganic ions, being an important step in the study of the electrolyte solutions. The history of the ion-pair concept starts in 1887 with Arrhenius, who structured the theory of electrolytic dissociation. Debye and Hückel defined in 1923 the activity coefficients and deduced the homonym equation that allows the assessment of those coefficients in aqueous solutions of electrolytes [25]. In 1926, Bjerrum introduced an association constant in the Debye-Hückel equation and demonstrated that ion-pair equilibrium is dependent on the dielectric constant of the solvent, on the temperature and on the size of the ions. For his theory, he considered spherical, nonpolarizable interacting ions [26]. Thus, the existence of ion pairs was accepted in low dielectric constant solvents. Subsequently, studies were performed in order to elucidate the existence of solvent molecules in the ion-pair structure [27]. Most of the experimental data used to confirm theoretical studies on ion pairing were conductivity measurements.

The subsequent development of organic synthesis and the physical-chemical study of association of more complex molecules, concomitant with the development of new analysis methods (spectrophotometry, chromatography), indicated that, when forming an ion pair, the interacting ions cannot be considered as rigid and spherical [28]. In 1967, Higuchi et al., considering the volume and charge distribution over the ions, studied how a contact ion pair can be solvated in various solvents. For the ion pairs formed between a large lipophilic cation and a small anion, the high negative charge per unit area, lead to the solvation with electrophilic molecules, such as chloroform, phenols and alcohols. The high negative charge on the surface of the ion pairs formed between a small cation and a large lipophilic anion induce the solvation with nucleophilic molecules, such as ethers, ketones, amides and phosphate esters. For the ion pairs formed between two large ions, no significant solvation was observed [29].

Hydrophobic interaction, typical for large unhydrated (hydrophobic) univalent ions, was proposed as a mechanism of ion-pair formation in aqueous solution by Diamond [30]. The driving force for the ion pairing is the water molecule preference to interact with itself by hydrogen bonding. The equilibrium is named *water structure-enforced ion pairing* and the complexes formed accordingly—*water structure-enforced ion pairs*. Thus, starting from this point, the existence of the ion pairs in water became an accepted fact.

When the interaction between the oppositely charged ions is strictly electrostatic, no new electronic bands appear in the absorption spectrum [28]. Spectral changes reported in the studies indicate that, for the ion pairs formed by organic ions, additional interactions (aromatic stacking, charge transfer, hydrogen bonding) might exist.

Aromatic stacking is indicated by a hypochromic effect in the absorption spectra and was demonstrated by thermodynamic studies for the interaction between organic species containing aromatic structures [31]. Considered to be a result of a non-classical hydrophobic effect, the *stacking* of the *aromatic* rings is determined by the interaction between the partial charges (positive and negative) that exists on the atoms situated in adjacent aromatic rings.

The redistribution of the charge between the ions (charge transfer) is identified spectrophotometrically by a hypsochromic (blue shift) or bathochromic (red shift) effect in the UV-Vis region, depending on the medium polarity. This type of interaction can be predicted by theoretical calculations, based on charge density and molecular orbital theory [32, 33]. The ionic associations based on such interaction have been named *ion-pair charge-transfer complexes* [32].

Similarly, it was proved that ion pairs can be formed also by the interaction between an acid and a base by *proton transfer* [34].

Thus, nowadays, it is generally accepted that electrostatic interactions, hydrophobic interactions and proton transfer are the main mechanisms involved in the ion-pair formation and that the ion pair stability depend on the structure and size of the ions, on their acid-base and hydrophobic properties and on the solvent nature as well.

The methods currently used for the study of the ion-pair equilibrium are spectrophotometry (molecular absorption, resonance light scattering and fluorescence), conductometry [35], chromatography [23, 24, 36] and capillary electrophoresis [37].

The development of computational chemistry makes possible simulation of associations between complex molecules. Thus, in silico investigations became a valuable tool in the study of the ionic association equilibrium. Such studies can explain the formation of a certain complex, or predict it, and are commonly validated by spectrophotometric methods.

2.2. Types of ion pairs

The formation mechanism and the structure of the ion pairs were established concurrently with the elucidation of the interaction types of the ions in solution. Considering the solvation, ion pairs exist in the *tight* (*contact, intimate*) form (no solvent molecule is involved in the ion pair) and in the *loose* form (one or more solvent molecules are included in the ion pair). Depending on the number of the solvent molecules involved, ion pairs are of *solvent-sharing* type (a single solvent molecule is included) and *solvent-separated* ion pairs (when more than one solvent molecule is involved) [1].

Considering the structure of the ions involved in the ion-pair equilibrium, an overview of the literature published allowed the identification of three categories: (a) inorganic ion pairs (both ions are inorganic), (b) ion pairs formed between an organic molecule in ionized form and an inorganic ion and (c) organic ion pairs (both ions are organic substances in ionized form). The inorganic ions can be included in an ionic association in the free form or as inorganic complex. The organic substances are transformed in the ionic form based on their acid-base properties, by selecting the optimum pH, or after a complexation reaction with an inorganic ion. The inorganic ion pairs are intensively studied by physical chemistry, for the theoretical background of the mechanisms of ion pairing. The ion pairs that contain an organic ion are more used in the pharmaceutical field.

The solubility of the ion pairs in the selected solvent can be also a criterion to classify the ion pairs. Thus, two categories can be discerned: insoluble and soluble ion pairs. Insoluble ion pairs are used in the assay of the pharmaceuticals by gravimetric methods [17, 18, 38] and atomic absorption spectrometry [39, 40] and also in pharmaceutical technology for drug release systems [5]. The applications that rely on the formation of soluble ion pairs are most numerous.

As an ion pair is electrically neutral, the number of ions involved depends on their charge. Frequently encountered in literature are binary ion pairs, formed between ions with the same charge, and ternary ion pairs, which contain one divalent ion and two monovalent counterions.

3. Spectrophotometric applications of ion pairing in pharmaceutical analysis

3.1. Molecular absorption spectrophotometry

Most of the ion-pair spectrophotometric assay methods published in pharmaceutical field are based on UV-Vis molecular absorption. These methods are robust, easy to perform, sensitive, accurate and precise. In association with an organic dye, pharmaceutical substances with no characteristic visible spectrum can be detected in this region.

Widely used are the extractive spectrophotometric methods: the ion pairs are extracted in an organic solvent, and the extract is further analyzed. For the quantitative extraction in the selected organic solvent, the optimum pH value, concentration of the reagents and ionic strength must be established.

The selection of the counterion should consider that bulky, univalent and having the charge distributed over the whole ion reagent has the best capacity to form ion pairs. With respect to the geometry of the counterion, planar types of organic dyes are appropriate for developing ion-pair absorption spectrophotometric methods [41]. Computational chemistry is a useful tool to evaluate the volume, geometry and charge density of the studied substances. By correlating these data with the results of the studies on the solvation of different types of ion pairs [29], the selection of the optimum solvent for the extraction is simplified.

The formation of the ion pair can be revealed spectrophotometrically by a shift of the absorption peak of the chromophore. As an example, the spectral changes that appeared at the formation of terbinafine-methyl orange (TBF-MO) ion pair in chloroform were used for the assay of terbinafine by ion-pair absorption spectrophotometry by Florea et al. [42]. MO is a planar dye containing aromatic rings, and the formation of TBF-MO ion pair is accompanied by a blue shift (from 502 to 408 nm) and hypochromic effect for the visible peak of MO (**Figure 1**). These spectral changes indicate the stabilization of the ion pair by aromatic stacking [31].

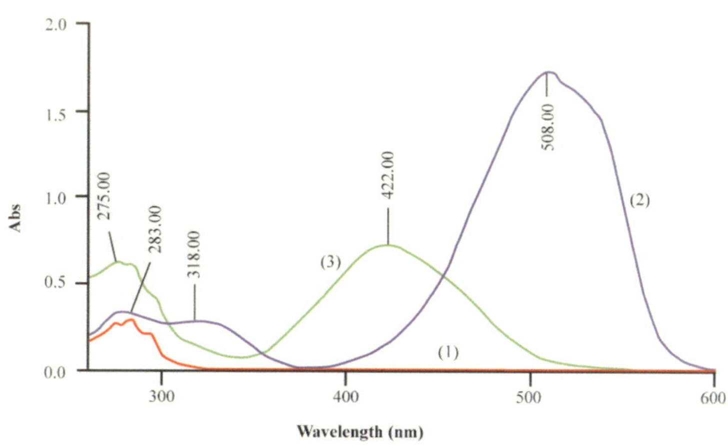

Figure 1. Molecular absorption spectra of TBF (1), MO (2) and TBF-MO ionic association (3) (from Ref. [42], with permission).

Bromocresol purple (BCP) [43] and alizarin red [44] were also used as counterions in the assay of TBF using extraction methods.

As counterions, the chain-type reagents having long alkyl groups are also useful. They are bulky and univalent, but their charge is not distributed over the whole ion. Even so, the main limitation arises mostly from the fact that they are colourless; therefore, they can be used as ion-pair reagents for the assay of coloured substances.

Hexadecyltrimethylammonium bromide (CTAB) was used to develop an extractive spectrophotometric method for the assay of nimesulide (NS) by Florea et al. [45]. As CTAB is a chain-type reagent, with no aromatic rings in the structure and no characteristic spectrum in UV-Vis region, NS in its ionized form is the reagent having a peak in visible range. Therefore, when the ion pair is formed, a red shift and hyperchromic effect appeared (**Figure 2**).

Sulfonephthalein dyes, such as bromocresol green (BCG), bromocresol purple (BCP) and brilliant blue G [46], were also used as counterions in the assay of NS using extraction-free methods. A comparison of the experimental data indicated a larger linearity range for the method based on the ion pair formed with CTAB.

Because extraction is a laborious procedure, the trends are to develop non-extractive (extraction-free) ion-pair–based spectrophotometric methods in nonaqueous or aqueous solutions. Generally, by dissolving the substances in the organic solvents, the ion pairs are formed mainly by a proton transfer mechanism.

Limitations in the development of non-extractive methods arise mainly from the physical-chemical properties of the reagents, namely, their solubility in the appropriate solvents.

Literature data generally resulted in narrower linearity ranges for the non-extractive methods compared with the extractive ones. Some examples are presented in **Table 1**.

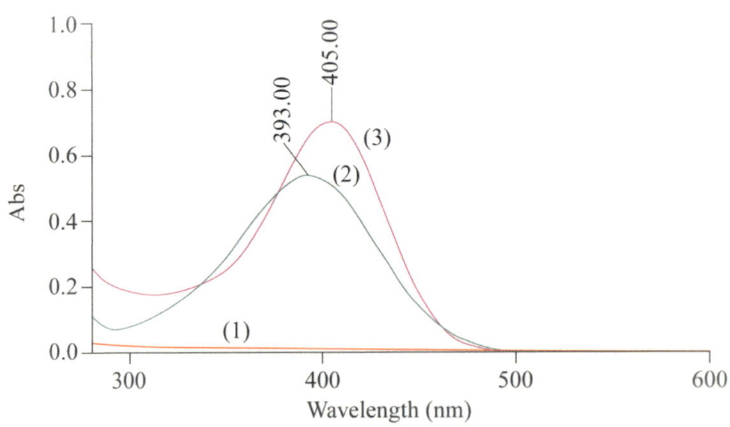

Figure 2. Molecular absorption spectra of CTAB (1), NS (2) and CTAB-NS (3) (from Ref. [45], with permission).

Analyte	Method	Counterion	Solvent	Linearity range (μg/mL)	Reference
Desloratadine	NE	Eosine	H_2O	0.31–2.81	[47]
	E	$[Co(SCN)_4]^{2-}$	$CHCl_3$	0.5–3	[48]
Losartan	NE	Eosine	H_2O	2.5–20	[47]
	E	Calmagite	$CHCl_3$	10–100	[49]
		Orange II	$CHCl_3$	10–100	[49]
Levofloxacin	NE	BCG	CH_2Cl_2	1–20	[50]
	E	Bromophenol blue	$CHCl_3$	1.85–31.5	[51]
		BCG	$CHCl_3$	1.85–25	[51]
Ampicillin	NE	Pyrocatechol violet	H_2O	0.2–28	[52]
	E	$[Mo(SCN)_6]^-$	CH_2Cl_2	1.5–77.5	[53]
Amoxicillin	NE	BCG	$(CH_3)_2SO$	1–13	[54]
	E	Methylene blue	$CHCl_3$	3.5–90	[55]

Table 1. Examples of extractive (E) and non-extractive (NE) ion-pair spectrophotometric methods for the assay of pharmaceutical substances.

3.2. Fluorescence spectroscopy

Among spectrophotometric methods, fluorimetry distinguishes itself by high sensitivity and specificity. In pharmaceutical sciences, fluorescence spectroscopy is an irreplaceable tool in the study of biochemical processes occurring at the cellular level. Substances having intrinsic fluorescence, named *fluorophores*, have characteristic structural features (rigid, plane structure with conjugated double bounds) and exhibit specific excitation (absorption) and emission (fluorescence) wavelengths, thus explaining the high specificity of the method [56]. Various interactions of the fluorophore with the surroundings can lead to a decrease of the fluorescence intensity. This effect is called quenching and can be used for quantification purposes, primarily for the determination of anions [57]. Molecular mechanisms such as the interaction with electron-deficient molecules (quenchers) in the excited state of the fluorophore (collisional quenching) or in the ground state (formation of non-fluorescent complexes with quenchers), together with different non-molecular effects, can be involved in the quenching process [56].

Ion-pair fluorescence assay methods are generally based on quenching. In the ion-pair structure, if one of the ions is a fluorophore, the counterion can act as a quencher. For a certain concentration range, the decrease of the fluorescence intensity is proportional with the analyte concentration. The development of these methods takes into consideration the same experimental conditions presented at Section 3.1, to obtain a quantitative ion-pair equilibrium (pH, ionic strength, solvent), but it is conditioned by the selection of an optimum fluorophore. Organic substances with native fluorescence that can be used as counterions are few; therefore, there are not many published applications. Literature data on ion-pair fluorescence methods for the assay of pharmaceutical substances are summarized in **Table 2**.

Fluorescent reagent	Analyte	Reference
Extractive methods		
Erythrosine B	Erythromycin	[58]
9,10-Dimethoxyanthracene-2-sulphonate	Imipramine	[59]
	Desipramine	
	Amitriptyline	
	Nortriptyline	
	Clomipramine	
	Doxepin	
Nonextractive methods		
Eosine	Astemizole, terfenadine, flunarizine as chelates with Pb^{2+}	[60]
	Amitriptyline	
	Clomipramine	[61]
	Rosiglitazone	[62]
	Pioglitazone	[63]
	Albendazole	[64]
Eosine (as chelate with Pd^{2+})	Ciprofloxacin	[65]
	Norfloxacin	
Safranin T	Meloxicam	[66]
4,5-Dibromofluorescein	Ceftazidime	[67]
	Ceftriaxone	
	Cefoperazone	

Table 2. Examples of extractive and non-extractive ion-pair fluorescence methods for the assay of pharmaceutical substances.

Berberine, an isoquinoline alkaloid, is a pharmacologically active fluorophore, with potential therapeutic effect in various diseases (Alzheimer's disease, cancer, viral infections, etc.). In order to get deeper insights into the details of its biological activity, the effect of the ion pairing on its fluorescence properties was studied using chloride and perchlorates anions [68]. Nanoparticles containing berberine-tetraphenylborate ion pair were prepared, and their ability to cross the cell membrane of cancer cells was studied by Soulié et al. [69].

Lately, the research in nanoscience opened an even wider pathway in fluorescence studies involving ion pairing. Quantum dots (QDs) are prone to exchange electrons with their complementary partners (acceptors or donors) upon excitation that can be transduced into detectable fluorescent signals [70]. Thus, sensitive assay methods can be developed by using QDs capped with different ligands in ionized form. An example is the determination method developed for albendazole, using glutathione-capped cadmium telluride QDs (GSH-CdTe QDs).

Ion-pair equilibrium takes place between albendazole in cationic form and anionic sites at the QD surface, and the effect was a decrease of the fluorescence intensity of capped CdTe QDs [70].

Various applications based on ion-pair equilibrium with fluorescence properties were developed along the time for the characterization of biomolecules in complex biological matrices by flow cytometry [71] and also for kinetic studies using fluorescence microscopy [72].

3.3. Light scattering spectrometry

Light scattering was observed by the Irish physicist John Tyndall in the late 1860s. The eminent British physicist Lord Rayleigh (John Strutt) developed the theoretical basis of electromagnetic wave interaction with particles smaller than the wavelength in the following decades (1870–1899). Now, the scattering of light by particles in a suspension is accepted to be elastic (without change in the wavelength of the incident light) and inelastic (the incident wavelength and the scattered one are different). Rayleigh scattering theory was developed for wavelengths much higher than the dimensions of the scattering particles.

Light scattering is widely used since the 1950s in chemical analysis; turbidimetric and nephelometric methods were developed for the analysis of polydisperse systems. Also, ion-pair–based turbidimetric methods were developed [73–75]. With the development of laser technology, first Raman scattering was separated and developed as an independent technique, allowing the analysis of vibrational and rotational states of a molecule.

Resonance light scattering, also known as resonance Rayleigh scattering (RLS), or enhanced Rayleigh scattering, is a simple, rapid and sensitive method for the study of aggregation of molecules. It was first predicted by Placzek in the mid-1930s and later studied as resonance-enhanced Rayleigh scattering (RERS) for diphenylpolyenes [76], for a series of coumarin dyes [77] and for aggregates containing porphyrins [78].

Starting with the 2000s, a series of studies underlined the utility of the method in the assay of pharmaceuticals as ion pairs with organic dyes [79–81] or using a counterion attached to nanoparticles [82, 83] but also for unravelling of their interaction mechanisms with macromolecules of biological interest (transport proteins, DNA) [84–86]. Recent studies have highlighted the potential of this technique to elucidate the action mechanisms of pharmaceutical substances at the molecular level: the mechanism of interaction of oridonin (natural substance with anticancer effect) with DNA macromolecule was revealed [87]; also, the molecular mechanism by which quercetol affects the bioavailability of propranolol was explained [88].

The ion-pair–based assay methods were developed according to the technique proposed by Pasternack et al. [78]. The RLS spectra are registered using a steady-state spectrofluorometer through synchronous scanning of both monochromators. Near or within the range of the absorption band, an enhancement of the scattered signal is observed, which no longer obeys Rayleigh's law. The effect was largely attributed to a scattering-absorption-re-scattering process. For a certain concentration range, increments in the scattering intensity are directly proportional to the concentration of the analyte.

The majority of the substances determined are hydrophilic organic molecules, largely hydrated in water. The ion pairs are formed mostly by an experimentally conducted hydrophobic ion pairing. An increased ionic strength determines the chemical species involved in the ion pairing to become more hydrophobic because the solvent molecules in their hydration shell are attracted in competitive solvation equilibria of the inorganic ions. Generally, the optimum pH and increased ionic strength are obtained using Britton-Robinson buffer. Molecular absorption spectra are used as a previous step in developing RLS methods. By monitoring changes in the absorption spectra, the optimum counterion is selected, and the experimental conditions for quantitative ionic association equilibrium (pH, ionic strength, reaction time) are established. For example, in the case of streptomycin (STR) assay in ionic association with Congo red (CR) [89], for the Britton-Robinson buffer (pH value 5.5), a maximum blue shift (from 497 to 487 nm) and hypochromic effect were obtained, indicating the quantitative formation of the STR-CR ion pair. In these experimental conditions, maximum scattering intensity was obtained (**Figure 3**).

Resonance light scattering methods have many advantages such as great sensitivity and selectivity, simple experimental procedure and the use of accessible equipment (classical spectrofluorimeter) [90], but no validated methods have been published yet. RLS signals suffer from fluctuations caused by many variable factors such as environmental conditions in the reaction medium (pH, ionic strength, temperature and polarity), reagent concentration and the incident light intensity [91].

The challenge for the analysts is to improve the technique in order to obtain reproducible results. Resonance light scattering ratiometry was proposed and applied to the study of the interaction between porphyrins and heparin, in order to solve the problems correlated with the single wavelength measurement. The method provides precise data by taking the intensity ratio at two suitable wavelengths [91].

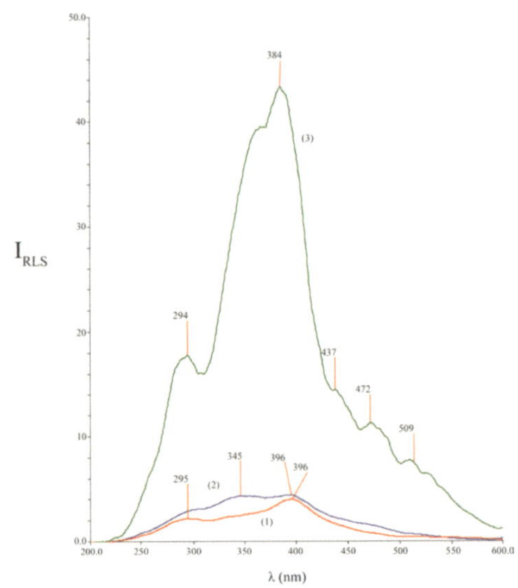

Figure 3. RLS spectra of CR (1), STR (2) and STR-CR ion pair (3) (from Ref. [89], with permission).

From our experience, slight variations of the ionic strength cause changes in the RLS signal intensity. In routine analysis, it is difficult to obtain identical values of this parameter, and therefore it is difficult to obtain reproducible results. A favourable effect on ion pairing may be obtained by adding small quantities of methanol or ethanol. They have strong water-structuring effect [92, 93], so the hydrophobic interactions for the ion pair can be enforced by engaging water and alcohol molecules in hydrogen bonds, thus dehydrating the substances of interest.

3.4. Challenges and perspectives in IP spectrophotometry

Among the permanent challenges in ion-pair spectrophotometry applied in the pharmaceutical field, one can number the increase of the sensitivity, enabling a more comprehensive study of the mechanisms underlying biochemical processes based on ion-pair equilibrium and finding appropriate conditions to obtain ion pairs for novel pharmacologically active substances.

In terms of increasing the sensitivity of the ion-pair–based methods, the best perspectives are offered by the RLS and fluorimetry, especially when the counterions fixed at the surface of QDs (capped QDs) are used. Using post-column ion pairing, RLS method has been incorporated as a detection technique in high-performance liquid chromatography [94] and capillary electrophoresis [95]. Studies are needed to obtain reproducible results of RLS and to validate the assay methods.

Ion pairing is a fundamental interaction in biological systems. Molecular recognition and protein function are biochemical processes based on ion pairing, and obtaining experimental evidence on the dynamics of macromolecules is a challenge. First experimental data on ion-pair dynamics at protein-DNA interfaces, obtained using nuclear magnetic resonance spectroscopy, were published by Anderson et al. [96].

Polyphenols, an important group of pharmacologically active substances, have not been characterized in terms of the ability to form pairs. Perspectives are opened by recently published study [97], which evaluates the photodynamic therapeutic effect of the curcumin on breast cancer cells using curcumin-methylene blue ion-pair–based nanoparticles. There are numerous substances in this class to be studied.

4. Conclusions

The present work underlined the existence of ion-pair spectrophotometry as a distinct group of methods largely used in the pharmaceutical field. Its evolution was dynamic and was correlated with the elucidation of ion-pair formation mechanisms and the development of computational chemistry. In medicines control, ion-pair molecular absorption spectrometry has the most numerous applications. Generally, organic solvents were used as reaction media. With the development of resonance light scattering techniques, the number of the applications of the ion pairs formed in aqueous solution has increased significantly. Fluorimetry, more sensitive, is also used as an assay method but mostly for biochemical purposes.

If one single feature has to be emphasized, the importance of ion-pair spectrophotometric methods in the pharmaceutical field consists in their versatility. Substances with or without characteristic absorption in UV-Vis range or intrinsic fluorescence, hydrophilic or hydrophobic and organic or inorganic, can be determined as ion pairs in bulk or complex matrices using rapid, sensitive and simple procedures.

Author details

Marinela Florea* and Mihaela Ilie*

*Address all correspondence to: florea.marinela@gmail.com; mihaela.ilie@umfcd.ro

Faculty of Pharmacy, Analytical Chemistry Department, "Carol Davila" University of Medicine and Pharmacy, Bucharest, Romania

References

[1] IUPAC. Compendium of Chemical Terminology. 2nd ed. (the "Gold Book"). Compiled by McNaught AD and Wilkinson A. Oxford: Blackwell Scientific Publications; 1997. XML on-line corrected version: http://goldbook.iupac.org; created by Nic M, Jirat J, Kosata B; updates compiled by Jenkins A. [Internet] 2006. DOI: 10.1351/goldbook. Available from https://goldbook.iupac.org/I03231.html [Accessed: 25 February 2017]

[2] Le Butto R, Kazakevich Y. Reversed phase HPLC. In: Kazakevich Y, Le Butto R, editors. HPLC for Pharmaceutical Scientist. 1st ed. Hoboken, NJ: John Wiley & Sons, Inc.; 2007. p. 200

[3] Meyer DJ, Manning MC. Hydrophobic ion pairing: Altering the solubility properties of biomolecules. Pharmaceutical Research. 1998;**15**(2):188-193. DOI: 10.1023/A:1011998014474

[4] Wang X, Jiang Y, Wang YW, Huang MT, Ho CT, Huang Q. Enhancing anti-inflammation activity of curcumin through O/W nanoemulsions. Food Chemistry. 2008;**108**:419-424. DOI: 10.1016/j.foodchem.2007.10.086

[5] Tang C, Zhang E, Li Y, Yang L. An innovative method for preparation of hydrophobic ion-pairing colistin entrapped poly(lactic acid) nanoparticles: Loading and release mechanism study. European Journal of Pharmaceutical Sciences. 2017;**102**:63-70. DOI: 10.1016/j.ejps.2017.02.036

[6] Jimenez-Kairuz AF, Allemandi DA, Manzo RH. The improvement of aqueous chemical stability of a model basic drug by ion pairing with acid groups of polyelectrolytes. International Journal of Pharmaceutics. 2004;**269**(1):149-156. DOI: 10.1016/j.ijpharm.2003.09.008

[7] Jimenez-Kairuz AF, Allemandi DA, Manzo RH. Equilibrium properties and mechanism of kinetic release of metoclopramide from carbomer hydrogels. International Journal of Pharmaceutics. 2003;**250**:129-136. DOI: 10.1016/S0378-5173(02)00525-2

[8] Sancilio FD, Stowell GW, Whittall L, White D, Whittle R. Preparation of narcotic-NSAID ion pairs. U.S. Patent No. US20050203115 A1. [Internet]; 2005. Available from http://google.co.id/patents/US20050203115 [Accessed: 25 February 2017]

[9] Jonkman JH, Hunt CA. Ion pair absorption of ionized drugs—Fact or fiction? International Journal of Clinical Pharmacy. 1983;**5**(2):41-48. DOI: 10.1007/BF01960074

[10] Song IS, Choi MK, Shim WS, Shim CK. Transport of organic cationic drugs: Effect of ion-pair formation with bile salts on the biliary excretion and pharmacokinetics. Pharmacology and Therapeutics. 2013;**138**(1):142-154. DOI: 10.1016/j.pharmthera.2013.01.006

[11] Kong L, Liu Z, Liu S, Wang D. Interaction of vancomycin with DNA and determination of DNA via resonance Rayleigh scattering and resonance nonlinear scattering. Analytical Methods. 2012;**4**(12):4346-4352. DOI: 10.1039/C2AY26050F

[12] Wang YT, Zaho FL, Li KA, Tong SY. Microdetermination of proteins by enhanced Rayleigh scattering spectroscopy with Morin. Fresenius Journal of Analytical Chemistry. 1999;**364**:560-564. DOI: 10.1007/s002160051386

[13] Kim SK, Sessler JL. Ion pair receptors. Chemical Society Reviews. 2010;**39**(10): 3784-3809. DOI: 10.1039/C002694H.

[14] Florea M, Monciu CM, Aramă CC. Pharmaceutical applications of ionic associations. Note I. Farmacia. 2007;**55**(6):605-612

[15] Constantinescu IC, Florea M, Aramă CC, Monciu CM. Assay of nimesulide by ion association titration. Farmacia. 2009;**57**(3):267-271

[16] Abdulrahman SAM, Basavaiah K. Application of ion-association titration for the assay of bupropion hydrochloride in pharmaceuticals. Chemical Industry and Chemical Engineering Quarterly. 2011;**17**(3): 299-306

[17] Nedelcu A. New methods for cinnarizine assay using the reaction with picrolonic acid [Rom.]. Farmacia. 1999;**47**(4): 31-39

[18] Nedelcu A. New methods for cinnarizine assay using the reaction with Reinecke salt [Rom.]. Farmacia. 2001;**49**(2):37-43

[19] Sun XX, Aboul-Enin HY. Internal solid contact sensor for the determination of doxycycline hydrochloride in pharmaceutical formulation. Talanta. 2002;**58**:387-396. DOI: 10.1016/S0039-9140(02)00292-8

[20] Salem AA, Barsoum BN, Izake EL. Potentiometric determination of diazepam, bromazepam and clonazepam using solid contact ion-selective electrodes. Analytica Chimica Acta. 2003;**498**:79-91. DOI: 10.1016/j.aca.2003.08.070

[21] Vinay KB, Revannasiddappa HD, Rajendraprasad N, Ramesh PJ, Xavier CM, Basavaiah K. Use of two sulfonthalein dyes in the extraction-free spectrophotometric assay of tramadol in dosage forms and in spiked human urine based on ion-pair reaction. Drug Testing and Analysis. 2012;**4**(2):116-122. DOI: 10.1002/dta.247

[22] Abdulrahman SAM, Basavaiah K. Spectrophotometric determination of dothiepin hydrochloride in pharmaceuticals through ion-pair complexation reaction. Chemical Industry and Chemical Engineering Quarterly. 2012;**18**(2):339-347

[23] Florea M, Aramă CC, Monciu CM, Determination of terbinafine by ion-pair reversed phase liquid chromatography. Farmacia. 2009;**58**(1):82-88

[24] Florea M, Monciu CM, Ilie M. Determination of nimesulide by ion pair high-performance liquid chromatography using tetrabutylammonium as the counterion. Analytical Letters. 2015;**48**(2):328-339. DOI: 10.1080/00032719.2014.940529

[25] Cecchi T. Ion-Pair Chromatography and Related Techniques. Boca Raton: Taylor&Francis Group; 2010. p. 21

[26] Quintanar-Guerrero D, Allemann E, Fessi H, Doelker E. Applications of the ion-pair concept to hydrophilic substances with special emphasis on peptides. Pharmaceutical Research. 1997;**14**(2):119-127

[27] Winstein S, Clippinger E, Fainberg AH, Robinson GC. Salt effects and ion-pairs in solvolysis. Journal of American Chemical Society. 1954;**76**(9):2597-2598.

[28] Ishchenko AA, Shapovalov SA. Heterogeneous association of the ions of dyes in solutions. Journal of Applied Spectroscopy. 2004;**71**(5):605-629. DOI: 10.1023/B:JAPS.0000049618.42857.0a

[29] Higuchi T, Michaelis A, Tan T, Hurwitz A. Ion pair extraction of pharmaceutical amines. Role of dipolar solvating agents in extraction of dextromethorphan. Analytical Chemistry. 1967;**39**(8): 974-979. DOI: 10.1021/ac60252a051

[30] Diamond RM. The aqueous solution behavior of large univalent ions. A new type of ion-pairing. Journal of Physical Chemistry. 1963;**67**(12):2513-2517. DOI: 10.1021/j100806a002

[31] Newcomb LF, Gellman SH. Aromatic stacking interactions in aqueous solution: Evidence that neither classical hydrophobic effects nor dispersion forces are important. Journal of American Chemical Society. 1994;**116**(11):4993-4994. DOI: 10.1021/ja00090a057

[32] Ito F, Nagamura T. Photochemical and photophysical properties of ion-pair charge transfer complexes for all-optical information processing. Journal of Photochemistry and Photobiology C: Photochemistry Reviews. 2007;**8**:174-190. DOI: 10.1016/j.jphotochemrev.2007.12.002

[33] Cai Y, Feng YP. Review on charge transfer and chemical activity of TiO_2: Mechanism and applications. Progress in Surface Science. 2016;**91**:183-202. DOI: 10.1016/j.progsurf.2016.11.001

[34] Dena ASA, Hassan WMI. Experimental and quantum mechanical studies on the ion-pair of levocetirizine and bromocresol green in aqueous solutions. Spectrochimica Acta Part A: Molecular and Biomolecular Spectroscopy. 2016;**163**:108-114. DOI: 10.1016/j.saa.2016.03.030

[35] Bračko S, Špan J. Conductometric investigation of dye–surfactant ion pair formation in aqueous solution. Dyes and Pigments. 2000;**45**(2):97-102. DOI: 10.1016/S0143-7208(00)00016-4

[36] Daignault LG, Jackman DC, Rillema DP. Behavior of simple salts on silica and C18 columns: Retention dynamics of cations, anions and ion pairs. Journal of Chromatography A. 1989;**462**:71-84. DOI: 10.1016/S0021-9673(00)91337-2

[37] Taga A, Uegaki K, Yabusako Y, Kitano A, Honda S. Simultaneous determination of the association constants of oligosaccharides to a lectin by capillary electrophoresis. Journal of Chromatography A. 1999;**837**(1-2):221-229. DOI: 10.1016/S0021-9673(99)00104-1

[38] Constantinescu CI. Contributions to the study of clemastine complex with the sodium tetraphenylborate. Note I [Rom.]. Farmacia. 2002;**L**(4):46-53

[39] Ragab GH, Amin AS. Atomic absorption spectroscopic, conductometric and colorimetric methods for determination of fluoroquinolone antibiotics using ammonium reineckate ion-pair complex formation. Spectrochimica Acta A: Molecular and Biomolecular Spectroscopy. 2004;**60**(4):973-978. DOI: 10.1016/S1386-1425(03)00327-5

[40] El-Brashy A.M., El-Sayed MM, El-Sepai FA. Spectrophotometric and atomic absorption spectroscopic determination of some fluoroquinolone antibacterials by ion-pair complex formation with bismuth (III) tetraiodide. Journal of the Chinese Chemical Society. 2005;**52**(2):253-262. DOI: 10.1002/jccs.200500038

[41] Tôei K. Ion-association reagents. A review. Analytical Sciences. 1987;**3**:479-488

[42] Florea M, Mociu CM. Spectrophotometric determination of terbinafine through ion-pair complex formation with methyl orange. Farmacia. 2008;**56**(4):393-401

[43] Qarah NAS, Basavaiah K, Swamy N. Ion-pair extractive spectrophotometric assay of terbinafine hydrochloride in pharmaceuticals and spiked urine using bromocresol purple. Journal of Applied Spectroscopy. 2016;**83**(4):694-702. DOI: 10.1007/s10812-016-0350-x

[44] Qarah NAS, Kanakapura B, Nagaraju S, Udigere C. Assay of terbinafine hydrochloride by extractive-spectrophotometry with alizarin red S.A modified approach. Eurasian Journal. 2015;**10**(1):34-45

[45] Florea M, Monciu CM, Andritoiu ML, Bacanu LG. Spectrophotometric determination of nimesulide through ion-pair complex formation with hexadecyltrimethylammonium bromide. Farmacia. 2008;**56**(6):639-646

[46] El-Henawee MM, Ragab GH, Amin El-Sayed A, Sultan AF. Spectrophotometric determination of nimesulide in pure and in pharmaceutical formulations using ion-associate complex formation. Asian Journal of Pharmaceutical Analysis and Medicinal Chemistry. 2014;**2**(4):240-248

[47] Abd El-Hay SS, El-Mammli MY, Shalaby AA. Determination of clemastine hydrogen fumarate, desloratadine, losartan potassium and moxepril HCl through binary complex formation with eosin. Arabian Journal of Chemistry. 2016;**9**:S541-S547. DOI: 10.1016/j.arabjc.2011.06.021

[48] Rele RV, Gurav PJ. A simple extractive spectrophotometric determination of loratadine, desloratadine and rupatadine from pharmaceutical formulations. International Journal of Pharma and Bio Sciences. 2012;**3**(2):89-95

[49] Prabhakar AH, Giridhar R. A rapid method for the determination of Losartan in bulk and in synthetic mixture for solid dosage form. Journal of Pharmaceutical and Biomedical Analysis. 2002;**27**:861-866. DOI: 10.1016/S0731-7085(01)00536-2

[50] El-Sayed AM, Metwally MES, El-Sepai AF. Spectrophotometric determination of some fluoroquinolone antibacterials through charge-transfer and ion-pair complexation reactions. Bulletin of the Korean Chemical Society. 2004:**25**(3):365-372

[51] Ashour S, Al-Khalil R. Simple extractive colorimetric determination of levofloxacin by acid-dye complexation methods in pharmaceutical preparations. Il Farmaco. 2005;**60**:771-775. DOI: 10.1016/j.farmac.2005.06.007

[52] Amin AS. Pyrocatechol violet in pharmaceutical analysis. Part I. A spectrophotometric method for the determination of some β-lactam antibiotics in pure and in pharmaceutical dosage forms. Il Farmaco. 2001;**56**(3):211-218. DOI: 10.1016/S0014-827X(01)01078-3

[53] Mohamed GG. Spectrophotometric determination of ampicillin, dicluxacillin, flucloxacillin and amoxicillin antibiotic drugs: Ion-pair formation with molybdenum and thiocyanate. Journal of Pharmaceutical and Biomedical Analysis. 2001;**24**(4):561-567. DOI: 10.1016/S0731-7085(00)00463-5

[54] Keskar MR, Jugade RM. A new spectrophotometric method for determination of amoxicillin using bromocresol green. World Journal of Pharmacy and Pharmaceutical Sciences. 2014;**3**(2):1340-1348

[55] Prasad BB, Gupta S. Extraction-spectrophotometric determination of certain β-lactam antibiotics with methylene blue. Indian Journal of Pharmaceutical Sciences. 2000;**62**(4): 261-266

[56] Lakowicz JR. Principles of Fluorescence Spectroscopy. 3rd ed. New York: Springer; 2006. pp. 63-93

[57] Moffat AC, Osselton MD, Widdop B, editors. Clarke's Analysis of Drugs and Poisons in Pharmaceuticals, Body Fluids and Postmortem Material. Vol.1, 4th ed. London: The Pharmaceutical Press; 2011. pp. 510-511

[58] Sanz A, Tomas V, Martinez-Lozano C, Perez-Ruiz T. Extraction-spectrofluorimetric method for the determination of erythromycin and its esters in pharmaceutical formulations using manual and flow-injection procedures. Analyst. 1993;**118**(5):567-571. DOI: 10.1039/AN9931800567

[59] Acedo-Valenzuela MA, Galeano-Diay T, Mora-Diez N, Silva-Rodriguez A. Response surface methodology for the optimisation of flow injection analysis with in situ solvent extraction and fluorimetric assay of tricyclic antidepressants. Talanta. 2005;**66**:952-960. DOI: 10.1016/j.talanta.2004.12.044

[60] Kelani K, Bebawy LI, Abdel-Fattah L. Determination of astemizole, terfenadine, and flunarizine hydrochloride by ternary complex formation with eosin and lead (II). Journal of Pharmaceutical and Biomedical Analysis. 1999;**18**:985-992. DOI: 10.1016/S0731-7085(98)00107-1

[61] Kaur K, Malik AK. Study on the fluorescence quenching reaction of amitriptyline and clomipramine hydrochlorides with eosin y and its analytical application. Journal of Fluorescence. 2013;**23**(3):533-542. DOI: 10.1007/s10895-013-1185-y

[62] Zonghui Q, Bing X, Rong T, Xiangqing G, Bangjiang W. Study on the interaction of rosiglitazone maleate and eosin y bymolecular spectroscopy and its analytical applications. Chemistry Bulletin/Huaxue Tongbao. 2010;**73**(12):1131-1138

[63] Qin Z, Tan R, Xie B, Zhao X, Wan B. Study on the interaction between pioglitazone hydrochloride and eosin Y by fluorescence and resonance Rayleigh scattering method. Chemistry Bulletin/Huaxue Tongbao. 2010;**73**(2):166-172

[64] Fengling T, Wei H, Yang J, Qin L. Study on the interaction between albendazole and eosin Y by fluorescence, resonance Rayleigh scattering and frequency doubling scattering spectra and their analytical applications. Spectrochimica Acta Part A: Molecular and Biomolecular Spectroscopy. 2014;**126**:135-141. DOI: 10.1016/j.saa.2014.01.126

[65] El Walily AFM, Belal SF, Bakry RS. Spectrophotometric and spectrofluorimetric estimation of ciprofloxacin and norfloxacin by ternary complex formation with eosin and palladium(II). Journal of Pharmaceutical and Biomedical Analysis. 1996;**14**(5): 561-569. DOI: 10.1016/0731-7085(95)01662-7

[66] Hassan EM. Spectrophotometric and fluorimetric methods for the determination of meloxicam in dosage forms. Journal of Pharmaceutical and Biomedical Analysis. 2002;**27**(5):771-777. DOI: 10.1016/S0731-7085(01)00530-1

[67] Fu S, Liu Z, Liu S, Liu J, Yi A. Study on the resonance Rayleigh scattering spectra of the interactions of palladium (II)-cephalosporins chelates with 4,5-dibromofluorescein and their analytical applications. Analytica Chimica Acta. 2007;**599**(2):271-278. DOI: 10.1016/j.aca.2007.07.071

[68] Megyesi M, Biczók L. Effect of ion pairing on the fluorescence of berberine, a natural isoquinoline alkaloid. Chemical Physics Letters. 2007;**447**:247-251. DOI: 10.1016/j.cplett.2007.09.046

[69] Soulié M, Frongia C, Lobjois V, Fery-Forgues S. Fluorescent organic ion pairs on berberine: Counter-ion effect on the formation of particles and on the uptake by colon cancer cells. RSC Advances. 2015;**5**:1181-1190. DOI: 10.1039/C4RA09993A

[70] Li Q, Tan X, Fu L, Liu Q, Tang W. A novel fluorescence and resonance Rayleigh scattering probe based on quantum dots for the detection of albendazole. Analytical Methods. 2015;**7**:614-620. DOI: 10.1039/C4AY02289K

[71] Nalefski EA, Shaw KTY, Raos A. An ion pair in class II major histocompatibility complex heterodimers critical for surface expression and peptide presentation. The Journal of Biological Chemistry. 1995;**270**(38):22351-22360. DOI: 10.1074/jbc.270.38.22351

[72] Nagai H, Miwa N, Segawa M, Wakida S, Chayama K. Quantification of Ag(I) and kinetic analysis using ion-pair extraction across a liquid/liquid interface in a laminar flow by fluorescence microscopy. Journal of Applied Physics. 2009;**105**:102015-1-102015-7. DOI: 10.1063/1.3116092

[73] Farhadi K, Savojbolaghi AK, Farajzadeh M, Maleki R. Development of turbidimetric methods for the determination of some N-substituted phenothiazine derivatives using sodium dodecyl sulphate and mercury (II) chloride. Analytical Letters. 2003;**36**(10):2183-2198. DOI: 10.1002/jccs.200300021

[74] Basavaiah K, Charan VS. Spectrophotometric and turbidimetric determination of methdilazine using bromophenol blue. Indian Pharmacist. 2003;**2**(18):96-100

[75] Basavaiah K, Charan V. Ion-pair complexometric determination of cyproheptadine hydrochloride using bromophenol blue. Science Asia. 2004;**30**(2):163-170

[76] Bauer DR, Hudson B, Pecora R. Resonance enhanced depolarized Rayleigh scattering from diphenylpolyenes. Journal of Chemical Physics. 1975;**63**:588-589

[77] Stanton SG, Pecora R, Hudson BS. Resonance enhanced dynamic Rayleigh scattering. Journal of Chemical Physics. 1981;**75**:5615-5626

[78] Pasternack RF, Bustamante C, Collings PJ, Giannetto A, Gibbs EJ. Porphyrin assemblies on DNA as studied by a resonance light-scattering technique. Journal of American Chemical Society. 1993;**115**:5393-5399. DOI: 10.1021/ja00066a006

[79] Li J, Yang X, Yang J, Lai L. Resonance Rayleigh scattering and nonlinear scattering methods for the determination of nicardipine hydrochloride using eosin Y as a probe. RSC Advances. 2016;**6**(31):25887-25893. DOI: 10.1039/C5RA25851K

[80] Yang J, Wang E, Zhou S, Yang Q. Effects of (R)- and (S)-propranolol hydrochloride enantiomers on the resonance Rayleigh scattering spectra with erythrosine B as probe and their analytical applications. Talanta. 2015;**134**:757-760. DOI: 10.1016/j.talanta.2014.12.030

[81] Han W, Qu F, Shang J, Lu W, Yang J, Ma Q. Determination of Amikacin in different pharmaceutical formulations using a resonance Rayleigh scattering method with pontamine sky blue. Current Pharmaceutical Analysis. 2014;**10**(2):105-111. DOI: 10.2174/1573412910999140214094944

[82] Li Q, Tan X, Li J, Pan L, Liu X. Glutathione-capped CdTe nanocrystals as probe for the determination of fenbendazole. Spectrochimica Acta-Part A: Molecular and Biomolecular Spectroscopy. 2015;**141**:10-15. DOI: 10.1016/j.saa.2015.01.026

[83] Tan X, Li Q, Shen Y, Wu H, Zaho Y, Yang J. Chiral recognition of tyrosine enantiomers based on decreased resonance scattering signals with silver nanoparticles as optical sensor. Chirality. 2015;**27**(3):194-198. DOI: 10.1002/chir.22410

[84] Quiao M, Li C, Shi Y, Liu S, Liu Z, Hu X. Study on interactions of aminoglycoside antibiotics with calf thymus DNA and determination of calf thymus DNA via the resonance Rayleigh scattering technique. Luminescence. 2015;**30**(7):1159-1166. DOI: 10.1002/bio.2876

[85] Xiao X, Huang M, Wang Y, Gan L, Zhu G, Xue J. Spectroscopic study on the interaction of pyronine y with nucleic acids and its analytical application. Spectroscopy Letters. 2012;**45**(8):569-574. DOI: 10.1080/00387010.2011.652335

[86] Cui ZP, Liu SP, Liu ZF, Zheng HZ, Hu XL, Xue JX, Tian J. Interaction of proteins with aluminum(III)-chlorophosphonazo III by resonance Rayleigh scattering method. Luminescence. 2014;**29**(7):728-737. DOI: 10.1002/bio.2614

[87] Chen Z, Wang Z, Chen J, Chen X, Wu J, Wu Y, Liang J. Resonance light scattering technique as a new tool to determine the binding mode of anticancer drug oridonin to DNA. European Journal of Medicinal Chemistry. 2013;**66**:380-387. DOI: 10.1016/j.ejmech.2013.05.042

[88] Mohseni-Shahri FS, Housaindokht MR, Bozorgmehr MR, Moosavi-Movahedi AA. The influence of the flavonoid quercetin on the interaction of propranolol with human serum albumin: Experimental and theoretical approaches. Journal of Luminescence. 2014;**154**:229-240. DOI: 10.1016/j.jlumin.2014.04.033

[89] Florea M, Monciu CM, Ilie M. Resonance Rayleigh scattering study of streptomycin-congo red ionic association in view of analytical application. Farmacia. 2014;**62**(2):318-328

[90] Lu W, Fernández Band BS, Yu Y, Li QG, Shang JC, Wang C, Fang Y, Tian R, Zhou LP, Sun LL, Tang Y, Jing SH, Huang W, Zhang JP. Resonance light scattering and derived techniques in analytical chemistry: Past, present and future. Microchimica Acta. 2007;**158**:29-58. DOI: 10.1007/s00604-006-0670-2

[91] Huang CZ, Pang XB, Li YF, Long YJ. A resonance light scattering ratiometry applied for binding study of organic small molecules with biopolymer. Talanta. 2006;**69**:180-186. DOI: 10.1016/j.talanta.2005.09.022

[92] Laaksonen A. Three-dimensional structure in water-methanol mixtures. The Journal of Physical Chemistry A. 1997;**101**:591-591. DOI: 10.1021/jp970673c

[93] Mejía SM, Mills MJL, Shaik MS, Mondragon F, Popelier PLA. The dynamic behavior of a liquid ethanol-water mixture: A perspective from quantum chemical topology. Physical Chemistry Chemical Physics. 2011;**13**:7821-7833. DOI: 10.1039/c0cp02869j

[94] Liao Q, Li WH, Luo LG. Determination of heparin in plasma by HPLC coupled with resonance light scattering detection. Chromatographia. 2013;**76**:1677-1682. DOI: 10.1007/s10337-013-2526-3

[95] Qi L, Han ZQ, Chen Y. Incorporation of flow injection analysis or capillary electrophoresis with resonance Rayleigh scattering detection for inorganic ion analysis. Journal of Chromatography A. 2006;**1110**(1-2):235-239. DOI: 10.1016/j.chroma.2006.01.053

[96] Anderson KM, Esadze A, Manoharan M, Brüschweiler R, Gorenstein DG, Iwahara J. Direct observation of the ion-pair dynamics at a protein-DNA interface by NMR spectroscopy. Journal of American Chemists Society. 2013;**135**:3613-3619. DOI: 10.1021/ja312314b

[97] Hosseinzadeh R, Khorsandi K. Methylene blue, curcumin and ion pairing nanoparticles effects on photodynamic therapy of MDA-MB-231 breast cancer cell. Photodiagnosis and Photodynamic Therapy. 2017;**18**:284-294. DOI: 10.1016/j.pdpdt.2017.03.005

Application of Flow-Injection Spectrophotometry to Pharmaceutical and Biomedical Analyses

Bruno E.S. Costa, Henrique P. Rezende,
Liliam Q. Tavares, Luciana M. Coelho,
Nívia M.M. Coelho, Priscila A.R. Sousa and
Thais S. Néri

Additional information is available at the end of the chapter

http://dx.doi.org/10.5772/intechopen.70160

Abstract

The discovery of new drugs, especially when many samples have to be analyzed in the minimum of time, demand the improvement or development of new analytical methods. Various techniques may be employed for this purpose. In this context, this chapter gathers the collection of paper and represents the review of past work on spectrophotometric technique coupled to a continuous flow system to determine low concentrations of several chemical species in different kinds of pharmaceutical and biological samples. A short historical background of the flow-injection analysis technique and a brief discussion of the basic principles and potential are presented. Part of this chapter is devoted to describing the sample preparation techniques, principles, and figures of merit of analytical methods. Representative applications of flow-injection spectrophotometry to pharmaceutical and biomedical analysis are also described.

Keywords: pharmaceutical, biomedical samples, flow-injection, spectrophotometry

1. Introduction

The monitoring of chemical species in pharmaceutical and biomedical samples is a field in which analytical chemistry plays an important role, contributing new procedures of analysis and instrumentation. Many methods have been developed for pharmaceutical and biomedical analysis including chromatographic, electrophoretic, and spectrophotometric methods. However, there are inherent difficulties associated with the types of samples involved. The most practical difficulty encountered is the preservation and integrity of the species during sampling, storage, and sample pretreatment. In the medical area, the main matrices are blood, serum, and urine, while in the pharmaceutical industry, there are many types of samples and variations in their compositions. This hinders the application of analytical techniques for the fast and accurate monitoring of pharmaceutical and biomedical species in real samples.

Spectrophotometry is the technique most commonly employed in chemical analysis, and it provides advantages in terms of the availability of instruments, simplicity of procedures, speed, precision, accuracy, and applicability to a wide range of bio-medically important substances. Due to recent advances, increasing attention is being given to the coupling of a spectrometer to a continuous flow system to determine low concentrations of several chemical species in different kinds of pharmaceutical and biological samples. The flow-injection analysis technique has found wide application, which can be mainly attributed to its versatility, ease of automation, high sampling frequency, and the requirement for minimum sample treatment prior to injection into the system.

This chapter draws attention to some of the important and unique aspects of the applications of flow-injection spectrophotometry, addressed within the context of pharmaceutical and biomedical analysis. A short historical background of the flow-injection analysis technique and a brief discussion of the basic principles and potential are presented.

A notable feature of this chapter is the large number of papers on chemiluminescence discussed herein. In addition, considerable attention is given to sample preparation techniques and the characteristics of analytical methods such as precision, accuracy, and sampling frequency. Representative applications of flow-injection spectrophotometry to pharmaceutical and biomedical analysis are also described.

2. FIA origin and development

The process of flow-injection analysis (FIA) was initially proposed in the 1970s by Prof. Dr. Jaromir Ruzicka of the Technical University of Denmark and was subsequently consolidated as a state-of-the-art technology for the automation and mechanization of chemical systems. At that time, the cited researcher spent a year in Brazil advising on the installation of the Laboratory of Analytical Chemistry at the Center of Nuclear Energy in Agriculture of the University of São Paulo (CENA/USP), where pioneering work was carried out that led to the FIA process becoming very well established. Since its introduction, more than 20,000 papers have been published reporting the development of advanced instrumentation methods in the context of chemical analysis, which are available for environmental, food, and clinical services involving pharmaceutical and biomedical samples [1, 2].

The FIA process involves the insertion of the sample into a carrier fluid that transports it to a suitable detection system. During this process, the sample can be brought into contact with reagents that are also inserted by propulsion, resulting in a controlled dispersion of the sample. The processes that characterize FIA systems have gained great prominence in contemporary analytical chemistry since several limitations have been overcome in the development of improved analytical procedures.

The origin and development of flow analysis systems was strongly influenced by the work of Skeggs, who proposed an approach called continuous flow analysis (CFA) [3]. For approximately 20 years, it was accepted that segmentation by dividing the flow into small regular compartments separated by air bubbles was the best strategy to avoid contamination and the widening of the discrete zone of the sample along the course, which is known today as dispersion [3–5]. It was only in the mid-1970s that it was widely accepted that segmentation could be omitted following an innovative proposal for a method employing the continuous flow of the samples and reagents with adequate dimensions and flow rates. The system was subsequently simplified, increasing the frequency of samples analyzed per unit of time, referred to as the analytical frequency. Due to the advantage of good reproducibility offered by FIA systems, it has also become possible to quantify the analyte even before the reaction between the sample and the reagent reaches equilibrium since the interval between the injection and detection is the same for the standard solutions and the sample.

In general, the FIA process consists of fluid propulsion, usually performed by a peristaltic pump that operates at constant flow, sample injection, reaction promoted in a homogenizing mixing coil with suitable geometry and a compatible detection technique, such as molecular spectrometry and atomic, chromatographic, and electroanalytical techniques.

Initially, Ruzicka used a hypodermic syringe to promote the injection of the sample, which gave rise to the name of the process [6]. Since then, other devices have been proposed for the insertion of the sample into the loader fluid, such as the proportional commutator injector and the rotary valve. The FIA systems have thus evolved and independent injections by multicomutation can be performed, enabling binary sampling [7]. In recent years, FIA systems have evolved in ways that have led to the development of sequential injection analysis (SIA) systems. In this case, the injection of the sample and the contact with the reagent flow occur through the selection of the port of a central selector valve in which the mixture is provided with bi-directional movement, alternating the propulsion direction occurring in a single line, in the absence of confluences [2, 8]. Due to these characteristics, SIA systems can be considered as differentiated from conventional FIA systems.

The classification of FIA systems has become necessary considering the variety of analytical procedures available. This can be based on the way in which the sample is introduced (continuous or intermittent) and on the flow characteristic (segmented, monosegmented, nonsegmented). **Figure 1** shows a classification scheme for flow analysis methods.

FIA systems have become commonly used and a number of variations in the configurations have been proposed in order to minimize the consumption of sample and reagents and to enhance the sensitivity of detection and the selectivity and precision of the analytical measurements.

The simplest configuration is the single-line flow diagram, where the loading fluid is the reagent itself, and the mixing occurs exclusively by dispersion. When the ratio between the volumes of the injected sample aliquot and its pathway is inappropriate, the addition of reagents by confluence may provide a more effective reaction where inert solutions, such as carriers, are employed. In order to overcome the excessive waste of reagents, which are continuously consumed, the system of flow injection through coalescing zones was proposed,

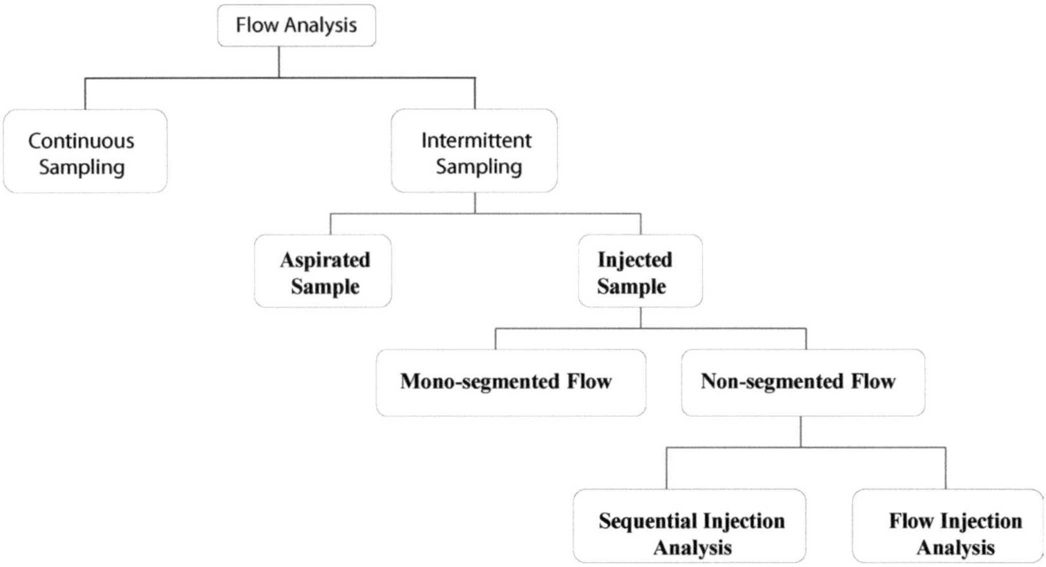

Figure 1. Schematic diagram of classification of flow analysis procedures.

thus minimizing the amount of waste generated. Since then, FIA systems have proven to be highly versatile and robust, making it possible to obtain strategically various arrangements and configurations that have been satisfactorily employed for extractive, separation, and pre-concentration purposes. In addition, there are a number of approaches through which clinical formulation products can be efficiently monitored for the certification of their quality.

3. Spectrophotometric flow-injection procedures for pharmaceutical samples

Spectrophotometric methods are the most commonly used techniques in chemical analysis due to the availability of instruments, simplicity of procedures, precision, and wide applicability. Based on the laws governing absorption and emission phenomena, it is possible to determine the concentrations of compounds in solutions, notably those of biological, chemical, or pharmaceutical interest [9].

Drug analysis, involving the pharmaceutical preparations or the raw materials used for their production, and the determination of drugs together with metabolites in biological samples (serum, plasma, saliva, urine, and some secretions) constitutes a large part of the activities carried out by pharmaceutical and clinical laboratories.

Spectrophotometry is the most commonly used technique for the determination of drugs, and it is based on chromogenic reactions or light absorption by the analyte. Chromogenic reactions for drugs include metal-ion complexes, redox reactions, and the formation of charge-transfer complexes.

Some of the flow-injection spectrophotometry procedures for the quantification of pharmaceutical samples [10–14] are detailed in **Table 1**. An important observation is the choice of carrier, aiming to avoid matrix effects and even clogging of the flow channels due to precipitation [14, 15].

Flow techniques, characterized by great flexibility, versatility, and ease of automation, allow the development and implementation of many analytical systems, which are compatible with a wide range of sample manipulation techniques, under highly reproducible conditions. Some flow-injection procedures are based on oxidation-reduction reactions. For the determination of N-acetyl-L-cysteine [16], this procedure involves the oxidation of the analyte of interest by Fe(III). The Fe(II) produced can be determined using 1,10-phenanthroline, and the chromophore formed is analyzed at 510 nm. On-line oxidation by Ce(IV) in acid medium—a procedure based on oxy-reduction—has been used to determine pyrazine. The colored free radical produced by the reaction was monitored at 510 nm [17].

Other procedures are based on the formation of a colored complex between the analyte of interest and metal-ions. For example, the determination of cimetidine with Cu(II) in acetate buffer (pH 5.9) can be carried out at a wavelength of 330 nm [18] and epinephrine with Fe(II) in amino acetic carbonate buffer (pH 8.3) at a wavelength of 530 nm [19].

A procedure for the determination of paracetamol (4-acetaminophen) has been described by Fatibello-Filho and Vieira [20]. The method is based on paracetamol oxidation by sodium hypochlorite, and the determination of the oxidant using o-toluidine dichloride as the chromogenic reagent at 430 nm. The analytical curve for paracetamol was linear in the concentration range of 8.50×10^{-6} to 2.51×10^{-4} mol L^{-1} with a detection limit of 5.0×10^{-6} mol L^{-1}. The relative standard deviation was less than 1.2% for a paracetamol solution of 1.20×10^{-4} mol L^{-1} ($n = 10$).

Analyte	Methodology	Detection limit	Ref.
Ketoprofen	Distilled water as a carrier for gels and citrate buffer, pH 6.5, for ampoules at 261 nm.	0.436 and 0.303 g ml^{-1} for gels and ampoules, respectively.	[10]
Lansoprazole	0.01 mol L^{-1} NaOH as carrier at flow rate of 1 ml min^{-1} and wavelength of 292 nm.	5.8×10^{-7} mol L^{-1}	[11]
Diazepan	0.1 mol L^{-1} HCl as carrier at flow rate of 6.8 ml min^{-1} and wavelength of 360 nm.	0.6 mg L^{-1}	[12]
Gemfibrozil	0.1 mol L^{-1} NaOH as carrier at flow rate of 1 ml min^{-1} and wavelength of 276 nm.	1.4 mg L^{-1}	[13]
Cefuroxime	Methanol:water (10:90 v/v) as carrier at flow rate of 1.0 ml min^{-1} and wavelength of 281 nm.	1.31×10^{-7} mol L^{-1}	[14]

Table 1. Flow-injection spectrophotometry procedures for the quantification of pharmaceutical samples.

Many methods for the determination of pharmaceuticals also involve the flow-injection procedure based on homogeneous reactions. For example, the indirect determination of olanzapine from the reaction with hexacyanoferrate (III) in which the measurement of the unreacted

Analyte	Methodology	Detection limit ($mg\ L^{-1}$)	Ref.
Levofloxacin	Oxidation with N-bromosuccinimide.	3.0	[23]
Benzylpenicillin	Derivatization with 4,6-dinitrobenzenofuroxane.	0.14	[24]
Isoproterenol	Oxidation by polyphenol oxidase immobilized on controlled-pore silica.	13.2	[25]
Paracetamol	Reaction with sodium hypochlorite followed by reaction with sodium salicylate.	0.4	[26]
Dipyrone	Reaction with ammonium molybdate to produce molybdenum blue.	32	[27]
Vitamin B complex (B1, B6, B12, and benfotiamine)	Multicomponent spectrophotometric analysis using Multivariate Curve Resolution Alternating Least Squares (MCR-ALS) algorithm.	0.0009–0.016	[28]
Promethazine and trifluoperazine	Bead injection spectroscopy-flow injection analysis (BIS-FIA) system and spectrophotometric detection.	0.00009–0.00014	[29]
Flutamide	Detection by electrospray ionization mass spectrometry.	0.001	[30]
Lansoprazole	Detection by electro spray ionization mass spectrometry.	0.0055	[31]

Table 2. FIA procedures using based on color-forming reactions.

oxidant is collected at 425 nm [21]. Diclofenac and mefenamic acid can also be oxidized in a flow system by hexacyanoferrate (III) and spectrophotometrically determined [22]. Other analytes that can be determined in homogeneous reactions are shown in **Table 2**.

4. Spectrophotometric flow-injection procedures for biomedical samples

The FIA technique can be coupled to various detection systems, such as a spectrophotometer, which allows a wide range of analytical devices to be combined [32].

A flow-injection analysis procedure using spectrophotometry was used to determine urea in blood plasma, employing the legume *Cajanus cajan* as a source of urease enzyme, in a mini-column coupled to the FIA collector. A confidence level of 90% and a relative standard deviation of 1.4% ($n = 12$) were obtained [33]. In another study, the development of a flow analysis procedure for the determination of total protein in a bovine blood plasma was carried out using the Biuret method. Samples of bovine plasma with 12.5 and 100.0 g L^{-1} of total protein were analyzed, and the analytical range was 2.5–20.0 g L^{-1}. The relative standard deviation of the procedure was 2.8%, and the analytical frequency was 76 determinations per hour. The results were compared with the traditional method of analysis (Biuret), and no statistically significant differences were observed at the 95% confidence level [34].

Sensors based on optical techniques are widely applied in different types of analysis, including biomedical sensing, and when connected to flow-injection analysis, a much faster analysis procedure is obtained [35]. In this context, a multicomutation flow system was used, which incorporates a sol-gel optical sensor (sensor: base catalyzed 4-(2-pyridylazo) resorcinol (PAR)) for the spectrophotometric determination of Cu(II) in urine samples using a photodiode detector with a maximum absorbance at 500 nm. The results were in agreement with those obtained by inductively coupled plasma mass spectrometry (ICP-MS), with a confidence level of 95% [35].

A FIA system was used to determine copper and zinc in water, pharmaceuticals, soils, and human hair samples. The product of the reaction with 2-carboxyl-2-hydroxy-5-sulfoformazylbenzene (Zincon) was introduced into a stream of carrier solution in the flow system. A sequential reaction of Cu(II) and Zn(II) was performed using Zincon, with the formation of two complexes and monitoring at 627 nm [36].

A flow procedure with spectrophotometric detection to determine bromopride in different matrices has been studied [37]. To increase the sensitivity of the reaction, a micellar medium (sodium dodecyl sulfate—SDS) was employed. Factorial planning was carried out to optimize the experimental parameters. The limit of detection was 1.07×10^{-7} mol L^{-1}. The method developed was satisfactorily applied in the determination of bromopride in pharmaceuticals and human urine, and recoveries were in the ranges 99.6–101.2 and 98.6–102.1%, respectively.

The application of a very sensitive and selective on-line flow-injection method for the determination of thorium(IV) after preconcentration in a minicolumn with N-benzoylphenylhydroxylamine-impregnated XAD-4 resin to biological samples has been described [38]. Sample rates of 40 and 11 h^{-1} were obtained at the 60 and 300 seconds preconcentration times, respectively; the preconcentration factors were 32 and 162, with detection limits of 0.76 and 0.150 μg L^{-1}, respectively [38].

Sarcosine has been investigated as a new marker for prostate cancer. A method for detecting sarcosine in biological samples (urine or blood plasma) has been proposed [39]. Ion exchange liquid chromatography with photometric detection at 570 nm was used as a separation method, which proved insufficient for the detection of sarcosine (70 μM). An off-line approach to the ninhydrin derivatization of the fractions collected was optimized, after which a known amount of ninhydrin was added followed by incubation of the mixture under the optimized temperature and time conditions. FIA system with electrochemical detection was used. In this case, 5 μL of sample was injected through a manual valve with a cell phase flow rate of 1 mL min^{-1} and spectrophotometric detection in the wavelength range of 450–800 nm. A detection limit of 1.7 μM was obtained for sarcosine [39].

5. On-line sample processing methods for in flow analysis

Flow analysis systems are widely used in analytical chemistry, contributing to increased reproducibility and accuracy of the methods. They also enable a reduction in the reagent consumption and the development of cleaner methods, meeting the requirements of "green" chemistry.

However, there are still limitations inherent to the procedures involved in preparing the samples, and these need to be suitable for each matrix.

Several procedures for analyzing pharmaceutical and biological samples have been developed where the sample preparation method performed on a laboratory bench is replaced by a flow procedure coupled directly to the instrument (spectrometer, chromatograph, electrophoresis unit, etc.). This increases the reliability of the method since it minimizes the potential for contamination inherent to the analysis, increases the reproducibility of the results, and increases the analytical frequency. These characteristics are due to the automation and processing of samples in closed systems under highly reproducible mixing and timing conditions. One of the factors that contributes to the success of this sample processing procedure carried out in closed systems is the combination of techniques and methodologies, known as hyphenation, which promotes faster analysis that is more efficient with less interference. The pretreatment of pharmaceutical and biological samples in flow is an important step in closed systems. Due to the complexity of these samples, the determination of chemical species presents significant challenges [40]. Thus, different types of procedures can be developed for the preparation of on-line samples for each matrix, according to its characteristics, such as solid phase extraction, solid phase microextraction, liquid-liquid microextraction, and chemical derivatization.

Some articles using on-line processes for the determination of compounds in pharmaceutical and biological samples are shown in **Table 3**. Two on-line procedures have been reported for the determination of ranitidine: chemiluminescence and UV-Vis detection [41, 42]. Several methodologies for the on-line preparation of saliva samples with detection by UV-Vis [43], ICP-OES [44], and AFS [45] are described in the literature.

Sample	Detection technique	Strategies for analysis	Ref.
Ranitidine and salbutamol	FIA-chemiluminescence	For $Ru(bipy)_3^{2+}$ chemiluminescence, a sulfuric acid carrier stream was employed into which $Ru(bipy)_3^{3+}$ and sulfuric acid was injected (20 μL), while a second stream delivered the analyte standard and sample solutions.	[41]
Ranitidine	UV-Vis	Injected samples were analyzed by spectrophotometry at 313 and at 615 nm after reaction with 3-methyl-2-benzothiazolinone and ferric chloride.	[42]
Sulfamethoxazole and trimethoprim	UV	Analytical procedure was solid phase extraction.	[46]
Saliva	UV-Vis	Analytical procedure involving extraction and preconcentration using 5-BrDMPAP.	[43]
Saliva	ICP-OES	The analytical procedure involved extraction by sorption and elution of the analytes.	[44]
Saliva	AFS	The flow system was equipped with a microwave and an ultraviolet photo-oxidation system.	[45]

Table 3. On-line treatment procedures for pharmaceutical and biological samples using spectrometric techniques.

The complexity of the matrices of the pharmaceutical and biomedical samples requires an efficient decomposition process, without losing the necessary characteristics for a precise quantification, maintaining the integrity of the analyte. The use of microwave radiation energy was found to be an efficient alternative to conventional sample preparation methods since the processing time is reduced, minimizing problems associated with the loss of the more volatile components. In spite of these advantages, the process requires the manual transfer of volumes, addition of reagents, and excessive dilutions, which are all potential sources of errors, for instance, contamination. The mechanization of the microwave sample preparation processes in a continuous stream decomposition system has contributed to improving the sample processing and, therefore, the analytical performance of the method.

The use of flow systems coupled to a microwave oven for the preparation of samples was first proposed by Burguera *et al.* [47], where urine samples were decomposed for further determination of lead. A volume of up to 100 μL of the sample was decomposed using a home microwave oven with a maximum power of 700 W and a 100 μL mixture of 0.4 M HNO_3 and 0.3 M HCl. The application of this system allowed an analytical frequency of 80 samples per hour. Since the first work of exploring the coupling of a microwave oven and a flow system, several systems have been developed and applied to a wide variety of samples, for instance, water, effluents, plants, food and minerals, along with biological fluids and tissues. The analysis of biological fluids is of great importance since it allows the diagnosis of various diseases, nutritional, and metabolic research, therapeutic monitoring involving the biological action of some metals, such as calcium, magnesium, iron, cobalt, zinc, and manganese, and the detection of some drugs (including cocaine and marijuana) [48, 49].

Coelho and collaborators developed an on-line decomposition system for urine samples using a microwave oven prior to the determination of calcium and magnesium by flame atomic absorption spectrometry (FAAS). The decomposition efficiency allowed a rapid treatment of the urine sample with an analytical frequency of 45 samples per hour. The system consisted of three decomposition coils inserted into the cavity of the microwave oven and a valve that allowed the interruption of the passage of the flow and confined the sample to the inside of the oven [50].

The preparation of flow samples in biomedical and pharmaceutical matrices employing spectroscopic techniques remains a challenge, and few studies have been reported in the literature when compared with the chromatographic methods. When spectrometric techniques are subjected to hyphenation, they are promising for the preparation of one or more samples, since the on-line detection systems cited in the literature favor a decrease in the use of batch procedures, thus minimizing the potential for contamination and automating the sample processing procedure.

6. Combination of FIA and other analytical systems

The advance of laboratory research has enabled the identification and quantification of analytes, individually or simultaneously.

One of the techniques that has contributed to the simultaneous determination of analytes is FIA system combined with other analytical systems, such as high performance liquid chromatography (HPLC), enzymatic reactions, gas chromatography, biosensors, electrophoresis, electrochemical, and immunoassays. According to Saurina [51], in most cases, these combinations enable analysts to detect and quantify up to three compounds simultaneously. The methods required to increase this number may not be compatible with the physical resources used in systems involving flow injection.

The combination of the FIA system with other analytical techniques enables reductions in the analysis time and the reagent/sample consumption and improved accuracy, sensitivity, selectivity, and sampling frequency. In addition, the analyst's contact with the sample is minimized, decreasing the potential for contamination. Thus, by combining the pretreatment (digestion, preconcentration, sample clean-up, and solvent-solvent extraction, etc.) with on-line sample introduction, the FIA system becomes a very efficient and advantageous technique [52–57].

In this context, the possibility of detecting multianalytes using various techniques involving the combination of the FIA system with other traditional analysis systems should be highlighted. Different separation and sample pretreatment procedures can be performed using detection techniques such as fluorescence, spectrophotometry, and electrochemistry, enabling the detection of innumerable analytes, including those present in samples involved in biological applications [58–63].

Several approaches to detection have been used, and electrochemical detectors are prominent in the scientific literature, notably in studies involving conventional amperometric detection coupled to an FIA system. The main characteristics of this combination are increased sensitivity, minimized contamination of the surface of the working electrode, the presence of negligible capacitive current and *in situ* measurement, etc. A limitation associated with this system is the instability of the electrochemical signal during the determination of some compounds, compromising the repeatability of the response and the reproducibility of the results [64].

Another example of combining FIA and an electrochemical system is found in the studies of Chaves *et al.* [65] in which three compounds were determined simultaneously: caffeine, ibuprofen, and paracetamol. The authors report results obtained by combining FIA with multiple pulse amperometry (MPA) using a wall-jet flow cell with a boron-doped diamond electrode. In this analysis, cyclic voltammetry (50 mVs^{-1}) was used.

According to Llorent-Martinez *et al.* [66] and Oliveira *et al.* [67], most of the methods involving detection by UV-Vis using flow procedures offer many advantageous of this combination, being simple, fast and direct methods offering good selectivity and sensitivity in the separation and/or preconcentration steps.

Vidal *et al.* [68] address the simultaneous determination of a mixture of three analytes that are often combined in pharmaceutical formulations: two analgesics (paracetamol and propyphenazone) and a stimulant drug (caffeine). The quantification was performed by separating the three compounds using an FIA system combined with a precolumn containing C18 silica gel to avoid spectral overlap of the compounds under analysis. The detection was conducted with a spectrophotometric detector through UV absorbance measurements. The results were satisfactory, since the compounds were quantified at low concentration ranges, that is, 25–350 μg mL^{-1} for paracetamol, 5–75 μg mL^{-1} for caffeine, and 15–150 μg mL^{-1} for propylphenazone. Also, the proposed method provided low detection limits ranging from 0.65 to 7.5 μg mL^{-1}.

Pistonesia *et al.* [69] carried out the simultaneous analysis of levodopa and benserazide in tablets of pharmaceutical formulations. The samples were not subjected to pretreatment, and the reaction mixture containing the sample and potassium periodate was directed to a flow cell (8 μL inner volume) inserted in a spectrophotometer with a UV-Vis diode array detector. The concentrations used for the construction of the calibration curves analyzed were 4.1×10^{-4} to 2.03×10^{-3} M for levodopa and 8.5×10^{-5} to 4.25×10^{-4} M for benserazide. During the analysis, the FIA system variables (flow velocity, reactor length, and injected volumes) were optimized and the validation of the method (considering the robustness, repeatability, reproducibility, and accuracy) was studied. The kinetic-spectrophotometric data provided by the diode array detection were treated applying partial least squares (PLS) multidimensional regression. Samples were injected into the HPLC system using preoptimized conditions. The results obtained using the HPLC analysis (197 and 49 for levodopa and benserazide, respectively) and the FIA/PLS system (200 and 48 for levodopa and benserazide, respectively) showed no discrepancy. Thus, a simple, fast, and direct method was obtained through the implementation of a flow-injection system coupled to UV-visible diode spectrophotometry.

Regarding the analysis of biological samples, Reis and Luca [70] described a procedure for the determination of total protein in bovine blood plasma using a FIA system. The use of the FIA system enabled the in-line dilution of bovine plasma samples containing between 12.5 and 100.0 g L^{-1} total protein. The conditions for the flow analysis were optimized and the results, when compared to those obtained with the traditional method (Biuret), did not indicate significant statistical differences (*t*-paired test) at the 95% confidence level. The proposed method provided fast results, low reagent consumption, and minimization of the sample handling, as well as an analytical frequency of 76 determinations per hour.

In addition to the combination of the FIA system with innumerous detectors, the coupling of flow analysis with pretreatment and separation systems has been employed in some studies, especially FIA combined with capillary electrophoresis (CE). The first studies involving this coupling were described in 1997 by Kuban *et al.* [71] and Fang *et al.* [72]. They described this as an advantageous combination, capable of overcoming certain limitations presented by CE, such as low sensitivity, precision, and analytical frequency. An important feature is that the sample pretreatment step in the FIA-CE system is performed online, minimizing the potential for sample contamination.

Numerous studies involving FIA-CE have been reported in the literature notably: Kuban and Karlberg [73] carried out the determination of small anions through dialysis; Chen and Fang [74] performed the preconcentration of samples; Arce *et al.* [75] determined cations and anions; Chen and Fang [74] monitored multi-components in drugs; Kuban *et al.* [76] determined trimethoprim and sulfamethoxazole in drug samples; and Kuban and Karlberg [77] determined pseudoephedrine in human plasma.

An example of an FIA-CE system is also described in a paper by Liu *et al.* [78], which proposes a method developed through the combination of flow injection and CE for the separation and determination of paracetamol (Par), pseudoephedrine hydrochloride (Pse), dextromethorphan, potassium bromide (Dex), and chlorphenamine hydrogen maleate (Chl) using uncoated fused silica capillaries. Detection was performed on a UV detector at 214 nm. During the analysis, a flow-injection analyzer was used to transport the background electrolytes and the samples. The system consisted of a double piston, a 16-way automatic switching valve with three sample loops, and a peristaltic pump. The limits of detection (LOD) values were 0.22, 0.29, 0.42, and 0.70 $\mu g\ ml^{-1}$ for the compounds Dex, Chl, Pse, Par, respectively. The low LOD values, the separation of the baseline of the peak of each analyte and the low cost of this FIA-CE system are characteristics that indicate that the proposed system is suitable for the identification and quantification of the compounds investigated.

Kuban *et al.* [76] described the determination of small inorganic cations (K^+, Na^+, Mg^{2+}, and Ca^{2+}) in blood, milk, or plasma samples by electrokinetic injection using an FIA-CE system. Since the undesirable adsorption of proteins onto the capillary wall during electrophoresis was inhibited, pretreatment of the samples was not necessary, and they could be injected directly into the system. In the initial stage, two injection modes were tested for all electrolyte and standard solutions: electrokinetic (EK) and prehydrodynamic (HD). The results indicated that EK injection was the better option because it showed high sensitivity and low matrix effects, with good repeatability of the cation migration times, mainly in the case of human plasma samples. In addition, a better performance was observed for the FIA-CE system when compared to the commercial CE system.

Other researchers have reported the determination of multianalytes using the FIA system combined with other systems of separation, identification, or quantification (detectors). These include the following: an immunoassay system using detection by chemiluminescence [79]; an electrochemiluminescence immunosensor for the detection of tumor markers [80]; biosensors with the use of enzymes [81]; the quantification of carbohydrates with amperometric biosensors [82]; and the analysis of pharmaceutical formulations combining FIA with HPLC or CE [83]. Thus, it is clear that FIA coupling with other analytical techniques allows the detection/quantification of multianalytes in pharmaceutical and biological samples, etc., either for the development of methods in laboratory research or in routine analysis.

Tzanavaras and Themelis published a review on the application of flow injection to pharmaceutical analysis that covers the topics of spectrophotometric determination of active pharmaceutical ingredients [84]. According to Tzanavaras and Themelis [84], the discovery of new drugs, especially when many samples have to be analyzed in the minimum of time, demand the improvement or development of new analytical methods.

7. Conclusions and future prospects

Many methods aimed at the monitoring of chemical species in pharmaceutical and biomedical samples have been developed and investigated in recent decades. This is a field in which analytical chemistry plays an important role, contributing new analysis procedures and instrumentation. However, methods for the determination and monitoring of pharmaceuticals are still scarce.

Although some progress has been made in the development of methodologies for the monitoring of chemical species in pharmaceutical and biomedical samples, some important points still need to be addressed, such as the sample pretreatment.

In this context, a further challenge has emerged for scientists, which is the development of new clean environmentally acceptable technologies with commercial feasibility. Thus, laboratory researchers need to improve the techniques for the identification and quantification of analytes, individually or simultaneously, with a focus on this challenge.

Acknowledgements

The authors are grateful for financial support from the Brazilian governmental agencies Conselho Nacional de Desenvolvimento Científico e Tecnológico (CNPq) and Coordenação de Aperfeiçoamento de Pessoal de Nível Superior (CAPES), the MG state government agency Fundação de Amparo à Pesquisa do Estado de Minas Gerais (FAPEMIG), and the GO state government agency Fundação de Amparo á Pesquisa do Estado de Goiás (FAPEG).

Abbreviations

AFS	atomic fluorescence spectrometry
BIS-FIA	bead injection spectroscopy-flow injection analysis
CE	capillary electrophoresis
CENA	Center of Nuclear Energy in Agriculture
CFA	continuous flow analysis
FAAS	flame atomic absorption spectrometry
FIA	flow injection analysis
HPLC	high performance liquid chromatography
ICP-MS	inductively coupled plasma mass spectrometry
ICP-OES	inductively coupled plasma optic emission spectrometry
MCR-ALS	multivariate curve resolution-alternating least squares
MPA	multiple pulse amperometry
PAR	4-(2-pyridylazo) resorcinol
PLS	partial least squares
SDS	sodium dodecyl sulfate
SIA	sequential injection analysis

USP	University of São Paulo
UV	Ultraviolet
Vis	Visible

Author details

Bruno E.S. Costa[1], Henrique P. Rezende[1], Liliam Q. Tavares[2], Luciana M. Coelho[2], Nívia M.M. Coelho[1]*, Priscila A.R. Sousa[2] and Thais S. Néri[1]

*Address all correspondence to: nmmcoelho@ufu.br

1 Institute of Chemistry, Federal University of Uberlândia, Uberlândia, Brazil

2 Department of Chemistry, Federal University of Goiás, Catalão, Brazil

References

[1] Siddiqui MR, Alothman ZA, Rahman N. Analytical techniques in pharmaceutical analysis: A review. Arabian Journal of Chemistry. 2017;**10**:S1409-S1421

[2] Trojanowicz M, Kotacinska K. Recent advances in flow injection analysis. Analyst. 2016; **141**:2085-2139

[3] Skeggs LT. New dimensions in medical diagnoses. Analytical Chemistry. 1966;**38**:31A-44A

[4] Ruzicka J. From continuous flow analysis to programmable Flow Injection techniques. A history and tutorial of emerging methodologies. Talanta. 2016;**158**:299-305

[5] Oliveira HM, Fonseca AJM, Cabrita ARJ. Flow analysis as an analytical tool for soil monitoring: From wet chemistry assays to nanomaterials based sample preparation. New Pesticides and Soil Sensors. 2017;**1**:483-517

[6] Ruzicka J, Hansen J. Flow injection analysis. Part I: A new concept of fast continuous flow analysis. Analytica Chimica Acta. 1975;**78**:145-157

[7] Morales-Rubio Á, De La Guardia M, Reis BF. Multi-commutation in spectrometry. TrAC Trends in Analytical Chemistry. 2009;**28**:903-913

[8] Mesquita RBR, Rangel AOS. A review on sequential injection methods for water analysis. Analytica Chimica Acta. 2009;**648**:7-22

[9] Skoog DA, Holler FJ, Crouch SR. Principles of Instrumental Analysis. 7th ed. Cengage Learning; Boston, USA, 2016

[10] Ozlu C, Basan H, Santana E, Ertas N, Goger NG. Quantitative determination of Ketoprofen in gels and ampules by using flow-injection spectrophotometry and HPLC. Journal of Pharmacy and Biomedical Analysis. 2005;**39**:606-661

[11] Yenicelli D, Dogrukol-Ak D, Tuncel M. Determination of lansoprazole in pharmaceutical capsules by flow injection analysis using UV-detection. Journal of Pharmacy and Biomedical Analysis. 2014;**36**:145-148

[12] Liawruangrath S, Makchit J, Liawruangrath B. A simple flow injection spectrophotometric procedure for the determination of diazepam in pharmaceutical formulation. Journal of Analytical Sciences. 2006;**22**:127-130

[13] Tzanavaras PD, Themelis DG. Validated flow injection spectrophotometric assay for the quality and stability control of gemfibrozil tablets. Analytical Letters. 2005;**38**:2165-2173

[14] Can NO, Altiokka G, Aboul-Enein HY. Determination of cefuroxime axetil in tablets and biological fluids using liquid chromatography and flow injection analysis. Analytica Chimica Acta. 2006;**576**:246-252

[15] Yenicelli D, Dogrukol-Ak D, Tuncel M. Determination of leflunomide in pharmaceutical tablets by flow-injection analysis. Journal of Liquid Chromatography and Related Technologies. 2005;**28**:1693-1701

[16] Fornazari ALT, Suarez WT, Vieira HJ, Fatibello-Filho O. Flow injection spectrophotometric system for N-acetyl-L-cysteine determination in pharmaceuticals. Acta Chimica Slovenica. 2005;**52**:164-167

[17] Misiuk W, Halaburda P. Flow injection spectrophotometric determination of perazine. Journal of Trace and Microprobe Techniques. 2003;**21**:95-102

[18] Garcia MS, Albero MI, Sanchez-Pedreno C, Abuherba MS. Spectrophotometric determination of cimetidine in pharmaceuticals and urine using batch and flow-injection methods. Journal of Pharmaceutical and Biomedical Analysis. 2003;**32**:1003-1010

[19] Solich P, Polydorou CK, Koupparis MA, Efstathiou CE. Automated flow-injection spectrophotometric determination of catecholamines (epinephrine and isoproterenol) in pharmaceutical formulations based on ferrous complex formation. Journal of Pharmaceutical and Biomedical Analysis. 2000;**22**:781-789

[20] Fatibello-Filho O, Vieira HJ. Spectrophotometric flow injection procedure to indirect determination of paracetamol in pharmaceutical formulations using o-tolidine as reagent. Eclética Química. 2008;**33**:47-54

[21] Jasinska A, Nalewajko E. Batch and flow-injection methods for the spectrophotometric determination of olanzapine. Analytica Chimica Acta. 2004;**508**:165-170

[22] Garcia S, Sanchez-Pedreno C, Albero I, Garcia C. Flow-injection spectrophotometric determination of diclofenac or mefenamic acid in pharmaceuticals. Microchimica Acta. 2001;**136**:67-71

[23] Al-Momani IF. Flow injection spectrophotometric determination of the antibacterial levofloxacin in tablets and human urine. Analytical Letters. 2006;**39**:741-750

[24] Evgenev MI, Garmonov SY, Shakirova LS. Flow-injection determination of benzylpenicillin novocaine salt in penicillin preparations with spectrophotometric detection. Journal of Analytical Chemistry. 2001;**56**:572-576

[25] Omuro-Lupetti K, Cruz-Vieira I, Fatibello-Filho O. Flow injection spectrophotometric determination of isoproterenol using an avocado (*Persea americana*) crude extract immobilized on controlled-pore silica reactor. Talanta. 2002;**57**:135-143

[26] Lavorante AF, Pires CK, Reis BF. Multicommuted flow system employing pinch solenoid valves and micro-pumps. Spectrophotometric determination of paracetamol in pharmaceutical formulations. Journal of Pharmaceutical and Biomedical Analysis. 2006;**42**:423-429

[27] Marcolino Jr LH, Sousa RA, Fatibello-Filho O, Moraes FC, Teixeira MFS. Flow-injection spectrophotometric determination of dipyrone in pharmaceutical formulations using ammonium molybdate as cromogenic reagent. Analytical Letters. 2005;**38**:2315-2326

[28] Hagazy MA, Abdelwahab NS, Fayed AS. A novel spectral resolution and simultaneous determination of multicomponent mixture of Vitamins B1, B6, B12, Benfotiamine and Diclofenac in tablets and capsules by derivative and MCR-ALS. Spectrochimica Acta Part A: Molecular and Biomolecular Spectroscopy. 2015;**140**:524-533

[29] Rama MJR, Medina AR, Diaz AM. Bead injection spectroscopy-flow injection analysis (BIS-FIA): An interesting tool applicable to pharmaceutical analysis: Determination of promethazine and trifluoperazine. Journal of Pharmaceutical and Biomedical Analysis. 2004;**35**:1027-1034

[30] Khan N, Nasser H, Yan JY, Chung FT, Wu HF. Detection of flutamide in pharmaceutical dosage using higher electrospray ionization mass spectrometry (ESI-MS) tandem mass coupled with Soxhlet apparatus. Analytical Chemistry Research. 2015;**3**:89-97

[31] Wu GL, Zhou HL, Shentu JZ, He QJ, Yang B. Determination of lansoprazole in human plasma by rapid resolution liquid chromatography–electrospray tandem mass spectrometry: Application to a bioequivalence study on Chinese volunteers. Journal of Pharmaceutical and Biomedical Analysis. 2008;**48**:1485-1489

[32] Keay PJ, Wang Y. Applications of flow injection analysis to analytical biotechnology. Trends in Biotechnology. 1997;**15**:76-81

[33] Luca GC, Reis BF. Sistema em fluxo para determinação espectrofotométrica de uréia em plasma de sangue animal empregando leguminosa como fonte natural da enzima urease. Química Nova. 2001;**24**:191-194

[34] Luca GC, Reis BF. Spectrophotometry of total protein in bovine blood plasma by flow analysis. Scientia Agricola. 2001;**59**:251-256

[35] Ojeda CB, Rojas FS. Recent development in optical chemical sensors coupling with flow injection analysis. Sensors. 2006;**6**:1245-1307

[36] Jeronimo PCA, Araújo AN, Montenegro MCBSM, Pasquini C, Raimundo JIM. Direct determination of copper in urine using a sol-gel optical sensor coupled to a multicommutated flow system. Analytical and Bioanalytical Chemistry. 2004;**380**:108-114

[37] Lima LS, Weinert P, Pezza L, Pezza HR. Sensitive flow-injection spectrophotometric analysis of bromopride. Spectrochimica Acta Part A: Molecular and Biomolecular Spectroscopy. 2014;**133**:597-604

[38] Ali A, Shahida S, Khan MH, Saeed MM. Online thorium determination after preconcentration in a minicolumn having XAD-4 resin impregnated with N-benzoylphenylhydroxylamine by FI-spectrophotometry. Journal of Radioanalytical and Nuclear Chemistry. 2011;**288**:735-743

[39] Cernei N, Zitka O, Ryvolova M, Adam V, Massari M, Hubalek J, Kizek R. Spectrometric and electrochemical analysis of sarcosine as a potential prostate carcinoma marker. International Journal of Electrochemical Science. 2012;**7**:4286-4301

[40] Peng J, Tang F, Zhou R, Xie X, Li S, Xie F, Yu P, Mu L. New techniques of on-line biological sample processing and their application in the field of biopharmaceutical analysis. Acta Pharmaceutica Sinica B. 2016;**6**:540-551

[41] Barnett NW, Hindson BJ, Lewis SW. Determination of ranitidine and salbutamol by flow injection analysis with chemiluminescence detection. Analytica Chimica Acta. 1999;**384**:151-158

[42] Orsine EMA, Martins JLS. Determination of ranitidine hydrochloride in pharmaceutical preparations by ultraviolet and visible spectrophotometry. Analytical Letters. 1993;**26**:1933-1941

[43] Luconi M, Silva MF, Olsina RA, Fernandez L. Flow injection spectrophotometric analysis of lead in human saliva for monitoring environmental pollution. Talanta. 2001;**54**:45-52

[44] Menegário AA, Giné MF. Micro-scale flow system for on-line multielement preconcentration from saliva digests and determination by inductively coupled plasma optical emission spectrometry. Spectrochimica Acta Part B: Atomic Spectroscopy. 2001;**56**:1917-1925

[45] Angeli V, Biagi S, Ghimenti S, Onor M, D'ulivo A, Bramanti E. Flow injection-chemical vapor generation atomic fluorescence spectrometry hyphenated system for organic mercury determination: A step forward. Spectrochimica Acta Part B: Atomic Spectroscopy. 2011;**66**:799-804

[46] Fernández de Córdova ML, Barrales PO, Torné GR, Díaz AM. A flow injection sensor for simultaneous determination of sulfamethoxazole and trimethoprim by using Sephadex SP C-25 for continuous on-line separation and solid phase UV transduction. Journal of Pharmaceutical and Biomedical Analysis. 2003;**31**:669-677

[47] Burguera JL, Burguera M, La Cruz OL, Naranjo OR. Determination of lead in the urine of exposed and unexposed adults by extraction and flow-injection/atomic absorption spectrometry. Analytica Chimica Acta. 1986;**186**:273-277

[48] Bradley CR, Rechnitz GA. Comparison of oxalate oxidase enzyme electrodes for urinary oxalate determinations. Analytical Letters. 1986;**19**:151-162

[49] Massom M, Townshend A. Determination of cholesterol by flow injection analysis with immobilized cholesterol oxidase. Analytica Chimica Acta. 1985;**174**:293-297

[50] Coelho LM, Arruda MAZ. On-line mechanised microwave biological fluid decomposition and off-line calcium and magnesium determinations by flame atomic absorption spectrometry. Química Analítica. 2002;**20**:243-249

[51] Saurina J. Flow-injection analysis for multi-component determinations of drugs based on chemometric approaches. Trends in Analytical Chemistry. 2010;**29**:1027-1037

[52] Ruzicka J, Hansen EH. Retro-review of flow-injection analysis. Trends in Analytical Chemistry. 2008;**27**:390-393

[53] Mervartova K, Polasek M, Calatayud JM. Recent applications of flow-injection and sequential-injection analysis techniques to chemiluminescence determination of pharmaceuticals. Journal of Pharmaceutical and Biomedical Analysis. 2007;**45**:367-381

[54] Economou A, Panoutsou P, Themelis DG. Enzymatic chemiluminescent assay of glucose by sequential-injection analysis with soluble enzyme and on-line sample dilution. Analytica Chimica Acta. 2006;**572**:140-147

[55] Kyaw T, Fujiwara T, Inoue H, Okamoto Y, Kummaru T. Reversed micellar mediated luminol chemiluminescence detection of iron (II, III) combined with on-line solvent extraction using 8-quinolinol. Analytical Sciences. 1998;**14**:203-207

[56] Miró M, Frenzel W. Automated membrane-based sampling and sample preparation exploiting flow-injection analysis. Trends in Analytical Chemistry. 2004;**23**:624-636

[57] Calleri E, Temporini C, Perani E, Stella C, Rudaz S, Lubda D, Mellerio G, Veuthey JL, Caccialanza G, Massolini G. Development of a bioreactor based on trypsin immobilized on monolithic support for the on-line digestion and identification of proteins. Journal of Chromatography A. 2004;**1045**:99-109

[58] Miranda CES, Carrilho E, Gervasio AP, Giné MF. Flow injection analysis-capillary electrophoresis hyphenated systems (FIA-CE): Challenges, applications and perspectives. Química Nova. 2002;**25**:412-419

[59] Hartwell SK, Grudpan K. Flow-based systems for rapid and high-precision enzyme kinetics studies. Journal of Analytical Methods in Chemistry. **2012** (2012);Article ID 450716:10

[60] Perrez-Ruiz T, Lozano CM, Tomas V, Ruiz E. Flow injection fluorimetric determination of L-dopa and dopamine based on a photochemical inhibition process. Microchimica Acta. 2007;**158**:299-305

[61] Takayanagi T, Yamashita H, Motomizu S, Musijowski J, Trojanowicz M. Preconcentration and decomposition of perfluorinated carboxylic acids on an activated charcoal cartridge with sodium biphenyl reagent and its determination at μg L^{-1} level on the basis of flow injection-fluorimetric detection of fluoride ion. Talanta. 2008;**74**:1224-1230

[62] Shitanda I, Takamatsu S, Watanabe K, Itagaki M. Amperometric screen-printed algal biosensor with flow injection analysis system for detection of environmental toxic compounds. Electrochimica Acta. 2009;**54**:4933-4939

[63] Saini S, Suri CR. Design and development of microcontroller based auto flow assembly for biosensor application. International Journal of Computer Applications. 2010;**6**:21-27

[64] Santos WTP, Gimenes DT, Richter EM, Angnes L. Flow injection analysis with multiple pulse amperometric detection: Potentialities and applications. Química Nova. 2011;**34**:1753-1761

[65] Chaves SC, Aguiar PNC, Torres LMFC, Gil ES, Luz RCS, Damos FS, Munoz RAA, Richter EM, Santos WTP. Simultaneous determination of caffeine, ibuprofen, and paracetamol by flow-injection analysis with multiple-pulse amperometric detection on boron-doped diamond electrode. Electroanalysis. 2015;**27**:2785-2791

[66] Llorent-Martinez EJ, Ootega-Barrales P, Fernández de Cordova ML. Trends in flow-based analytical methods applied to pesticide detection: A review. Analytica Chimica Acta. 2011;**684**:30-39

[67] Oliveira MC, Nogueira RFP, Neto JAG, Jardim WF, Rohwedder JJR. Flow injection spectrophotometric system for hydrogen peroxide monitoring in photo-fenton degradation processes. Química Nova. 2001;**24**:188-190

[68] Vidal AD, Barrales PO, Díaz AM. Simultaneous determination of paracetamol, caffeine and propyphenazone in pharmaceuticals by means of a single flow-through UV multi-parameter sensor. Microchimica Acta. 2003;**141**:157-163

[69] Pistonesia M, Centurion ME, Fernandez BS, Damianib PC, Alejandro C, Olivieri AC. Simultaneous determination of levodopa and benserazide by stopped-flow injection analysis and three-way multivariate calibration of kinetic-spectrophotometric data. Journal of Pharmaceutical and Biomedical Analysis. 2004;**36**:541-547

[70] Reis BF, Luca GC. Spectrophotometry of total protein in bovine blood plasma by flow analysis. Scientia Agricola. 2001;**59**:251-256

[71] Kuban P, Engstrom A, Olsson JC, Thorsen G, Tryzell R, Karlberg B. New interface for coupling flow-injection and capillary electrophoresis. Analytica Chimica Acta. 1997;**337**:117-124

[72] Fang ZL, Liu ZS, Shen Q. Combination of flow injection with capillary electrophoresis. Part I. The basic system. Analytica Chimica Acta. 1997;**346**:135-143

[73] Kuban P, Karlberg B. Preconcentration and decomposition of perfluorinated carboxylic acids on an activated charcoal cartridge with sodium biphenyl reagent and its determination at μg L^{-1} level on the basis of flow injection-fluorimetric detection of fluoride ion. Talanta. 1998;**45**:477-484

[74] Chen HW, Fang ZL. Combination of flow injection with capillary electrophoresis: Part 4. Automated multicomponent monitoring of drug dissolution. Analytica Chimica Acta. 1998;**376**:209-220

[75] Arce L, Rios A, Valcarcel M. Flow injection–capillary electrophoresis coupling to automate on-line sample treatment for the determination of inorganic ions in waters. Journal of Chromatography A. 1997;**791**:279-287

[76] Kuban P, Oldhoff O, Karlberg B. Direct determination of small cations in proteinaceous samples using a flow injection–capillary electrophoresis system. Journal of Chromatography A. 1999;**857**:321-326

[77] Kuban P, Karlberg B. On-line monitoring of kraft pulping liquors with a valveless flow injection–capillary electrophoresis system. Analytica Chimica Acta. 2000;**404**:19-28

[78] Liu X, Liu L, Chen H, Chen X. Separation and determination of four active components in medicinal preparations by flow injection-capillary electrophoresis. Journal of Pharmaceutical and Biomedical Analysis. 2007;**43**:1700-1705

[79] Fu Z, Yan F, Liu H, Yang Z, Ju H. Channel-resolved multianalyte immunosensing system for flow-through chemiluminescent detection of α-fetoprotein and carcinoembryonic antigen. Biosensors Bioelectronics. 2008;**23**:1063-1069

[80] Zhang Y, Liu W, Ge S, Yan M, Wang S, Yu J, Li N, Song X. Multiplexed sandwich immunoassays using flow injection electrochemiluminescence with designed substrate spatial-resolved technique for detection of tumor markers. Biosensors Bioelectronics. 2013;**41**:684-690

[81] Suwansa-Ard S, Kanatharana P, Asawatreratanakul P, Limsakul C, Wongkittisuka B, Thavarungkul P. Semi-disposable reactor biosensors for detecting carbamate pesticides in water. Biosensors Bioelectronics. 2005;**21**:445-454

[82] Vargas E, Gamella M, Campuzano S, Guzmán-Vázquez de Prada A, Ruiz MA, Reviejo AJ, Pingarrón JM. Development of an integrated electrochemical biosensor for sucrose and its implementation in a continuous flow system for the simultaneous monitoring of sucrose, fructose and glucose. Talanta. 2013;**105**:93-100

[83] Dantan N, Frenzel W, Kuppers S. Flow injection analysis coupled with HPLC and CE for monitoring chemical production processes. Chromatography. 2001;**54**:187-190

[84] Tzanavaras PD, Themelis DG. Review of recent applications of flow injection spectrophotometry to pharmaceutical analysis. Analytica Chimica Acta. 2007;**588**:1-9

Chapter 11

Factorial Design and Machine Learning Strategies: Impacts on Pharmaceutical Analysis

Marwa S. Elazazy

Additional information is available at the end of the chapter

http://dx.doi.org/10.5772/intechopen.69891

Abstract

Pharmaceutical analysis is going through an expeditious progress as the perception of 'multivariate data analysis' (MVA) becomes gradually more assimilated. Pharmaceutical analysis comprises a range of processes that covers both chemical and physical assessment of drugs and their formulations employing different analytical techniques. With the revolution in instrumental analysis and the huge amount of information produced, there must be an up-to-date data processing tool. The role of chemometrics then comes up. Multivariate analysis (MVA) has the capability of effectively drawing a complete picture of the investigated process. Moreover, MVA reproduces the arithmetic influence of variables and their interactions through a smaller number of trials, keeping both efforts and capitals. Spectrophotometry is among the most extensively used techniques in pharmaceutical analysis either direct (single component) or derivative (multicomponent). In addition to these recognized benefits, using chemometrics in conjunction with spectrophotometry affects three vital characteristics: accuracy, precision and robustness. The impact of hyphenation of spectrophotometric analytical techniques to chemometrics (experimental design and support vector machines) on analytical laboratory will be revealed. A theoretical background on the different factorial designs and their relevance is provided. Readers will be able to use this chapter as a guide to select the appropriate design for a problem.

Keywords: chemometrics, experimental design, machine learning strategies, support vector machines, pharmaceutical analysis, spectrophotometry

1. Introduction

Nowadays, an enormous amount of information is being generated by the state-of-the art analytical instrumentations, an issue that necessitates the presence of a potent data processing approach. Chemometry, a division of science that has seen a major progress in the past few decades, depends on eliciting data and the development of a mathematical model that describes the relationship between the response signal and the process variables [1–3]. In simple words, chemometrics is the term that is used to describe the case when chemistry, biology and other branches of science meet with mathematics and computer science [4]. As a multidisciplinary science, chemometrics can be used to resolve many problems beyond the boundaries of chemistry, including medicine, pharmacy, environment and other domains of natural and applied sciences [5, 6].

Chemometric techniques, including both multivariate data analysis (MVA) and factorial designs, play a vital role in analysing systems that are both large and multidimensional, an issue that adds to the power of this methodology. Moreover, the growing in complexity from the conventional univariate data analysis (one-variable and a single response at a time) to multivariate data analysis (more than one factor and a single or multiple responses) is greatly reflected on the imperative analytical outcomes, for example, sensitivity and selectivity [7, 8]. Additionally, being a versatile approach, application of chemometry can offer several more advantages. At the simple level (first order, vector data), samples that cannot be signalled using the existent calibration setting can now be effectively modelled. At more sophisticated levels (second- or higher orders), and in addition to the accurate determination of the calibrated analyte, not only new sample constituents can be identified but also their impact on the entire response can be adequately modelled.

Pharmaceutical analysis is experiencing an expeditious growth as the concept of 'multivariate data analysis' becomes progressively integrated. As being known, pharmaceutical analysis encompasses both chemical and physical evaluation of drugs and their dosage forms using different analytical strategies. Yet, the common routine in most of analytical laboratories is to meditate only one-variable and one response at time. Measuring the impact of this variable on the analytical signal is the only source of any generated data [1]. Nevertheless, quality of collected information would be significantly improved if the impact of more than one-variable, their linear, second- and third-order interactions on a single or multiple responses was defined through an arithmetic model [9].

Incorporation of 'design of experiments' (DOE) in any (or all) of the phases of drug development would be of a great effect, not only on the quality of data produced, but also on the analytical process itself in terms of better understanding and usage of generated data, as well as resources preservation.

This chapter focuses on the impact of using hyphenated chemometric-spectroscopic techniques in pharmaceutical analysis. Experimental designs as well as machine learning strategies, as essential parts of chemometrics, will be the main topic of the chapter. The reader does not need to be familiar with the complicated mathematical concepts. Rather, and for practicality and reader's advantageousness, a brief on the simple hypotheses needed to get DOE straightforward will be revealed.

Distinctive application of chemometrics in the field of drug analysis will be shown as we go forward. Material presented throughout the chapter will be of interest to students, chemometricians, drug manufacturers, quality control chemists and pharmacists.

2. Experimental design

Design of experiments (DOE) is a fundamental part of multivariate analysis techniques. However, DOE is comprehended to deal with a limited number of factors (determined according to the design used) in comparison to the other multivariate techniques.

Moreover, multivariate methods either bilinear such as partial least squares (PLS) and principal component analysis (PCA), or multi-way models such as Tucker-3 and parallel factor analysis (PFA), are commonly deemed as supplementary methodologies to DOE. Factors that were not considered in the initial set-up of DOE, as well as their effect, can now be recognized by the subsequent multivariate techniques [6, 10–12].

The typical scenario for setting DOE starts with deciding upon the experimental objective as well as the number of factors to be investigated. The most common objectives can be summarized as follows [13–16]:

- *Screening goal:* where all factors that might contribute to the response are considered and labelled as the main effects. Only factors proved to be significant will be considered for the second stage, which is known as optimization or fine tuning. In this phase, levels for each factor are adjusted to a narrower range to get the optimum response.
- *Response surface goal:* where main factors as well as factor-factor interactions (linear, quadratic, etc.) can be determined.
- *Optimization goal:* the experiment is designed in this case to get the best proportion for a factorial blend needed to get the optimum response (minimum or maximum).

Table 1 recaps the rules for selecting a design based on the number of factors and the envisioned goal of the experiment.

Up to now, the conventional approach for investigating the influence of several factors on a response depends on fixing the levels of all factors except the one to be investigated. This approach is known as one-variable at a time (OVAT). Although still being applied for analytical method development, OVAT usually confronts several difficulties.

One of the main limitations accompanying this rehearsal is the need for a big number of trials. Nevertheless, the resulting delineation of 'ideal conditions' and hereafter the system execution cannot be handled with a high extent of certainty. One reason for that is the absence of an evaluation for the variable-variable interactions in the paradigms premeditated using OVAT.

Number of factors	Screening goal	Response surface goal
Two to four factors	Full or fractional factorial designs (FFD)	Central composite (CCD) or Box-Behnken (BBD) designs
Five or more factors	Fractional factorial (FFD) or Plackett-Burman (PBD) designs	Preliminary assessment using the appropriate screening design is required to control the number of factors.

Table 1. Design selection rubric.

Multivariate data analysis (MVA) and its advantages mentioned earlier has the ability to replicate the arithmetical influence of the discrete factors and similarly their interactions through a reduced number of experimentations, saving both efforts and resources [16, 17].

The set-up of experimental design then can be viewed as 2–3 phases depending on the number of factors to be investigated and the objective of investigation: *screening, optimization* and *verification*.

2.1. Screening

Usually, a consecutive investigation process starts with testing a relatively large number of prospective variables. Screening designs then are factorial designs that can be used to get the few utmost substantial variables affecting the response, **Table 1**. Several designs can be used for this purpose, which are mentioned the following section.

2.1.1. *Two-level full factorial design (2^k-FFD)*

This design can be used when the number of variables (*k*) is between 2 and 15. Each variable is set at two levels: low (−1) and high (+1). Therefore, for three factors, for example, eight runs will be conducted excluding the central points and replicates. **Table 2** presents the design table when three factors X_1, X_2, and X_3 are investigated using the proposed two-level full factorial design (FFD). **Figure 1** shows the pattern of experiments in a design for three factors, arrows illustrate the direction of increase of the factors.

2.1.2. *Two-level fractional factorial design (2^{k-p})*

Even when the number of factors is small, many runs are needed if an FFD is to be used. For example, for five factors, $2^5 = 32$ experiments are needed in the base run only. In case replicates are needed and central points are added, the number of runs becomes large and the objective of using the DOE to save time and efforts becomes meaningless. The only way out for such a

Run order	X_1	X_2	X_3
1	−1	−1	−1
2	1	−1	−1
3	−1	1	−1
4	1	1	−1
5	−1	−1	1
6	1	−1	1
7	−1	1	1
8	1	1	1

Note: Runs are shown in standard order.

Table 2. A two-level, full factorial design table for three factors.

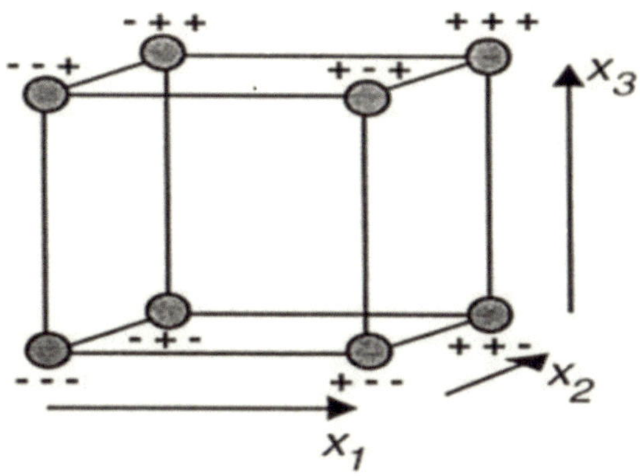

Figure 1. Pattern of experiments in a 2^3 FFD.

case is to cautiously select a fraction (*p*) of the original runs proposed by the two-level FFD. For the previous example (3 factors), instead of performing 16 experiments (8 × 2 replicates) and by using a ½ fraction, only 8 runs will be performed in the 2 replicates.

Figure 2 shows a comparison between a full (2^k) and a fractional (2^{k-p}) factorial designs used to investigate three factors. While eight runs are needed in the first set-up, only four runs will be performed in the second arrangement, where main effects are confounded with the two-way interactions.

2.1.3. Plackett-Burman design (PBD)

This design has run numbers that are multiple of 4. Using this design allows performing a number of trials $N = 4n$ in order to investigate a number of factors $f = 4\ (n - 1)$. PBD is an efficient approach when only main or large effects are of interest. In other words, this design can detect the most imperative factors affecting the experiment from a comparatively large number of factors (2–47) and without putting any concerns on interactions and non-linear effects. Minitab®, a commonly used software for this purpose, can generate a PBD for up to 47 factors.

PBD, in specific, is one of the commonly used approaches in robustness tests used in method validation compared to fractional factorial design, for example. The main reason for selecting PBD as a robustness test is that this design focuses only on the main effects, while factor-factor interactions are highly confounded with the large main effects, as previously mentioned [18–21].

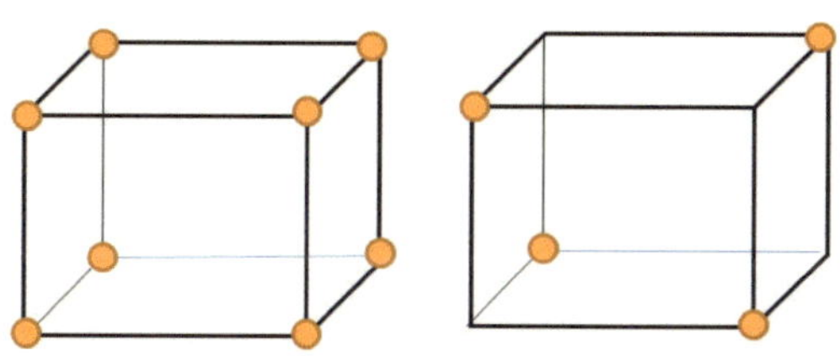

Figure 2. A 2^3 full factorial (left pane) and a 2^{3-1} fractional factorial designs (right pane) for three factors.

It is noteworthy to mention that, for any of the designs, identification of significant factors can be achieved using several tools. Pareto chart of standardized effects, normal and half-normal probability plots are among these tools.

2.2. Optimization

After selection of the most important factors from the previous screening process, levels of these factors need to be adjusted 'tuned' to identify the most suitable variable settings for optimizing a response. It is noteworthy to mention that significant factors can be also identified based on a former knowledge with the process under consideration. Another objective for this process is to assess the variable-variable linear interactions as well as the quadratic effects. This estimation gives an indication on how the response surface looks like. This approach is hence known as '*response surface methodology (RSM) designs*' [13].

Following the application of a response surface design, graphical representation of the developed polynomial mathematical model is assembled. Contour plots (2D) or response surface plots (3D) are used to graphically envisage the model.

2.2.1. Box-Behnken (BB) design

As a response surface design, BB design can capably determine the first- and second-order constants. BB design is simple, and independent with no contribution from a preceding factorial or fractional factorial design. Three levels for each factor are proposed; however, runs where all variables at their upper domains or all at lower domains are not included [22]. BB design is an economic choice since it involves less design points and hence a fewer number of runs compared to other RSM designs.

2.2.2. Central composite (CC) design

Unlike the BB design, CC designs usually contain in-built points from the factorial or fractional factorial designs (2^f trials) with added centre points that are enhanced with a group of axial points (2^f trials), **Figure 3**. Thus to scrutinize a number of factors = f, a number of experiments $N = 2f + 2f + 1$ will be conducted. The design in such a configuration allows the estimation of data curvature. Furthermore, due to inclusion of data points from a prior screening design, CC design can be used in a consecutive experimental set-up. Classification of CC designs depends on the value of alpha (α) or the distance between the axial points and the centre. Three types of CC design then exist: *circumscribed (CCC)*, *inscribed (CCI)* and *face-centred (CCF)* [1, 13, 23–26].

2.3. Statistical validation

Following the last step, generated models can be statistically assessed using conventional approaches such as 'analysis of variance' (ANOVA). In this approach, variances are used to decide whether the means are different. For ANOVA to be properly conducted, the response variable has to be continuous and at least one of the investigated variables is categorical. For a factor to be significant, the p-value is usually less than α of 0.05 [1, 23–26].

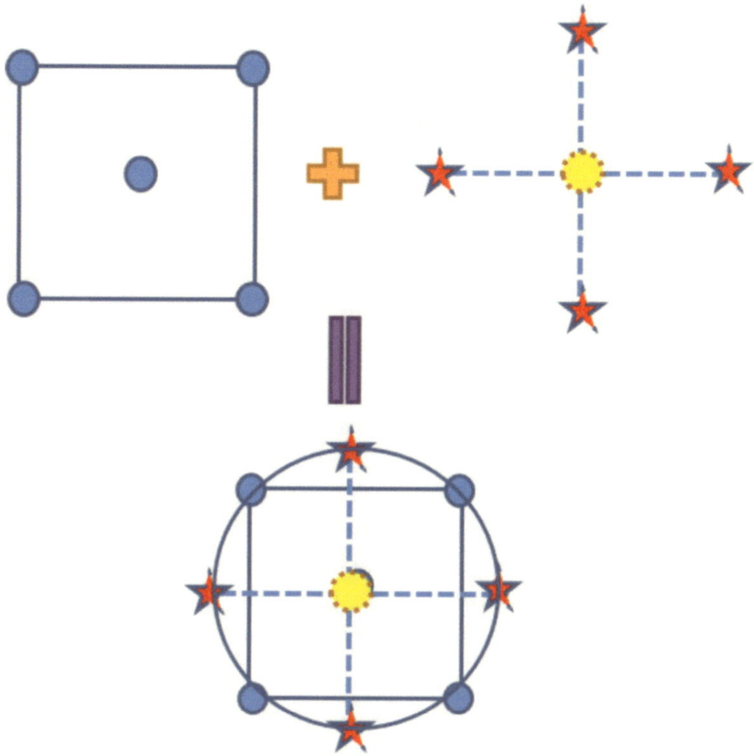

Figure 3. Central composite (CC) design for two factors.

Another model-fitting approach is the residual analysis. Residual plots are generally used to scrutinize the goodness of fit in regression and ANOVA. Examples of residual plots given by Minitab® include normal probability plots, residual versus fits, histograms and residuals versus order plots.

3. Support vector machines (SVMs)

SVM is a prevalent classification tool which was proposed by Vapnik [27]. As a kernel-based technique, support vector machines (SVMs) have seen a major development in the past few years. During such a short period, SVMs have found several applications in pharmacy, medicine and drug development industry. For example, SVMs have been used in finding the relation between drug structure and its activity 'structure-activity relationships (SAR)'. Moreover, SVMs with a capability of differentiating various drug substrates and classifying them as drugs or non-drugs are widely applied in drug design [28]. Fields of applications of SVMs extend to chemometrics, biosensors, computational biology and industrial modelling processes. Though being famous for the treatment of non-linear data, their application in handling linear models is still conceivable [27–32].

4. Pharmaceutical analysis and chemometrics

As mentioned earlier in this chapter, drug analysis covers all features related to both in- and after process (quality control) assay of drug substances. Details of these aspects include processes starting with drug synthesis, testing of physico-chemical properties, SAR and mechanism of drug action [28, 33, 34]. Quality control assays include stability testing of both raw and formulated drug materials, content homogeneity, solubility and dissolution properties. Nonetheless, drug assays are not circumscribed to the pure materials and the dosage forms, but the practice extends to include all complicated matrices (biological, foods, drinks, etc.). Moreover, analyses do not consider the active constituents only, but also look for the additives, degradation products and the impurities.

Different analytical techniques have been proposed for the determination of drugs (pure form, pharmaceutical formulations, biological fluids, etc.). For established drugs, standard analytical techniques can be obtained from compilations such as pharmacopoeias. The presence of almost daily new produces, however, requires constructing an appropriate analytical design. This design should inaugurate sufficient data on the analytical process and the product of concern. Data obtained should also be valid throughout the entire process of drug development and the procedure itself needs to be robust and applicable, when needed, in different laboratories.

These specifications do not mean that there is a need for a sophisticated technique such as chromatography. Yet, spectrophotometry might be an equivalent choice in the case being linked to an arithmetic backbone [16, 35–39]. Both single and multicomponent analyses (derivative spectrophotometry (DS)) can be readily linked to chemometry. Furthermore, analysis of a single response (e.g. absorbance) or multiple responses (at different wavelengths) can be better controlled using mathematical modelling [35–42].

Many challenges face the pharmaceutical analyst especially when trying to develop a new analytical method, inaugurate a drug stability study and establish automation into the laboratory. Handling these challenges using chemometrics will be revealed in the coming subsections.

Spectroscopic techniques have been used for long in pharmaceutical analysis. Ultraviolet and visible (UV-vis), infrared (IR), spectrofluorometry and near infrared (NIR) spectroscopy are among the most popular techniques in this concern. The application of techniques such as spectrophotometry in pharmaceutical analysis, though being simple, rapid, cost-effective and suitable for routine analysis, confronts many problems. A major problem that hinders the applicability of this technique is the lack of selectivity. Even in the analysis of a mixture of two or more components, the inability to select the most appropriate wavelength would have a negative impact on sensitivity, selectivity and reproducibility as well. Chromatography, though being a well-developed modern technique that is widely used in pharmaceutical analysis, suffers also from similar glitches. Inappropriate chemical deviations such as peaks from the matrix, alterations of mobile phase concentrations, baseline drift and shifts in retention times would greatly influence the cogency of the obtained results.

In both cases (and probably for other analytical techniques), the application of chemometrics to interpret the obtained data would be an ideal solution if the approach is able to account for all variations in the obtained data as well as get quantitative data from the tested samples. In addition, the used approach should be able to reduce the effects of these variations on the anticipated response.

In the coming subsections, we will consider the impacts of linking chemometry on pharmaceutical analytical techniques. More details will be given in the recent advances that have been made in this field and how spectrophotometry in specific has been affected.

4.1. Spectrophotometry

Spectrophotometric techniques are, as mentioned before, among the most widely used approaches in pharmaceutical analysis. Direct application of spectrophotometric analysis is only possible if the selected wavelength is not affected by another concomitant analyte. As an approach, application of spectrophotometry entails a study of a variety of factors affecting a single response or multiple responses [37–39].

With the advent of chemometrics, data processing programs and user-friendly software, the outdated OVAT approach is being gradually replaced with MVA in the analytical laboratories. In general, in addition to the known advantages of using chemometrics in conjunction with spectrophotometry, three crucial performance features are usually assessed with this hyphenation; accuracy, precision and robustness.

DOE and SVM are among the widely used chemometric approaches in spectrophotometric analysis of drugs and formulations. The main idea behind implementing these chemometric techniques is to establish the concept of thinking before doing, arrange and perform a controlled experiment, interpret the obtained results, and hence maximize the efficiency of used technique and obtained data. Generally, preservation of resources and conducting the fewest number of experiments are taken into consideration. This comprehensive knowledge and control of the running process are represented by a multi-aspect assembly of input variables together with method parameters, in other words, the 'design space'. The outcome of application of 'design space' is reflected on a pledge of quality as defined by International Conference on Harmonisation (ICH) tripartite rules [43].

As we mentioned earlier, DOE can be used in many stages of the pharmaceutical industry. For example, while screening designs can be used at the early stages of method development, optimization and testing of robustness are used just before the discharge of the finalized product [44].

Several other examples exist in the literature showing the application of DOE and SVM in the pharmaceutical industry. For instance, a two-level full factorial design (2^3-FFD) was used to decide upon the most substantial factors in the formulation of ascorbic acid tablets that are resistant to oxidative degradation using hydrophilic polymers. Measured responses were the tensile strength, disintegration time and the release features of these tablets [45]. In another application, Plackett-Burman design was employed to investigate the impact of seven factors on the release of theophylline from hydrophilic vehicles. According to the proposed model,

12 experiments were performed and a polynomial model was generated. Out of the seven variables, only two were proved to be significant [46].

In many cases of drug analysis, chemical pre-treatment of the analyte(s) prior to measurement of the anticipated response is sometimes needed. Usually, this preceding treatment would serve to correct for lack of sensitivity and selectivity encountered using direct spectrophotometry. Practices that are now ordinarily used in this concern are condensation, ion-pairing, charge transfer complexation, metal ion chelation, diazotization and redox reactions. With this pre-treatment, the process becomes technically more complicated and requires an investigation of a larger number of factors. A compelling solution in this case is provided by chemometrics. The literature now shows a huge amount of records on the hyphenation of factorial designs to spectrophotometric drug analysis, compared to the situation earlier.

For example, the Hantzsch condensation reaction was used for the derivatization of sodium alendronate, an inhibitor of bone resorption that is commonly used for management of osteoporosis, and which does not have any chromophore. Analysis of sodium alendronate was done both in its pure form and in oral solutions. Plackett-Burman screening design was used to investigate the effect of seven factors on the absorbance of the resulting condensation product. Only four factors were proved to be important and this finding was verified by ANOVA testing. Tuning of factors' levels was done using a circumscribed central composite design (CCCD). Moreover, data obtained from the CCCD including both variables and responses were treated with Statsoft® software employing artificial neuron network (ANN). A network of the multi-layer perceptron type (MLP) that has three hidden layer neurons gave the best results. Similarly, data from the CCCD were processed using different SVM kernels. Best results were obtained using a radial-basis function (RBF) kernel [37].

Chemical derivatization of midodrine hydrochloride both as per se and in formulations (tablets and oral drops) was performed using the Hantzsch reaction accompanied by a two-level 2^4-FFD. Variables proved to be significant ($p < 0.05$) were warily attuned utilizing a response surface methodology (RSM) with a face-centred central composite design. The suggested model represented a perfect example for probing the efficiency of factorial designs in optimizing the reaction conditions and maximizing the output [38]. Statistical validation of the proposed technique was performed by using ANOVA in two successive steps. Moreover, D-optimality design was chosen to minimalize the variance in the regression coefficients of the fitted model. **Table 3** shows the screened factors and the response domains employing the proposed screening design.

A suitable approach in finding the most significant variables for screening designs and the optimal locations following an optimization design is usually the graphical representation of the data or the generated model. This feature is usually implemented in chemometrics' software such as Statsoft® and Minitab®. The outcome of screening designs is customarily represented by the Pareto chart of standardized effects, where factors passing the reference line are considered significant. Similar conclusions can be drawn using normal and half-normal probability plots. **Figure 4** shows a Pareto chart showing the significant factors obtained after screening of all factors affecting the formation of a charge transfer complex between *p*-synephrine and *p*-chloranil employing a full factorial design.

Screened factor	Symbol	Level		Maximum absorbance of the product (Y)
		Low (−)	High (+)	
Temperature (°C)	X_1	25.0	100.00	0.602
Reaction time (min.)	X_2	5.00	30.00	0.495
Reagent volume (mL)	X_3	0.10	1.00	0.489
pH of acetate buffer	X_4	2.40	5.60	0.493
Response	Y	Target		

Table 3. Screened factors and response domains for a two-level (2^4) full factorial design (FFD) premeditated for Hantzsch reaction (reproduced from author's own work [38] with permission from the Royal Society of Chemistry).

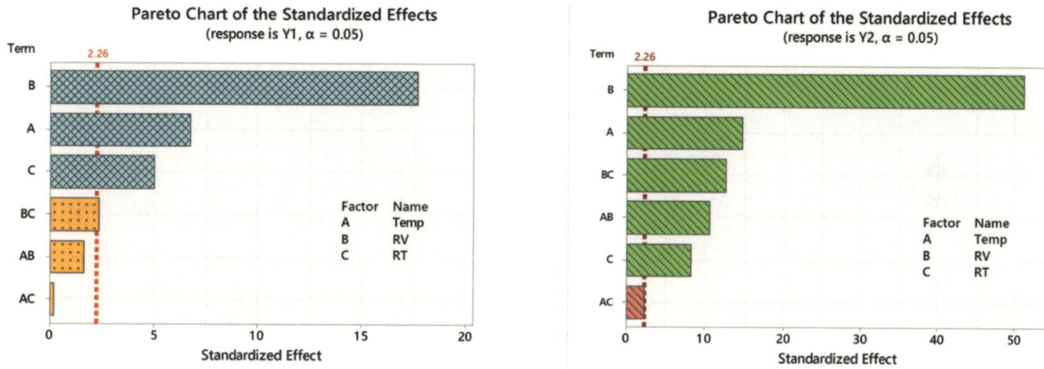

Figure 4. Pareto chart of standardized effects (reproduced from author's own work [39] with permission from the Royal Society of Chemistry).

Two types of graphs are commonly used to 'pinpoint' the optimal conditions; the response surface (3D) and contour (2D) plots. As shown in **Figure 5** [39], contour lines are produced when points that have the same absorbance are connected. On the other hand, 3D surface plots (figure is not shown) provide a stronger idea on interactions compared to contour plots. Both representations reveal a good matching with the obtained results, employing the polynomial equation.

Analysing one response is a simple task where analysis of each paradigm would merely identify zones of anticipated results. Conversely, concurrent optimization of two or more responses as a function of *n* variables is not that plausible. Different strategies are usually followed for this purpose; overlaid contour plots and global desirability function are among the commonly used approaches [39].

Overlaid contour plots are executed only if few responses are of concern (usually two responses). Simply, higher and lower bounds for each response are outlined. Contours for response boundaries versus variables under analysis are then displayed. A region that ensures both responses is recognized as the 'feasible' area [47, 48]. The plot usually shows the feasible regions where compromised optimum values for both responses meet. However, when more than one factor is involved and considering more than one response, a large number of graphs are requested, an issue that makes the procedure of pictorial observation tiresome.

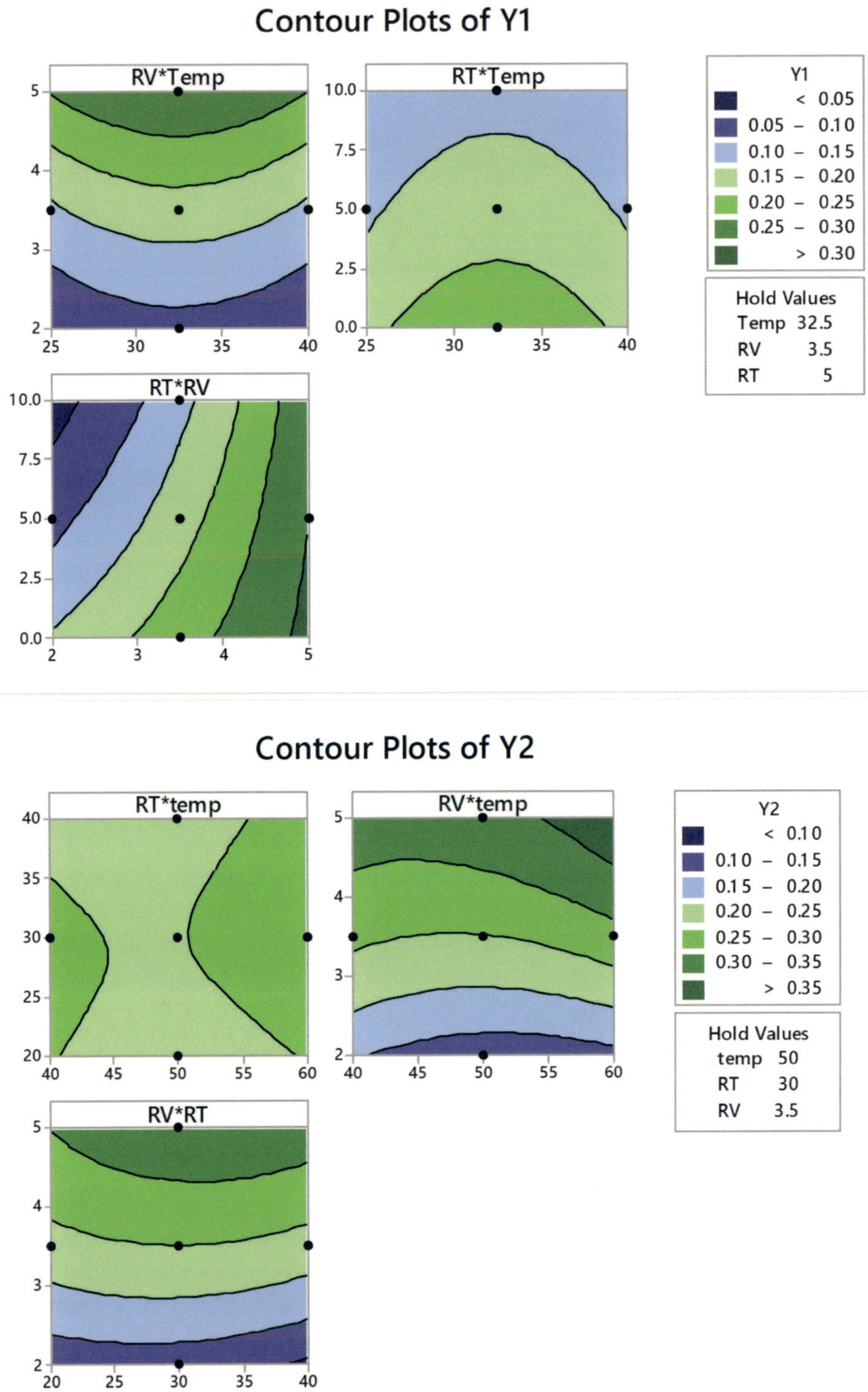

Figure 5. Two-dimensional contour plots for FCCD showing Y1 and Y2 as a function of different variable interactions (reproduced from author's own work [39] with permission from the Royal Society of Chemistry).

Additionally, the overlaying process is not that practicable as the best regions for each response are a bit far from each other.

Derringer function is another approach that can be used in this case. Individual desirability for each response is used to calculate the global desirability employing the following function:

$$D = (d_1^{r1}\ d_2^{r2} \ldots . d_m^{rm})^{1/\Sigma ri} = \left(\Pi_{i=1}^{n} d_1^{ri}\right)^{\frac{1}{\Sigma ri}} \quad (1)$$

where D is the overall desirability, d is the single desirability, r is the significance of each response compared to the other and m is the number of responses to be optimized [49, 50]. In general, as the value of D gets closer to 1.0000, the desirability of this variable arrangement on the proposed response gets higher. **Figure 6** shows the desirability function plot following the optimization employing an FCCD approach. The horizontal dashed lines represent current response values. The vertical solid lines show the optimal value for each variable.

A serious drawback that hinders drawing useful data, either assessable or qualitative, from spectrophotometry is the overlapping of absorption bands. This overlapping might be arising

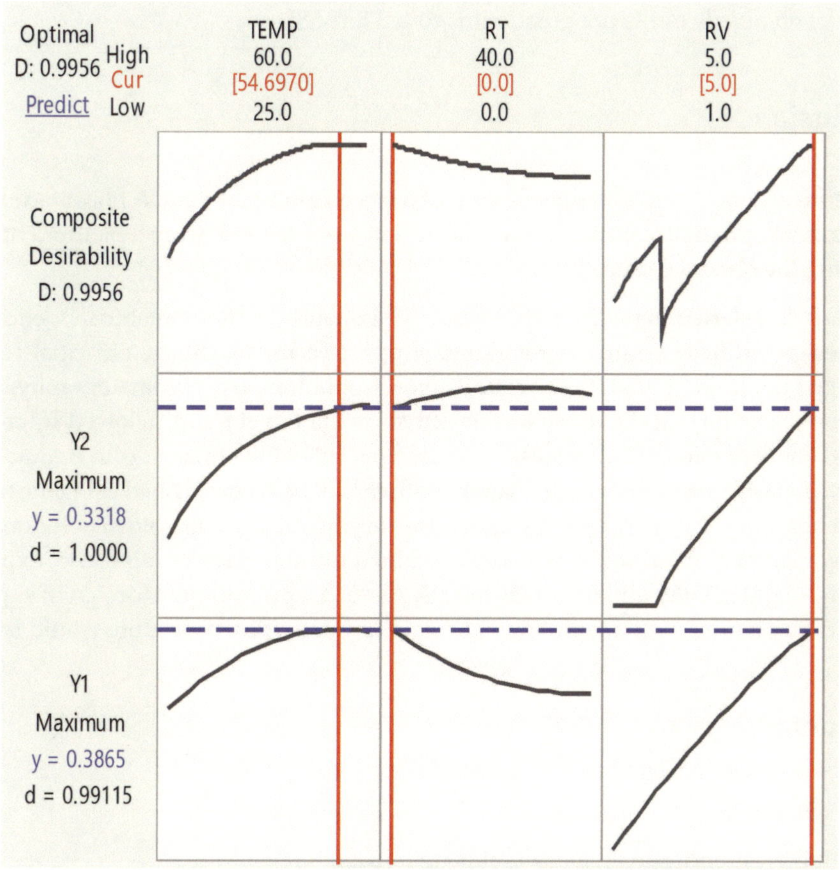

Figure 6. Desirability function plot for the FCC design (reproduced from author's own work [39] with permission from the Royal Society of Chemistry).

from the presence of drug or non-drug impurity, the presence of more than one component in the target formulation or due to the presence of degradation products. The presence of these components in one formulation at unequal concentration levels augments the problem. A compulsive solution to this problem is using derivative spectrophotometry (DS). This approach depends on differentiation of the regular absorption spectrum using arithmetical transformation into a first-order derivative or a higher order derivative. Several advantages are achieved using DS including but not limited to an improvement in resolution, reduction of noise level, elimination of interferences, augmentation of sensitivity and selectivity, and accordingly an improvement in separation efficiency [51–54].

The situation is not complicated if no chemical interaction among the components, and their spectra are only partially overlapped. In such a case, an acceptable resolution can be achieved employing first derivative spectra. Depending on the spectral characteristics of components to be analysed and the nature of interventions in multicomponent samples, chemometric algorithms have been proved to be a powerful tool in resolving binary (or more) mixture. Approaches such as principal component regression (PCR) and partial least squares (PLSs) have been widely applied both for zero- or higher- order spectra. A combination of MVA and derivative spectral data is highly beneficial where features such as easiness of application and reliability of obtained results are greatly improved [55–58].

5. Conclusion

Pharmaceutical analysis involves generation of a large amount of data. A pharmaceutical analyst then has an apparently intimidating task and needs to choose from a plethora of methods for handling the obtained data.

Chemometry has started to realize its potential. Assimilation of chemometric modelling (experimental design, artificial neuron networking, support vector machines, principal component analysis, etc.) to different analytical methods (spectrophotometry, chromatography, etc.) with the purpose of optimizing the analytical objectives is the novel trend followed by researchers nowadays. For every analytical process, the principal role of the analyst is to optimally obtain informative data. Unfortunately, best usage of data cannot be accomplished using the traditional univariate analysis. Multivariate analysis, in contrary, would be the golden solution, where a reasonable amount of information would be obtained through a fewer number of experiments, reduced effort and smaller amount of chemicals. As such, application of 'design of experiments (DOE)' becomes a need, and integration of DOE in any analytical procedure would be a must.

Author details

Marwa S. Elazazy

Address all correspondence to: marwasaid@qu.edu.qa

Department of Chemistry and Earth Sciences, College of Arts and Sciences, Qatar University, Doha, Qatar

References

[1] Massart DL, Vandeginste BGM, Deming SN, Michotte Y, Kaufman L. Chemometrics: A Textbook. Amsterdam: Elsevier; 1988. Chapter 2

[2] Wold S. Chemometrics, why, what and where to next? Journal of Pharmaceutical and Biomedical Analysis. 1991;**9**(8):589-596

[3] Mocák J. Chemometrics in Medicine and Pharmacy. Nova Biotechnologica et Chimica. 2012;**11**(1): 11-25

[4] Lopes JA, Costa PF, Alves TP, Menezes JC. Chemometrics in bioprocess engineering: Process analytical technology (PAT) applications. Chemometrics and Intelligent Laboratory Systems. 2004;**74**:269-275

[5] Krantz-Rülcker C, Stenberg M, Winquist F, Lundström I. Electronic tongues for environmental monitoring based on sensor arrays and pattern recognition: A review. Analytica Chimica Acta. 2001;**426**(2):217-226

[6] Singh I, Juneja P, Kaur B, Kumar P. Pharmaceutical applications of chemometric techniques. ISRN Analytical Chemistry. 2013;**2013**:1-13

[7] Miller CE. Chemometrics in process analytical chemistry. In: Bakeev KA, editor. Process Analytical Technology. Oxford, UK: Blackwell Publishing Ltd.; 2005

[8] Olivieri AC. Perspective analytical advantages of multivariate data processing. One, Two, Three, Infinity? Analytical Chemistry. 2008;**80**:5713-5720

[9] Fisher RA. The Design of Experiments. New York: Haffner Press; 1935

[10] Bro R. Multivariate calibration what is in chemometrics for the analytical chemist? Analytica Chimica Acta. 2003;**500**:185-194

[11] Martens H, Martens M. Multivariate Analysis of Quality: An Introduction. Chichester, UK: Wiley; 2000

[12] Huang J, Kaul G, Cai C, Chatlapalli R, Hernandez-Abad P, Ghosh K, Nagi A. Quality by design case study: An integrated multivariate approach to drug product and process development. International Journal of Pharmaceutics. 2009;**382**:23-32

[13] Box GEP, Draper NR. Response Surfaces, Mixtures, and Ridge Analyses. 2nd ed. Hoboken, NJ, USA: Wiley; 2007. ISBN 978-0-470-05357-7

[14] Bruns R, Scarmiano I, Neto B. Statistical Design — Chemometrics. 1st ed. Elsevier Science; Volume 25 (Data Handling in Science and Technology); Amsterdam: Elsevier; 2006

[15] Carlson R. Design and Optimization in Organic Synthesis. 3rd ed. Amsterdam: Elsevier; 1991

[16] Leardi R. Experimental design in chemistry: A tutorial. Analytica Chimica Acta. 2009; **652**:161-172

[17] Eiroa AA, Diévart P, Dagaut P. Improved optimization of polycyclic aromatic hydrocarbons (PAHs) mixtures resolution in reversed-phase high-performance liquid chromatography by using factorial design and response surface methodology. Talanta. 2010;**81**:265-274

[18] Plackett RL, Burman JP. The design of optimum multifactorial experiments. Biometrika. 1946;**33**:305-325

[19] Morgan E. Chemometrics: Experimental Design. Analytical Chemistry by Open Learning. Chichester: Wiley; 1991. pp. 118-188

[20] Box GEP, Hunter W, Hunter J. Statistics for Experimenters, An Introduction to Design, Data Analysis and Model Building. New York: Wiley; 1978. pp. 306-418

[21] Vander Heyden Y, Massart DL. Review of the use of robustness and ruggedness in analytical chemistry. In: Smilde A, de Boer J, Hendriks M, editors. Robustness of Analytical Methods and Pharmaceutical Technological Products. Amsterdam: Elsevier; 1996. pp. 79-147

[22] Box GEP, Behnken DW. Some new three level designs for the study of quantitative variables. Technometrics. 1960;**2**:455-475

[23] Dejaegher B, Vander Heyden Y. The use of experimental design in separation science. Acta Chromatographia. 2009;**21**:161-201

[24] Dejaegher B, Durand A, Vander Heyden Y. Experimental design in method optimization and robustness testing. In: Hanrahan G, Gomez FA, editors. Chemometric Methods in Capillary Electrophoresis. New Jersey: John Wiley & Sons; 2010. pp. 11-74

[25] Montgomery DC. Design and Analysis of Experiments. 4th ed. New York: John Wiley; 1997

[26] Lewis GA, Mathieu D, Phan-Tan-Luu R. Pharmaceutical Experimental Design. New York: Marcel Dekker; 1999

[27] Vapnik V. The Nature of Statistical Learning Theory. New York: Springer; 2000

[28] Korkmaz S, Zararsiz G, Goksuluk D. Drug/nondrug classification using support vector machines with various feature selection strategies. Computer Methods and Programs in Biomedicine. 2014;**117**:51-60

[29] Zararsiz G, Elmali F, Ozturk A. Bagging support vector machines for leukaemia classification. International Journal of Computer Science Issues. 2012;**9**:355-358

[30] Ivanciuc O. Applications of support vector machines in chemistry. In: Lipkowitz KB, Cundari TR, editors. Reviews in Computational Chemistry, Weinheim: Wiley-VCH; 2007. pp. 291-400

[31] Hamel L. Support vector machines. In: Larose DT, editor. Knowledge Discovery with Support Vector Machines. Hoboken, New Jersey, USA, John Wiley & Sons, Inc; 2009, pp. 89-132

[32] Naguib IA, Abdelaleem EA, Draz ME, Zaazaa HE. Linear support vector regression and partial least squares chemometric models for determination of hydrochlorothiazide and benazepril hydrochloride in presence of related impurities: A comparative study. Spectrochimica Acta Part A: Molecular and Biomolecular Spectroscopy. 2014;**130**:350-356

[33] Puzyn T, Leszczynski J, Cronin MT, editors. Recent Advances in QSAR Studies. Methods and Applications. Heidelberg, Germany: Springer; 2010. p. 414

[34] Merz KM, Ringe D, Reynolds CH. Drug Design: Structure- and Ligand-Based Approaches. New York: Cambridge University Press; 2010

[35] Hamel L. Support vector machines. In: Larose DT, editor. Knowledge Discovery with Support Vector Machines. Hoboken, New Jersey, USA, John Wiley & Sons, Inc; 2009, pp. 89-132

[36] Berridge JC, Jones P, Roberts-Mcintosh AS. Chemometrics in pharmaceutical analysis. Journal of Pharmaceutical and Biomedical Analysis. 1991;**9**:597-604

[37] Korany MA, Ragab MAA, Youssef RM, Afify MA. Experimental design and machine learning strategies for parameters screening and optimization of Hantzsch condensation reaction for the assay of sodium alendronate in oral solution. RSC Advances. 2015;**5**:6385-6394

[38] Elazazy MS. Determination of midodrine hydrochloride via Hantzsch condensation reaction: A factorial design based spectrophotometric approach. RSC Advances. 2015;**5**:48474-48483

[39] Elazazy MS, Ganesh K, Sivakumar V, Huessein YHA. Interaction of p-synephrine with p-chloranil: Experimental design and multiple response optimization. RSC Advances. 2016;**6**:64967-64976

[40] Boeris MS, Luco JM, Olsina RA. Simultaneous spectrophotometric determination of phenobarbital, phenytoin and methyl phenobarbital in pharmaceutical preparations by using partial least-squares and principal component regression multivariate calibration. Journal of Pharmaceutical and Biomedical Analysis. 2000;**24**:259-271

[41] Berridge JC. Chemometrics and method development in high-performance liquid chromatography. Part 1: Introduction. Chemometrics and Intelligent Laboratory Systems. 1988;**3**:175-188

[42] Berridge JC. Chemometrics and method development in high-performance liquid chromatography. Part 2: Sequential experimental designs. Chemometrics and Intelligent Laboratory Systems. 1989;**5**:195-207

[43] ICH, 2005. Q2 (R1), Validation of analytical procedures: text and methodology, ICH Harmonised Tripartite Guideline. International Conference on Harmonisation of Technical Requirements for Registration of Pharmaceuticals for Human Use, Chicago, USA, 2005

[44] Eriksson L, Johansson E, Kettaneh-Wold N, Wikström C, Wold S. Design of Experiments: Principles and Applications. 3rd ed. Umeå, Sweden: MKS Umetrics AB; 2008

[45] Odeniyi MA, Jaiyeoba KT. Optimization of ascorbic acid tablet formulations containing hydrophilic polymers. Farmacia. 2009;**57**:157-166

[46] El-Malah Y, Nazzal S. Hydrophilic matrices: application of Placket-Burman screening design to model the effect of POLYOX–carbopol blends on drug release. International Journal of Pharmaceutics. 2006;**309**:163-170

[47] Sivertsen E, Bjerke F, Almøy T, Segtnan V, Næs T. Multivariate optimization by visual inspection. Chemometrics and Intelligent Laboratory Systems. 2007;**85**:110-118

[48] Hasniyati MR, Zuhailawati H, Sivakumar R, Dhindaw BK. Optimization of multiple responses using overlaid contour plot and steepest methods analysis on hydroxyapatite coated magnesium via cold spray deposition coated magnesium via cold spray deposition. Surface Coatings and Technology. 2015;**280**:250-255

[49] Derringer G, Suich R. Simultaneous optimization of several response variables. Journal of Quality Technology. 1980;**12**:214-219

[50] Minitab 17 Statistical Software. Computer software; 2010. State College, PA: Minitab, Inc. (www.minitab.com)

[51] Talsky G. Derivative Spectrophotometry. 1st ed. Weinheim: VCH; 1994

[52] Sanchez Rojas F, Ojeda CB. Recent development in derivative ultraviolet/visible absorption spectrophotometry: 2009-2011 a review. Microchemical Journal. 2013;**106**:1-16

[53] Hasan NY, Abdel-Elkawy M, Elzeany BE, Wagieh NE. Stability indicating methods for the determination of aceclofenac. Il Farmaco. 2003;**58**:91-99

[54] El-Saharty YS, Refaat M, El-Khateeb SZ. Stability-indicating spectrophotometric and densitometric methods for determination of aceclofenac. Drug Development and Industrial Pharmacy. 2002;**28**:571-582

[55] De Luca M, Oliverio F, Ioele G, Ragno G. Multivariate calibration techniques applied to derivative spectroscopy data for the analysis of pharmaceutical mixtures. Chemometrics and Intelligent Laboratory Systems. 2009;**96**:14-21

[56] Bautista RD, Jiménez AI, Jiménez F, Arias JJ. Simultaneous determination of drugs in concentration ratios above 40 1 by application of multivariate calibration to absorbency and derivative spectrophotometric signals. Fresenius Journal of Analytical Chemistry. 1997;**357**:449-456

[57] Dinç E, Ustündağ O. Chemometric resolution of a mixture containing hydrochlorothiazide and amiloride by absorption and derivative spectrophotometry. Journal of Pharmaceutical Biomedical Analysis. 2002;**29**:371-379

[58] Brown C, Vega-Montoto L, Wentzell P. Derivative preprocessing and optimal corrections for baseline drift in multivariate calibration. Applied Spectroscopy. 2000;**54**:1055-1068